Friedrich Löffler
Hans Dietrich
Werner Flatt

**Dust Collection
with Bag Filters
and Envelope Filters**

Friedrich Löffler (editor)
Hans Dietrich
Werner Flatt

Dust Collection with Bag Filters and Envelope Filters

With 203 figures

Translated by Hermann Simon

Springer Fachmedien Wiesbaden GmbH 1988

Translator: Hermann Simon, Sydney, Australia

Vieweg is a subsidiary company of the Bertelsmann Publishing Group.

All rights reserved
© Springer Fachmedien Wiesbaden, 1988
Originally published by Friedr. Vieweg & Sohn Verlagsgesellschaft mbH, Braunschweig in 1988

No part of this publication may be reproduced, stored in a retrieval system or transmitted in any form or by any means, electronic, mechanical, photocopying, recording or otherwise, without permission of the copyright holder.

Typesetting: Vieweg, Braunschweig

ISBN 978-3-663-07901-9   ISBN 978-3-663-07900-2 (eBook)
DOI 10.1007/978-3-663-07900-2

Contents

Preface XII

1 **Fundamental Priniciples of Particle Separation in Fabric Filters** 1
Friedrich Löffler
  1.1 Introduction 1
  1.2 Particle Properties and Separation Efficiency 2
      1.2.1 Particle Size – Measures of Fineness 3
      1.2.2 Representation of Particle Size Distribution 7
          1.2.2.1 Cumulative Distribution $Q_r(x)$ 7
          1.2.2.2 Density Distribution $q_r(x)$ 8
          1.2.2.3 Distribution Parameters 10
          1.2.2.4 Conversion of Distributions 11
          1.2.2.5 Approximation Functions 12
      1.2.3 Characterization of Dust Collection 17
          1.2.3.1 Total Collection Efficiency 18
          1.2.3.2 Separation Function, Grade Efficiency 19
          1.2.3.3 Calculating Total Collection Efficiency 21
  1.3 Separation in Fibre Assemblies 23
      1.3.1 Collection Efficiency in Fibre Assemblies 23
      1.3.2 Collection Efficiency of a Single Fibre 24
          1.3.2.1 Transport Mechanisms 25
          1.3.2.2 Adhesion Efficiency 31
          1.3.2.3 Summarising Particle Separation on Single Fibres 35
      1.3.3 Dust Separation in Bag Filters 36
      1.3.4 Cleaning and its Effect on Dust Emission 39
  1.4 Pressure Loss in Cleanable Fabric Filters 41
      1.4.1 Approximations derived from Experimental Measurements 42
      1.4.2 Attempts at Filter Modelling 42
      1.4.3 The Effect of Cleaning Intensity on Residual Pressure Drop 46
      1.4.4 The Effect of Electrostatic Forces on Pressure Drop 49
  1.5 Design of Multichamber Fabric Filters 50
  Literature 53

2 **Filter Media** 55
Hans Dietrich
  2.1 The Nucleus 55
  2.2 Fibres as Building Blocks 56
      2.2.1 Cotton 56

	2.2.2	Wool	57
	2.2.3	Chemical and Mineral Fibres	58
		2.2.3.1 Introduction	58
		2.2.3.2 Percentage Share by Fibre Types	61
		2.2.3.3 Polyester (PES)	61
		2.2.3.4 Polyacrylonitrile (PAN)	61
		2.2.3.5 Aliphatic Polyamide (PA)	63
		2.2.3.6 Aromatic Polyamides (Aramides)	64
		2.2.3.7 Polypropylene (PP)	64
		2.2.3.8 Polytetrafluoroethylene (PTFE)	65
		2.2.3.9 Mineral Fibres (Natural and Man Made)	66
2.3	Textile Structures as Dust Filters		82
	2.3.1	Fabrics	82
	2.3.2	Needlefelts	88
	2.3.3	Wool Felt as The Forerunner	92
	2.3.4	Nonwovens	95
		2.3.4.1 Nonwovens bonded with Adhesive Binders	95
		2.3.4.2 Nonwovens that shrink	95
		2.3.4.3 Nonwovens through swelling	95
		2.3.4.4 Spunbondeds	95
2.4	Market and Technology		98
2.5	Quality Criteria		100
	2.5.1	Area Weight	101
	2.5.2	Density	101
	2.5.3	Air Permeability	101
	2.5.4	Bursting Strength	103
	2.5.5	Breaking Extension	103
	2.5.6	Abrasion Resistance	104
	2.5.7	Additional Quality Parameters	104
2.6	Filter Medium and Cleaning Methods		105
2.7	Selection Criteria		112
2.8	Filter Medium und Separation Process		114
2.9	Chemical Attack on Filter Media		125
2.10	Hot Gas Filtration		131
2.11	Filter Medium and Static Electricity		136
2.12	Flammability of Filter Media		142
2.13	Filter Media in Use		144
2.14	Warranties for Filter Media		160
2.15	Assembly, Accessories, Support Elements Maintenance and Care, Washing and Cleaning of Filter Media		165
2.16	Trends and the Future		172
Literature			175

Contents

3 Filtration Plants 180
Werner Flatt
- 3.1 Introduction 180
- 3.2 Summary 180
  - 3.2.1 History, Patents, Definitions 180
    - 3.2.1.1 Shaker Type Filters 184
    - 3.2.1.2 Back Blowing Type Filters 186
    - 3.2.1.3 Reverse Flow Filters 186
    - 3.2.1.4 Reverse Jet Filters 188
    - 3.2.1.5 Ring Jet Filters 190
    - 3.2.1.6 Direct Pulse Filters 192
  - 3.2.2 Trends 194
    - 3.2.2.1 Significant Developments since 1970 194
    - 3.2.2.2 Explosion Protection 195
    - 3.2.2.3 Higher Air to Cloth Ratios 195
    - 3.2.2.4 Ease of Maintenance 195
    - 3.2.2.5 Market Trends 196
  - 3.2.3 List of Manufacturers of Different Filtertypes 197
    - 3.2.3.1 Shaker Type with Reverse Gas Cleaning 197
    - 3.2.3.2 Blowring Filters 199
    - 3.2.3.3 Reverse Flow Filters 199
    - 3.2.3.4 Reverse Jet Filters with Venturis 201
    - 3.2.3.5 Ring Jet Filters 203
    - 3.2.3.6 Direct Pulse Filters (without Venturis) 203
- 3.3 Filter Configuration and its Effect on Operating Conditions 204
  - 3.3.1 Method of Measurement and Comparisons 204
    - 3.3.1.1 Permanent Test Installations 204
    - 3.3.1.2 Pressure Measurement in a Single Bag with Piezoelectric Sensor 204
    - 3.3.1.3 Determining Area Mass on a Single Bag 208
  - 3.3.2 Influence of Cleaning 209
    - 3.3.2.1 Shaker Filters 210
    - 3.3.2.2 Counter Current Cleaning 210
    - 3.3.2.3 Pulse Cleaning 211
    - 3.3.2.4 Bag Length 219
  - 3.3.3 Bag Diameter 221
  - 3.3.4 Flue Gas Distribution 223
  - 3.3.5 Summary and Conclusions 227
- 3.4 Designing the Filter Surface Area 227
  - 3.4.1 General 227
  - 3.4.2 Standard Design Guides 230
  - 3.4.3 System Characteristics 230
  - 3.4.4 Design Calculations using a Base Value, Indexes and Coefficients 231
  - 3.4.5 Base Values for Dusts and Fields of Application 234

		3.4.6 Creating a Data Bank 237
		3.4.7 Pilot Studies 237
		3.4.8 Design Fundamentals 237

- 3.5 Ease of Maintenance and Intervals 243
  - 3.5.1 Cleaning and Control Systems 243
  - 3.5.2 Filter Elements 244
  - 3.5.3 Wearing Parts, Spares and Consumables 245
  - 3.5.4 Special Maintenance Facilities 245
- 3.6 Explosion Protection 247
  - 3.6.1 Regulations and Definitions 248
  - 3.6.2 Construction and Tests 248
  - 3.6.3 Planning Based on Field Experience 252
- 3.7 Heat and Sound Insulation 256
- 3.8 The Cleaning Air 257
  - 3.8.1 Properties and Function 257
  - 3.8.2 Cleaning Air Production 258
    - 3.8.2.1 Quality of Cleaning Air 258
    - 3.8.2.2 Line Filters 259
    - 3.8.2.3 Air Driers 259
  - 3.8.3 Pressure Lines 260
- 3.9 Certification 261
  - 3.9.1 Filter Resistance 261
    - 3.9.1.1 Basics 261
    - 3.9.1.2 Measurement Techniques 263
  - 3.9.2 Filter Efficiency and Emission Limits 263
    - 3.9.2.1 Basics 263
    - 3.9.2.2 Measurement of Collection Efficiency 264
- 3.10 Investment Analysis and Operating Cost Comparison 265
  - 3.10.1 Criteria 265
  - 3.10.2 Cost Evaluation 266
  - 3.10.3 Comparison Table for Difference Offers 268
  - 3.10.4 Discussion 268
- Literature 268

Index 270

Preface

Technological requirements, legislative pressures, and last but not least a growing environmental concern by the population at large have led to substantial improvements in particle separation from gases during the past few years. It has been repeatedly documented that, despite rising production and increasing fuel consumption, the last 15 years have witnessed a reduction in particulate matter by more than 50 %. Changes in raw material and fuels, improvements in techniques and advances in rational implementation of separation methods have largely contributed to this.

Updated processing conditions, advances in measuring techniques, as well as more attention to environmental stresses make it even more important not to rest on our laurels. It has become increasingly evident that the assessment of a dust problem by quantity alone is not enough. A precise assessment of the quality of the particulate matter is of equal importance. Chemical composition and physical characteristics are also important parameters. Constituents and particle size distribution demand increased attention, since both influence the subsequent treatment of the dust. To make further reductions in environmental dust levels, and to increase the separation performance of filters, especially in the fine particle area, their fields of application must be extended widely even more.

Correctly engineered and well maintained bag and envelope filters have a proven record of performance for fine particle separation. Together with the recent advances in filter media and application techniques, bag and envelope filters have not only opened new fields, but also achieved a market share of over 40 % in dust collector sales.

Technical development has been empirical mostly and, as is often the case in process technology, preceeds fundamental basic research. Consequently, there are few systematic examinations of the problems surrounding bag and envelope filters.

The state of knowledge and technology — to the extent ot which it is accessible through the technical literature — is found in infrequent monograms or field reports. Textbooks frequently dismiss filtration in a single chapter. ("Air Filtration" by C. N. Davis published in 1973 covers only the fundamentals of deep bed filters.)

The present authors considered it judicious, therefore, to set out an up to date position of separation technology for bag and envelope filters. The current publication puts particular emphasis on the aspects of application technology. It addresses itself to those specialists involved in the planning, building, selection and running of filter installations, i.e. the "practitioner". The material offered will go beyond what would be required at a tertiary level; nevertheless, we commend our treatment of the subject to those scholars requiring or wanting a deeper understanding of filtration.

The book is divided into three chapters. Chapter 1 encompasses the results of theoretical and experimental research in the separation process and the concomitant differential pressure. Chapter 2 deals with the "core" of the filter — filter medium — and gives a detailed introduction into the problems of developing and manufacturing the filtration media. Filter installations is the subject matter in Chapter 3. In this chapter a systematic treatment is offered of the different filtration sytems, including detailed constructions and problems in in-use situations.

At this point the editor takes great pleasure in thanking his coauthors for their willingness to contribute their respective knowledge to this book. From its inception it was desired that the chapters on "Filter Media" and "Filter Installations" be written by authors with an intimate practical knowledge. It is much appreciated that these contributors, who are both actively engaged in the technology, have so generously given their time. It is probable that they grossly underestimated the time involvement in putting this publication together. That they "pulled through", despite the economic recessions over recent times, deserves special commendation.

Karlsruhe, June 1986 *Friedrich Löffler*

1 Fundamental Principles of Particle Separation in Fabric Filters

Friedrich Löffler

1.1 Introduction

Fibrous filter media have acquired a firm place in todays separation technology. They are the medium and technique of choice when efficient separation is required.

Depending on area of application, eventual layout, and mode of operation they fall into two large groups: *storage filters* and *cleaned filters* [1].

Storage filters are used for low levels of dust (mg/m^3); typically in the air conditioning and ventilation industry. The open fibre mats have *pore volumes* greater than 90%, or even exceeding 99%. The particles are retained within the interior of this medium. Once saturated, the mats are discarded; regeneration by washing or blowing off the dust is practiced occasionally. Some recent constructions provide for cyclic blow-back. Typical face velocities for these filters are 0.1 to 3 m/s.

In order to give a measure of the performance requirements for high efficiency filters DIN 24184 requires: a 99.97% efficiency for a particle range of 0.3 to 0.5 μm for a class S filter.

Filters requiring cleaning are used where dust levels reach the order of grams per m^3, and even up to several hundred grams per m^3. Such levels are common in industrial processing. For these purposes the fibre medium functions as a support base. Until recently woven fabrics were used exclusively, while today non-wovens or felts have become common. Their pore volume lies between 70 to 90%. The initial filtering phase takes place within the interior of the medium and then relocated to the surface. Thus, the filter cake deposited on the surface becomes the actual filter medium, which can reach separation efficiencies of more than 99.9%. As the layer of dust increases the filter resistance rises. Thus, periodic cleaning becomes necessary; in severe cases at minute intervals. Typical face velocities lie between 0.5 and 5 cm/sec, or 18 to 180 m/h, respectively.

This book deals with filters that are cleaned, and where the filter medium is in the form of bags or envelopes. Such filters have experienced a rapid development in recent years; they have become the nucleus for efficient separation of fine particles.

Design criteria for cleaned filters are largely empirical; i.e. experiments and experience are the foundations for their design and construction. Despite this, spectacular successes have been recorded.

In many instances separation efficiencies exceed 99% across a broad spectrum of measureable performances. Discharge concentration can be reliably kept below 5 mg/m^3.

Nevertheless, there are still situations where the residual emission is inadequate. Cause of failure and prevention must be determined in this case.

Besides separation efficiency the filter resistance is of significant — often critical — importance. The most important task in designing filters is to choose operating conditions and filter medium combinations that will maintain a preset differential pressure over a long period — often years.

Chapter 1 describes particles size distributions and separation; it then sets out the state of knowledge about theoretically and experimentally obtained results to explain the separation process and the filter resistance.

1.2 Particle Properties and Separation Efficiency

The behaviour of particles during separation is determined by such properties as size, shape, and density. To explain the separation and cleaning behaviour of fibrous media additional characteristics must be considered: Surface activity, agglomeration tendencies, flow properties, electrostatic behaviour, temperature effects and chemical properties. Some of these properties are in turn influenced by particle size.

An adequate description of particle systems requires detailed measurement and definition of the various particle characteristics. It is very important to consider how each of the characteristics are distributed; it is not sufficient to quote a mean value only. Our present knowledge has considerable gaps in this area. Much more attention must be paid to a comprehensive description of particle properties based on an in depth appreciation of particle behaviour, in order to arrive at a reliable design based on a comprehensive understanding of all the particle properties. The same can be said for extending the scope of current measuring techniques as well as a consideration whether such methods are still appropriate, or whether fundamentally new procedures are called for.

Those particle characteristics that determine adhesion or bouncing will be dealt with in more detail in subsequent sections. In this first section the significance of particle size and its distribution will be introduced. The units, symbols and representations for this are suggested in various DIN Standards [2, 3] which will form the basis for this exposition. Readers wishing to take their studies further are referred to the comprehensive specialized literature, e.g. [4, 5, 6]. (The very complex treatment of particle shape distribution is not covered in this text; here also extensive literature is available [7, 8, 9].)

It is a general fact that the size of all particles dispersed in a gas is distributed over a range which — depending on its origin — can extend over a wide range. The properties and behaviour of the dispersed phase is determined largely by this particle distribution. This may become more apparent, if it is remembered that particle weight varies with the cube of the diameter, while the flow forces are proportional to the first or second power only. To characterize a dispersed phase, therefore, it is essential to obtain and describe the particle distribution over as wide a range as possible, best across its entire range.

The definitions and terms introduced to describe particle distribution are necessary in order to understand separation functions introduced in the following sections.

1.2 Particle Properties and Separation Efficiency

1.2.1 Particle Size — Measures of Fineness

The particle assembly can be described by a distribution function. Each of these particle properties is assigned a dispersion measure. This dispersion measure must be a physical and measurable entity that can be quantified. The dispersion measures relating to particle size is termed *Fineness;* refer to DIN 66141 [2]. Here are some examples:

a) Geometrical Particle Measures

For regularly shaped particles (cubes, spheres) it is sufficient to give type of shape and main dimensions.

Non-uniform particles are more the rule than the exception. A unique, single parameter description, is not possible here, so that agreement must be reached on how measurement is to be done. Forinstance, by giving the main dimensions of a regularly shaped particle.

During microscopic analysis, e.g. automatic counting, statistical measures are generated (fig. 1.1). x_F is the Feret diameter: Distance between tangents perpendicular to measuring axis, x_M the Martin diameter: The chord bisecting the projected area parallel to the direction of measurement, x_{oe} is the longest chord parallel to the measuring direction — frequently determined by optical image analyzing methods.

Important: The value recorded depends on the angular position of the particle. Even with mono sized particles measured under conditions of random angular positioning a size distribution is measured.

The geometric equivalent diameter is a dimension of a regularly shaped particle whose geometric properties are equivalent to those under examination, e.g. diameter of the circle of an equal projection area, diameter of a sphere of equivalent surface area, or diameter of a sphere of equal volume.

b) Settling Velocity

The stationary settling velocity of a single particle in an infinite medium in gravity field (DIN 66111 [3]) is a frequently used measure to characterize particles. This property defines the movement of a particle. "Infinitely large" implies that the interaction between particles that are settling and the container wall is negligible. Since the settling velocity is a function of the type of medium, it must be specified in regard to density,

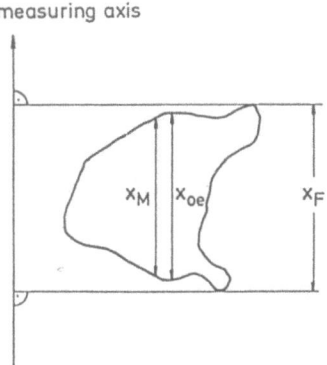

Fig. 1.1
Statistical Particle Measurement Parameters
XF = Feret diameter
XM = Martin diameter
X = Longest chord parallel to direction of measurement

viscosity and its temperature at the time of measurement. For instance, the settling velocity of a spherical particle of 10 μm dia. in air is $2.7 \cdot 10^{-1}$ cm/s, in water it is $5 \cdot 10^{-3}$ cm/s at 20 °C. This is based on a density difference $\Delta\rho$ of 1 g/cm³ between particle and medium.

For very fine particles (below 1 μm) the settling velocity is measured using the increased acceleration field generated by a centrifuge. These parameters are converted to characterize the particle distribution within a gravity field.

By assuming a spherical particle it is possible to derive the equivalent diameter of a sphere with the same settling velocity and the same density difference $\Delta\rho$. From the measured settling velocity it is possible to calculate the equivalent diameter of the settling velocity x_{w_f}.

For a sphere the following equation expresses the settling velocity:

$$w_f^2 = \frac{4}{3} \cdot \frac{1}{c_w (\mathrm{Re})} \cdot \frac{\Delta\rho}{\rho_f} g x \qquad (1)$$

Where

x = Particle diameter
$\Delta\rho = \rho_p - \rho_f$
ρ_p = Density of particle
ρ_f = Fluid density
g = Acceleration due to gravity
$c_w (\mathrm{Re})$ = Drag coefficient of a sphere
$\mathrm{Re} = \dfrac{w_f \cdot x \cdot \rho_f}{\mu}$ Reynold No
μ = Fluid viscosity

Fig. 1.2 shows the settling velocity as a function of particle diameter for various densities. The drag coefficient c_w is a function of the Reynolds No. For $\mathrm{Re} \leq 0.25$, i.e. within the "Stokes Law" region: –

$$c_w = \frac{24}{\mathrm{Re}}$$

so that

$$w_f = \frac{1}{18} \cdot \frac{\Delta\rho}{\mu} g x^2 \qquad (2)$$

In the region $\mathrm{Re} \leq 0.25$ the equivalent diameter for the same settling velocity is also called the "Stokes Diameter" x_{St}. Thus

$$x_{St} = \sqrt{\frac{18 \mu w_f}{g \Delta\rho}} \qquad (3)$$

Equation (2) and (3) are applicable for $\Delta\rho = 1$ g/cm³ up to approx $x \leq 50$ μm in air, -for water up to $x \leq 80$ μm.

1.2 Particle Properties and Separation Efficiency

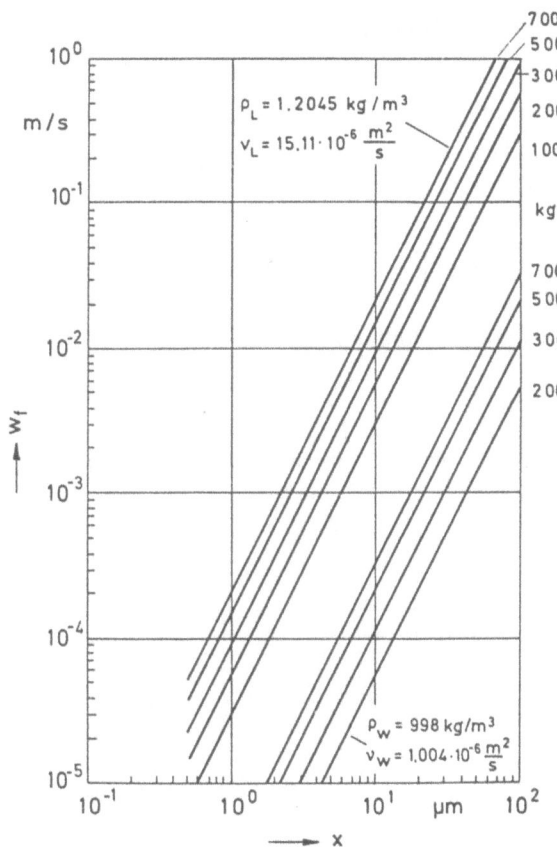

Fig. 1.2

Setting velocity of spheres in water, viz. air, as a function of particle size and density

In dust separation it is useful to use the concept of an "aerodynamic diameter" x_{ae}, which is derived from the settling velocity. For this purpose a density difference $\Delta\rho$ of 1 g/cm³ is used for an equivalent diameter sphere. This differs from DIN 66111 and should be kept in mind when making any comparisons. Equation (3) can be used for a conversion, i.e. —

$$x_{ae} = x_{St} \sqrt{\Delta\rho_{St}} \tag{4}$$

Here $\Delta\rho_{St}$ is the density difference on which the calculation of x_{St} is based. Cascade impactors are commonly calibrated in x_{ae}.

For a derivation of equation (2) and (3), respectively, it is assumed that the fluid phase in which the particles are suspended is a continuum. This implies that the mean free path of the gas molecules λ is much smaller than the particle x. In the case of fluids this condition holds invariably for all technically interesting cases. However, for dispersions of fine particle in gases the drag law requires adjustment. For standard temperature and pressure $\lambda \approx 0.06$ μm.

Stokes law can be corrected with factor Cu introduced by Cunningham:

Settling velocity $\quad w_{Cu} = w_{St}\, Cu$ (5)

Stokes dia $\quad x_{Cu} = x_{St}\sqrt{\dfrac{1}{Cu}}$ (6)

Now, according to Millikan [27]

$$Cu = 1 + 1.246\,\frac{2\lambda}{x} + 0.42\,\frac{2\lambda}{x}\exp\left[-0.87\,\frac{x}{2\lambda}\right]$$ (7)

For $x > 0.1\,\mu m$ it is sufficient to use a simplified approximation

$$Cu \approx 1 + 0.864\,\frac{2\lambda}{x}$$ (8)

Table 1.1 sets out some values for Cu that are derived from eq. 7, viz. 8. For a particle size of $1\,\mu m$ the correction factor exceeds already the 10% limit.

Table 1.1 Cunningham Correction Factor (Cu)

$x/\mu m$	0.01	0.1	0.5	1.0	2.0	5.0	10.0
Cu (eq. (8))	12.23	2.123	1.225	1.112	1.056	1.022	1.011
Cu (eq. (7))	22.31	2.899	1.328	1.162	1.082	1.032	1.016

c) Optical Measures of Fineness

The degree of light scattering that a particle produces can be used as a measure of particle size. By calibrating with reference particles of known dimensions an "equivalent light scattering diameter" x_{sc} can be derived. Another method of calibration is to use a reference method for measuring the size distribution of the same material. Such optical devices are used increasingly for calibration of impactors, viz. these are calibrated in aerodynamic diameters x_{ae}.

d) Electrical Measures of Fineness

In this type of measurement (Coulter counter or similar) fineness is determined by measuring the electrical resistance within a capillary tube through which the particles flow suspended in an electrolyte. The change in electrical resistance is proportional to the volume of particles. From this value the diameter of a sphere of equivalent volume x_v can be derived.

Important: When comparing particle distributions it is important to note how these distributions were obtained, because for irregularly shaped particles significant differences occur depending on the method employed. (Differences of the order of 2 to 3 fold are not uncommon.)

1.2 Particle Properties and Separation Efficiency

1.2.2 Representations of Particle Size Distribution

To adequately deliniate a particle size distribution it is necessary to give, besides particle size, information on the quantity increments. For this purpose the total quantity of all measured particles is grouped according to the chosen fineness parameter and then assigned to partial quantities.

A table is the simplest form of representation. A graphic form is much more discriptive and allows comparisons to be made. DIN 66141 requires the fineness parameter to be always on the abscissa; the relative quantities are plotted on the ordinate. Two types of distribution functions are used:

1.2.2.1 Cumulative Distribution $Q_r(x)$

The cumulative distribution is defined as the integral of all subsets between the minimum particle size x_{min} and a particular value x, expressed as a proportion of the total quantity in the range x_{min} to x_{max} (fig. 1.3). Thus, $Q_r(x_i)$ is the proportionate part — dimensionless — for which $x \leq x_i$ (the upper band limit is counted as part of the interval!).

The limiting values for Q_r are $Q_r(x_{min}) = 0$ and $Q_r(x_{max}) = 1$ i.e. $0 \leq Q_r(x) \leq 1$. $Q_r(x)$ always increases with x, it never decreases, but may remain constant.

To enable a reference to the type of measure of population used in determining a particular distribution an index r has been introduced. The table below gives a summary of these possible types; absolut numbers and mass are the most frequently used. $Q_3(x)$ is occasionally described in terms of $D(x) =$ undersize, viz. $R(x) =$ oversize, where $R(x) = 1 - D(x)$. These terms are borrowed from sieving analysis.

Type	Index	Distribution Function
Numer	0	$Q_0(x), q_0(x)$
Length	1	$Q_1(x), q_1(x)$
Area	2	$Q_2(x), q_2(x)$
Mass, Volume	3	$Q_3(x), q_3(x)$

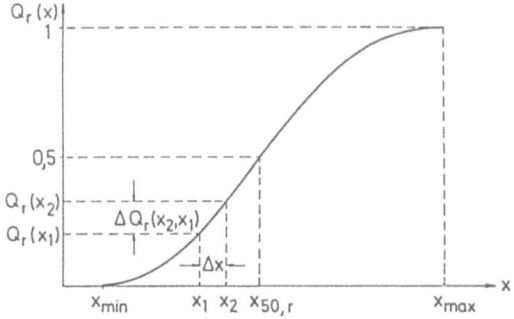

Fig. 1.3
Cumulative distribution curve $Q_r(x)$

1.2.2.2 Density Distribution $q_r(x)$

The density distribution is defined as the proportion of quantity in the range x and x + Δx divided by the width of the interval Δx, i.e.

$$q_r(x_1, x_2) = \frac{\Delta Q_r(x_1, x_2)}{\Delta x} = \frac{Q_r(x_2) - Q_r(x_1)}{x_2 - x_1} \tag{9}$$

if $Q_r(x)$ can be differentiated –

$$q_r(x) = \frac{dQ_r(x)}{dx} \tag{10}$$

$q_r(x)$ has the dimension (length)$^{-1}$, if x is expressed as a length.

Very important: The proportionate quantity is related to the width of the interval; therefore the term "density" is used.

The shaded area in fig. 1.4 between x_1 and x_2 expresses the proportion between these two limits; this is evident from eq. (9). The area under the density distribution $q_r(x)$ is the integral between x_{min} and x_{max}, i.e.

$$\int_{x_{min}}^{x_{max}} q_r(x)\, dx = 1 \quad \text{Standardisation condition}$$

and

$$\int_{x_{min}}^{x_i} q_r(x)\, dx = Q_r(x_i) \tag{11a}$$

viz.

$$\int_{x_1}^{x_2} q_r(x)\, dx = Q_r(x_2) - Q_r(x_1). \tag{11b}$$

Numeric example to determine $Q_r(x)$ and $q_r(x)$:

Using a counting method particle numbers ΔM_i are obtained. Since a counting method was employed, the index is zero.

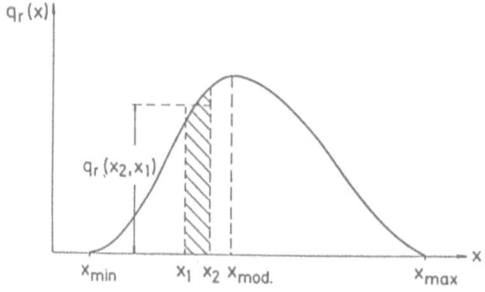

Fig. 1.4 Density distribution $q_r(x)$

1.2 Particle Properties and Separation Efficiency

For the total number M_{tot}, ΔM_i is the value for the ith interval and the ratio is —

$$\Delta Q_0(x_i) = \frac{\Delta M_i}{M_{tot}}$$

Cummulative distribution between x_{min} and x_i:

$$Q_0(x_i) = \sum_{x_{min}}^{x_i} \Delta M_i / M_{tot} = \sum_{x_{min}}^{x_i} \Delta Q_0(x)$$

Density distribution within the ith interval

$$q_0(\bar{x}_i) = \frac{\Delta Q_0(x_i)}{\Delta x_i}$$

with

$$\Delta x_i = x_i - x_{i-1} \quad \text{and} \quad \bar{x}_i = \frac{x_{i-1} + x_i}{2}$$

i.e.

x_i = upper limit of interval
$\bar{x}_i$ = mean value of interval

and

$$\Delta Q_0(x_i) = Q_0(x_i) - Q_0(x_{i-1}).$$

i	x_i μm	$\bar{x}_i$ μm	Δx_i μm	ΔM_i	$\sum_0^i \Delta M_i$	$Q_0(x_i)$	$q_0(\bar{x}_i)$ 1/μm
0	0						
1	2	1	2	0	0	0	0
2	4	3	2	2	2	0.069	0.0345
3	6	5	2	5	7	0.241	0.0862
4	8	7	2	12	19	0.655	0.207
5	10	9	2	6	25	0.862	0.1035
6	12	11	2	3	28	0.965	0.0517
7	14	13	2	1	29	1.0	0.0172
8	16	15	2	0	29	1.0	0

$M_{tot} = 29$

Note: For a graphic presentation $Q_0(x_i)$ is always plotted at the upper limit of the interval, while $q_0(\bar{x}_i)$ is plotted at the centre of the interval.

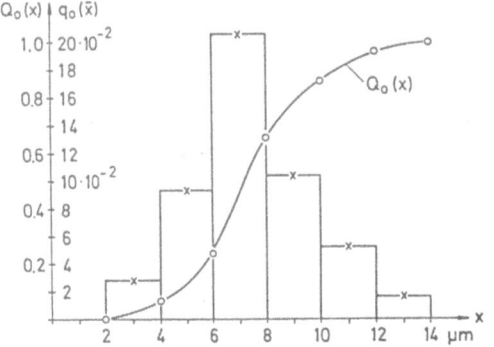

Fig. 1.5

Figurative example for the determination of $Q_0(x)$ and $q_0(x)$

Fig. 1.5 shows the histogram that is formed when $q_0(\bar{x}_i)$ is plotted. A smooth curve can be generated from the histogram for $q_0(x)$ by drawing a curve through the midpoints of the bars where the areas below the histogram and the smooth curve are equal. Graphic or numeric differentiation of the $Q_0(x)$ function represents a further possibility. The curve Q_0 in fig. 1.5 plotted through the upper limit of each interval is also a smoothened curve.

1.2.2.3 Distribution Parameters

Several parameters can be used to characterize measured distributions, e.g. from a graph it is possible to read off directly:

Median value $x_{50,r}$ = Particle size, for which $Q_r(x_{50,r}) = 0.5$
Modal value x_{mod} = Particle size, for which $q_r(x)$ is maximum.
(caution: some populations have more than one maximum)

It is possible to calculate several integral mean values from the distribution functions: These values are also called *moments* of the population. The universal definition for the kth moment of a q_r distribution is:

$$M_{k,r} = \int_{x_{min}}^{x_{max}} x^k q_r(x)\,dx \tag{12}$$

$M_{k,r}$ describes the mean x^k of the q_r distribution. From this the associated mean particle size $x_{k,r}$:

$$\bar{x}_{k,r} = \sqrt[k]{M_{k,r}}. \tag{13}$$

1.2 Particle Properties and Separation Efficiency

Several means may be calculated; two of these are worth mentioning:

a) $M_{1.0}$ is the weighted mean of the number distribution, i.e. the arithmetic mean, or the mean $\bar{x}$.

$$\bar{x} = M_{1.0} = \int_{x_{min}}^{x_{max}} x\, q_0(x)\, dx = \sum_{x_{min}}^{x_{max}} \bar{x}_i\, \Delta Q_0(x_i) \,. \tag{14}$$

b) the mean $\overline{x^3}$ of a population is

$$\overline{x^3} = M_{3.0} = \int_{x_{min}}^{x_{max}} x^3\, q_0(x)\, dx = \sum_{x_{min}}^{x_{max}} \overline{x_i^3}\, \Delta Q_0(x_i) \,. \tag{15}$$

$\overline{x^3}$ can be used to convert a particle number concentration c_n into a mass concentration c_m: The mean volume $\bar{v}$ per particle is:

$$\bar{v} = k_v\, \overline{x^3}\,;\ k_v = \text{shape factor} = \frac{\pi}{6}\ \text{for spheres} \tag{16}$$

so that

$$c_m = c_n\, \rho_s\, k_v\, \overline{x^3}$$

for additional moments and their relationshipes to one another refer to DIN 66141 and [4].

Furthermore, it is emphasized that a positioning of a parameter by itself does not uniquely define a population. Additional data about the width of the population is required. For this indeces such as x_{min} and x_{max}, or better still — because it can be more accurately determined — $x_{0.05}$ and $x_{0.95}$ etc.

An integrated measure of dispersion is the variance σ_r^2 that is commonly used in statistics:

$$\sigma_r^2 = \int_{x_{min}}^{x_{max}} (x - \bar{x}_r)^2 \cdot q_r(x)\, dx \,. \tag{17}$$

The square root of the variance is called the Standard Deviation.

1.2.2.4 Conversion of Distributions

It is frequently desireable to be able to convert a distribution curve obtained by on system of measurement into another. Such is the case, if a counting method (e.g. scatter or microscope) is used to determine frequency distribution $q_0(x)$; the mass distribution $q_3(x)$ is also of interest. Such a conversion is possible as long as for a fineness parameter a geometrical dimension has been measured, e.g. an equivalent diameter, which is usually the case.

The general equation to find $q_t(x)$, given $q_u(x)$ is:

$$q_t(x) = \frac{x^{t-u} q_u(x)}{\int_{x_{min}}^{x_{max}} x^{t-u} q_u(x) \, dx} \quad . \tag{18}$$

for a particular case:

$$q_3(x) = \frac{x^3 q_0(x)}{\int_{x_{min}}^{x_{max}} x^3 q_0(x) \, dx} \tag{19}$$

resp.:

$$q_0(x) = \frac{\dfrac{q_3(x)}{x^3}}{\int_{x_{min}}^{x_{max}} \dfrac{q_3(x)}{x^3} \, dx} \tag{20}$$

1.2.2.5 Approximation Functions

Measured distributions can frequently be expressed as mathematical functions. When plotted on special coordinate systems, they yield a straight line.

Please note: A measured population distribution *may* be a straight line, but not necessarily so.

The common approximation functions are two parameter distributions (size modulus and distribution modulus) and are normalised.

a) *Power distribution* according to Gaudin-Schuhmann (DIN 66143)

This function defines a mass distribution $Q_3(x)$ in the range $0 \leqslant x \leqslant x_{max}$:

$$D(x) = Q_3(x) = \left(\frac{x}{x_{max}}\right)^m \quad \text{for } x \leqslant x_{max} \tag{21}$$

$$Q_3(x) = 1 \quad \text{for } x \geqslant x_{max}$$

Distribution parameters are x_{max} and m.

When plotted on a double logarithmic grid (fig. 1.6) this power function yields a straight line with gradient m, which is measured either directly off the graph as the tangent angle; alternatively, if the line is displaced parallel to coincide with the so called pole (example 1), whence the gradient can be read off directly from the log-log paper. Sectioned approximations are also possible (example 2).

1.2 Particle Properties and Separation Efficiency

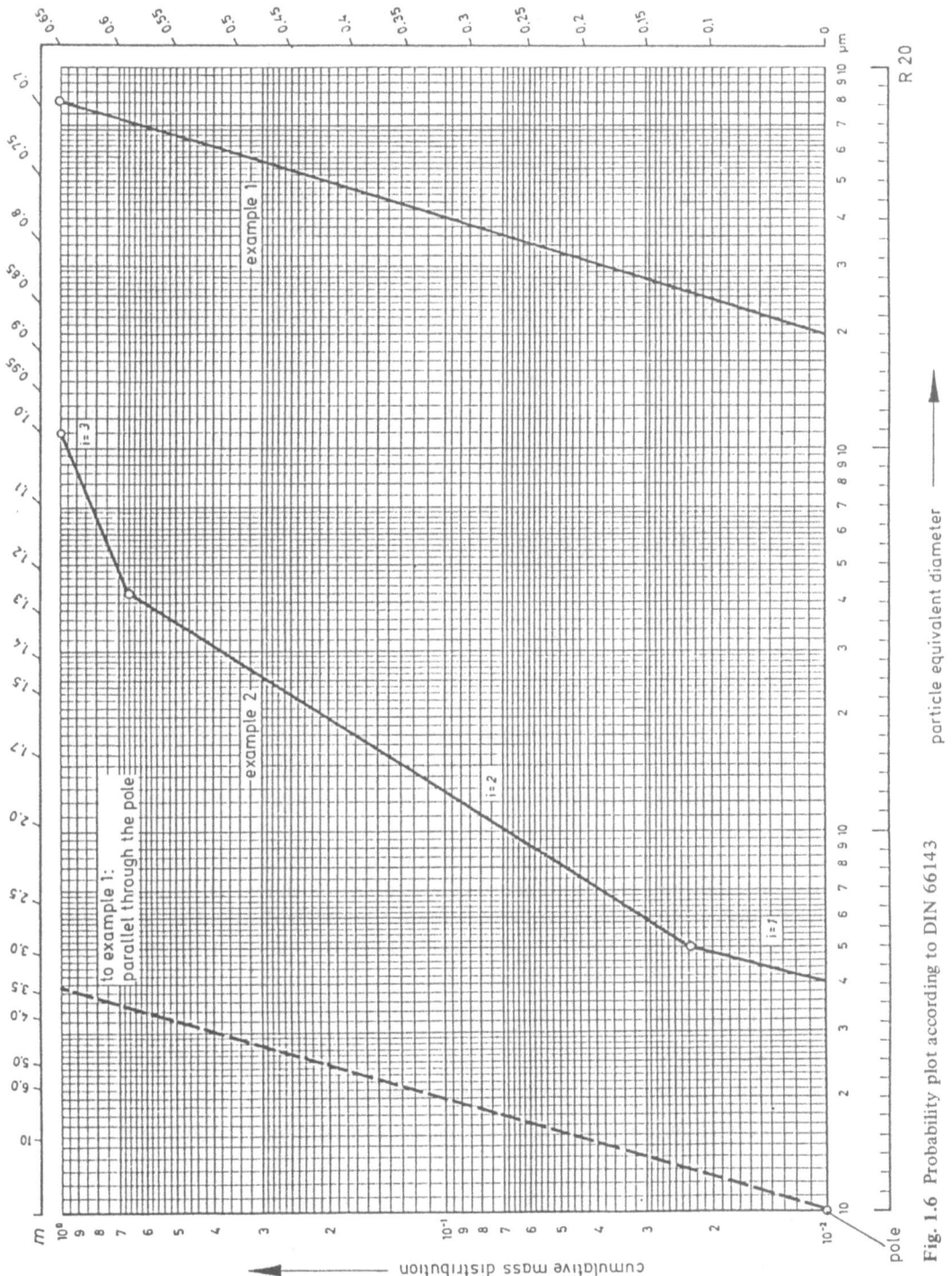

Fig. 1.6 Probability plot according to DIN 66143

b) *RRSB* – (Rosin-Rammler-Sperling-Bennet) *Distributions* (DIN 66145)
The RRSB distribution function has been defined by Bennet as:

$$Q_3(x) = 1 - \exp\left(-\left(\frac{x}{x'}\right)^n\right) \tag{22}$$

x' and n are the distribution parameters, where x' is the particle size – for $Q_3(x') = 0.632 = 63.2\%$, n = gradient of the straight line in the "RRSB-paper" (fig. 1.7).
When plotted on special graph paper with a $\log x$ abscissa and $\log\log 1/(1-Q_3(x))$ ordinate eq. (22) gives a straight line. The graphing technique (DIN 66145) includes two additional scales at the margin to read off the gradient n by displacing the line parallel until it passes through the pole; the specific surface S_v area can be determined also.

c) *The logarithmic normal distribution* (DIN 66144)
This is a universal distribution function that also can be used successfully for particle size distribution. It is applicable to all Q_r distributions. The logarithmic distribution is derived from the Gauss normal distribution as follows:

$$Q_r(x) = \frac{1}{\sqrt{2\pi}} \int_{-\infty}^{t'} \exp\left(-\frac{1}{2}t^2\right) dt \tag{23}$$

by substituting the abscissa:

$$t = \frac{1}{\sigma} \ln(x/x_{50,r}) . \tag{24}$$

In this instance the distribution parameters are the median $x_{50,r}$ and the standard deviation σ of the distribution. (DIN 66144 uses different symbols.)
The logarithmic normal distribution forms a straight line when plotted on probability paper (fig. 1.8).
A particular advantage of the log. normal distribution is the simple conversion of the cumulative distribution, i.e. determining Q_3 from Q_0 or vice versa. It can be shown from theory that a Q_r distribution derived from a normally distributed Q_t population is also normally distributed with the same standard deviation σ, i.e. it is simply displaced parallel. The new median is easily calculated from:

$$x_{50.3} = x_{50.0} \cdot \exp(3\sigma^2) \quad \text{(given } Q_0, \text{ find } Q_3) \tag{25}$$

or

$$x_{50.0} = x_{50.3} \cdot \exp(-3\sigma^2) \quad \text{(given } Q_3, \text{ find } Q_0) . \tag{26}$$

The standard deviation can be read off the scale at the margin or calculated. It is:

$$= \ln \frac{x_{84}}{x_{50}}, \quad \text{viz.} \quad \sigma = \ln \frac{x_{50}}{x_{16}} . \tag{27}$$

In the American literature and in aerosol physics the "geometric standard deviation" is also used.

1.2 Particle Properties and Separation Efficiency

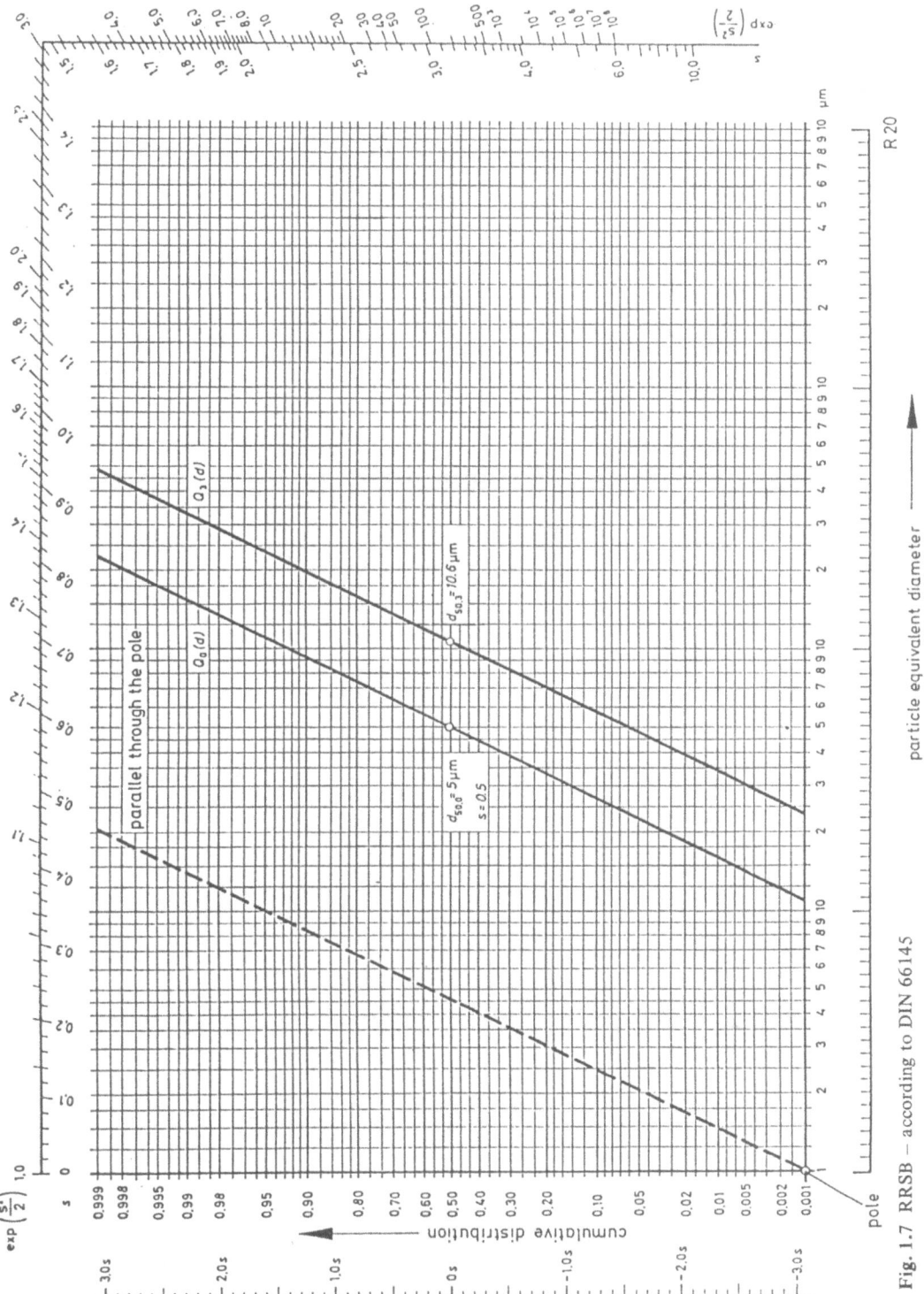

Fig. 1.7 RRSB – according to DIN 66145

16   1 Fundamental Principles of Particle Separation in Fabric Filters

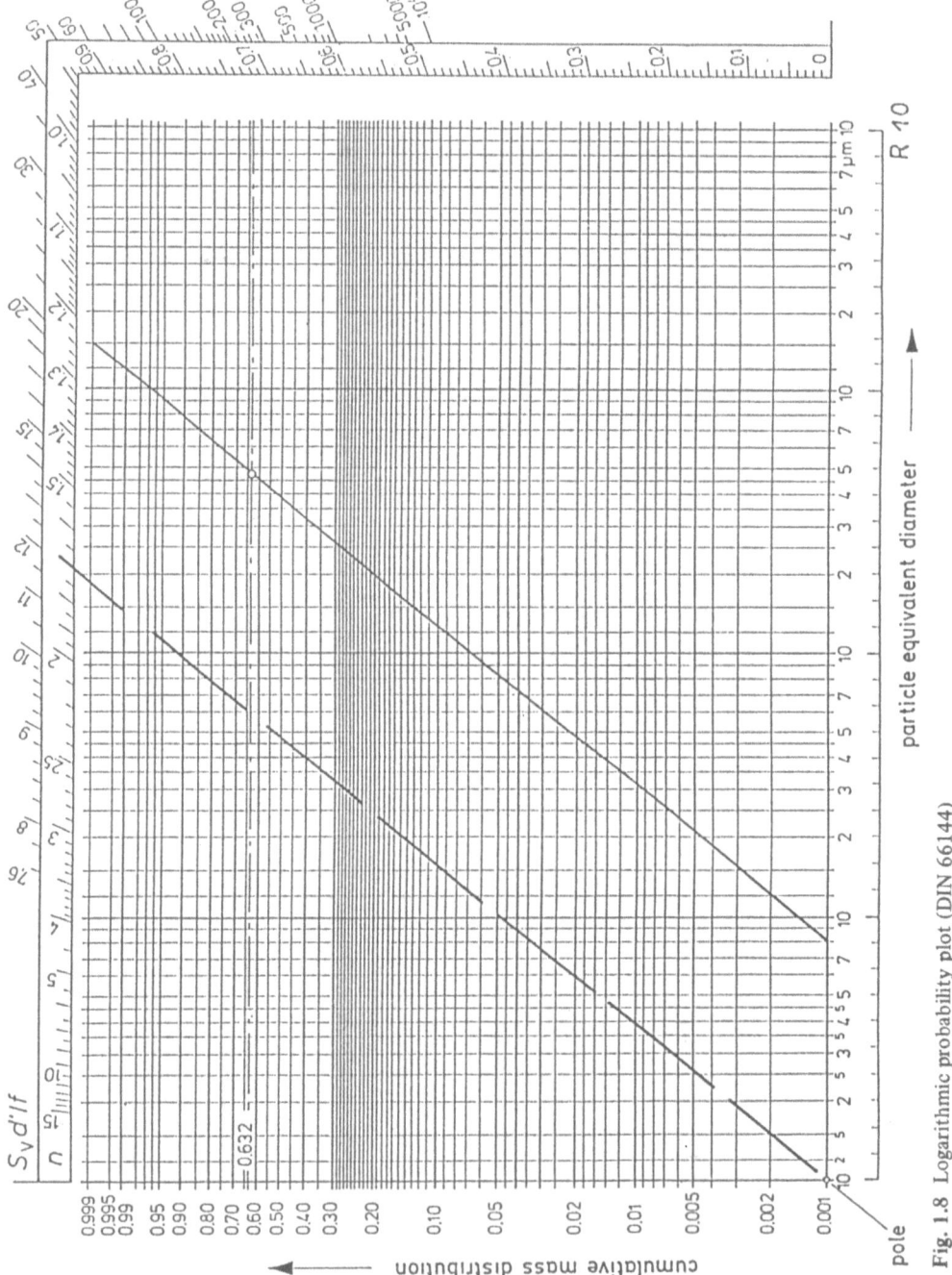

Fig. 1.8 Logarithmic probability plot (DIN 66144)

1.2 Particle Properties and Separation Efficiency

It is defined as —

$$\sigma_g = \frac{x_{84}}{x_{50}} = \sqrt{\frac{x_{84}}{x_{16}}}, \qquad (28)$$

i.e. $\sigma_g = \exp \sigma$ viz. $\sigma = \ln \sigma_g$. This should be remembered when making any comparisons. For $\sigma_g \leqslant 1.15$, viz. $\sigma \leqslant 0.14$ a particle distribution is regarded as "mono disperse" in aerosol language. It should be noted that many distributions cannot be expressed simply by a straight line approximation. In such cases an attempt can be made to a sectional approximation with varying gradients at different parts of the curve. This can be helpful in calculating the specific surface area [4].

1.2.3 Characterization of Dust Collection

For characterization of collection efficiency integral methods (for the total particle assembly) as well as differential methods (for each particle size class) can be used. Definitions as well as the relationships between these methods and conclusions that can be drawn therefrom are set out below.

For fig. 1.9:

$\dot{M}$ = Mass flow rate of particles
$q(x)$ (i.e. $Q(x)$) = Particle size distribution
$\dot{V}$ = Gas volume flow rate
$C = \dot{M}/\dot{V}$ = Dust content

For the coarse particle discharge $\dot{V}_c$ and C_c are placed in brackets, because these variables need only be defined where the part of the gas stream exits with the discharge. This does not normally occur with filters.

A more general description of these relationships as well as those variables that are of significance can be found in DIN 66142 [10]. Only one set of conditions will be treated below, where the dust component in the gas stream is separated into two components. The technical terms for this set of conditions have become part of the technical language commonly used in the dust collection industry.

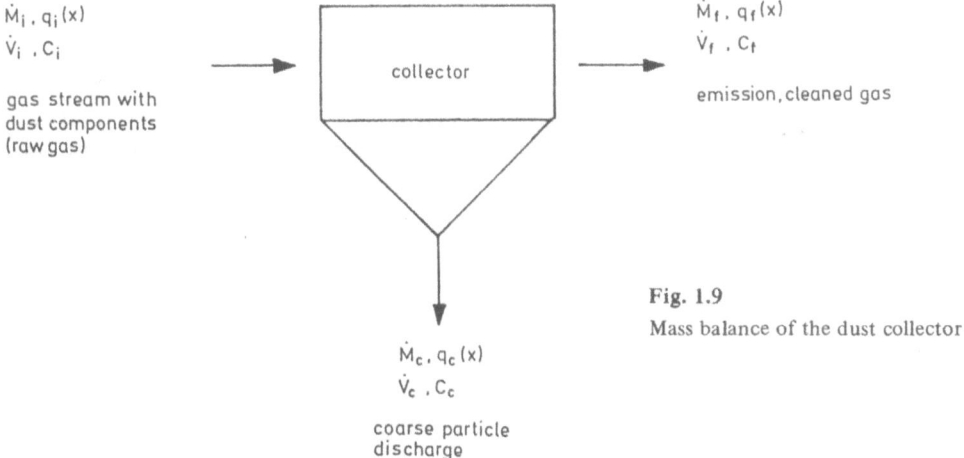

Fig. 1.9
Mass balance of the dust collector

1.2.3.1 Total Collection Efficiency

The integral mass balance is —

$$\dot{M}_i = \dot{M}_c + \dot{M}_f \tag{29}$$

Total collection efficiency ϕ is defined as —

$$\phi = \frac{\dot{M}_c}{\dot{M}_i} \tag{30}$$

viz. when combined with eq. (29) —

$$\phi = 1 - \frac{\dot{M}_f}{\dot{M}_i} . \tag{31}$$

Penetration P is defined as —

$$P = \frac{\dot{M}_f}{\dot{M}_i}$$

which when combined with eq. (33) becomes:

$$P = 1 - \phi . \tag{32}$$

The mass flow rates in the raw gas and in the cleaned gas can also be expressed by the product of gas flow rate and dust content: The following holds —

$$\phi = \frac{\dot{V}_i C_i - \dot{V}_f C_f}{\dot{V}_i C_i} \tag{33}$$

By assuming that $\dot{V}_i = \dot{V}_f$ eg. (33) becomes:

$$\phi = 1 - \frac{C_f}{C_i} \tag{34}$$

viz.

$$P = \frac{C_f}{C_i} . \tag{35}$$

Dust collectors frequently reach collection efficiencies above 99%. The significant difference in collection efficiency between filters of 99.9% and 99.99% efficiency, respectively, becomes more apparent when the penetrations are compared (ref. table 1.2). The same applies to an even greater extent to the decontamination factor DF that is used in some fields of dust control:

$$DF = \frac{1}{P} = \frac{\dot{M}_i}{\dot{M}_f} , \tag{36a}$$

by combining with eq. (35) —

$$DF = \frac{C_i}{C_f} . \tag{36b}$$

1.2 Particle Properties and Separation Efficiency

Table 1.2
Comparison between separation efficiency ϕ, discharge ratio P, and decontamination factor DF

ϕ	0.99	0.999	0.9999
P	0.01	0.001	0.0001
DF	100	1.000	10.000

Eq. (34) and (35) represent a model by which collection efficiency or penetration can be measured. The dust content in the cleaned and raw gas is measured gravimetrically from which ϕ, or P, respectively, is calculated. Similar principles apply when determining number concentrations C_{fo} and C_{io}. The values so obtained − (filter collection) efficiency differs of course from those obtained gravimetrically.

It should also be noted that total collection efficiency depends not only on the operating conditions of the dust collector, but also on the type of dust, and especially on the particle size distribution $Q_i(x)$. Quoting a total collection efficiency relates to a particular case only; this limits severely its applicability to a different set of circumstances.

1.2.3.2 Separation Function, Grade Efficiency

The effectiveness of dust separation depends on particle size. This becomes more significant for depth filters using a fibre medium. That separation efficiency is a function of particle size is also observed in cleanable filters − at least during some phases of the cyclic operation. For this reason it makes more sense − as will be shown − to define and determine collection efficiency in terms of its dependence on particle size. This leads up to grade efficiency (ref. DIN 66142); often termed "fractional efficiency".

A mass balance using eq. (29) must hold for particles of any size. Since q(x) dx is that portion in the interval x and x + dx, for each particle size the following must hold:

$$\dot{M}_i \, q_i(x) \, dx = \dot{M}_c \, q_c(x) \, dx + \dot{M}_f \, q_f(x) \, dx \tag{37}$$

by combining with eq. (30) and (31), it follows −

$$q_i(x) = \phi \, q_c(x) + P \, q_f(x) . \tag{38}$$

When eq. (38) is integrated between the limiting values x_{min} to x_{max}, the following is obtained −

$$\int_{x_{min}}^{x_{max}} q_i(x) \, dx = \phi \int_{x_{min}}^{x_{max}} q_c(x) \, dx + P \int_{x_{min}}^{x_{max}} q_f(x) \, dx . \tag{39}$$

Normalising conditions (eq. (11)) requires the integral sum across the entire density distribution to be equal to one, so that eq. (39) becomes −

$$1 = \phi + P . \tag{40}$$

This means that the mass balance, as required by eq. (29) is fulfilled. It is useful, therefore, when making a graphic representation of a separation process, to plot the density distributions $\phi \cdot q_c(x)$ and $P q_f(x)$. It is sufficient then, if two of the three possible curves q(x) and one mass ratio ϕ or P is known; the remaining third curve can be determined by a point by point plot using eq. (38). In such a presentation, as exemplified in fig. 1.10, the area under each of the curves (linear plotting) expresses the total separation

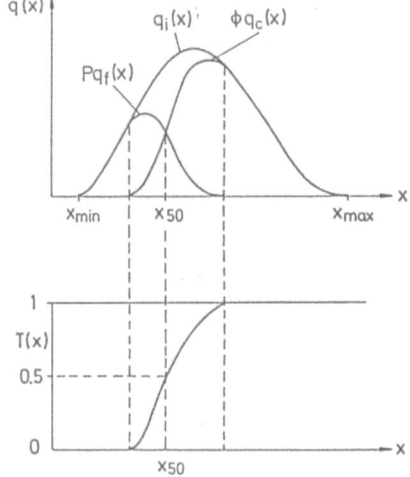

Fig. 1.10
Representation of a separation by means of distribution and separation functions

efficiency ϕ, viz. the penetration P. If eq. (38) is integrated from x_{min} to any particular particle size x, it follows that:

$$Q_i(x) = \phi Q_c(x) + PQ_f(x) . \tag{41}$$

Thus, what applies to sums of distribution applies equally to density distributions. The separation function T(x) gives the ratio, for each particle size x, of that quantity that is separated to the quantity of particles of the same size x in the inlet. Thus, by definition –

$$T(x) = \frac{d \dot{M}_c(x)}{d \dot{M}_i(x)} = \frac{\dot{M}_c q_c(x) dx}{\dot{M}_i q_i(x) dx} \tag{42}$$

so that

$$T(x) = \frac{\phi \cdot q_c(x)}{q_i(x)} . \tag{43}$$

T(x) normally covers the range of values from zero (no particles collected) to unity (all particles separated):

$$0 \leqslant T(x) \leqslant 1 .$$

The separation function is independent of the measure of the quantities (number or mass). With the help of eq. (43) the separation curve can be calculated, or it may be derived graphically from the distribution curve. By utilising eq. (38), eq. (43) can be rewritten as: —

$$T(x) = 1 - \frac{P q_f(x)}{q_i(x)} \tag{44}$$

or

$$T(x) = \frac{\phi q_c(x)}{\phi q_c(x) + P q_f(x)} \tag{45}$$

1.2 Particle Properties and Separation Efficiency

Equations (43), (44) and (45) suggest procedures for determining $T(x)$. In some instances it becomes important which of these equations is used for measurement and calculation; e.g. eq. (44) will give negative results, if fragmentation occurs within the separating apparatus, (e.g. atomising of droplets). Were eq. (45) to be used, this effect would not become evident, since only the material leaving the apparatus is examined.

Fig. 1.10 is an example in point; there are separators where $T(x)$ experiences a minimum and increases again as x decreases. Such is the case with very efficient fibre filters or electrostatic precipitators.

Practical Determination of $T(x)$

Equations (43) to (45) require that at least two of the three possible particle size populations, plus one mass ratio, must be known to calculate $T(x)$. A relatively simple method to determine $T(x)$ is to measure the distribution in the raw gas and cleaned gas with a light scattering particle analyser. The calculation is performed with eq. (44) as follows:
Since

$$q(\bar{x}_j) = \frac{\Delta Q(x_j)}{\Delta x_j}.$$

For a uniform interval width $\Delta x_{jf} = \Delta x_{ji}$

$$T(x) = 1 - \frac{P q_f(x)}{q_i(x)} = 1 - \frac{P \Delta Q_f(x_j)/t_f}{\Delta Q_i(x_j)/t_i}$$

where t_i and t_f is the recording time on the raw gas and cleaned gas side, respectively.

$$\Delta Q(x_j) = \Delta Q_0(x_j) = \frac{\Delta Z(x_j)}{Z_{tot}}$$

if $\Delta Z(x_j)$ is the number of particles in the jth interval and Z_{tot} is the total number of particles counted.

Now, $P = Z_{f\,tot}/Z_{i\,tot}$ so that

$$T(\bar{x}_j) = 1 - \frac{\Delta Z_f(x_j)\, t_i}{\Delta Z_i(x_j)\, t_f} \qquad (46)$$

for equal recording times, $t_i = t_f$

$$T(\bar{x}_j) = 1 - \frac{\Delta Z_f(x_j)}{\Delta Z_i(x_j)}. \qquad (47)$$

It is also possible to determine $T(x)$ from a particle count distribution curve q_i and q_f with the aid of membrane filter sampling using eq. (46) and (47), where the counting area is equal for both sets. Similarly, these equations can be used for the analysis of measurements from cascade impactors, whereby ΔZ is not the particle count, but the weighed fraction at each separation stage.

1.2.3.3 Calculating Total Collection Efficiency

The total or overal collection efficiency can be calculated given the raw gas particle size distribution and a known separation curve. From the definition of separation function it follows, that the fraction of each particle size (x) that is separated is: —

$$T(x) \cdot q_i(x) \cdot dx = T(\bar{x}_j)\, \Delta Q_i(x_j).$$

By summing across the entire particle population:

$$\phi = \int_{x_{min}}^{x_{max}} T(x) \, q_i(x) \, dx = \sum_{x_{min}}^{x_{max}} T(\bar{x}_j) \, \Delta Q_i(x_j) \, . \tag{48}$$

Also, the distribution of particles emitted into the cleaned gas $q_f(x)$ can be calculated from

$$q_i(x) = \phi \, q_c(x) + P \, q_f(x)$$
$$\phi + P = 1$$
$$\phi \, q_c(x) = T(x) \, q_i(x)$$

so that

$$q_f(x) = \frac{q_i(x) \, [1 - T(x)]}{1 - \int_{x_{min}}^{x_{max}} T(x) \, q_i(x) \, dx} \tag{49a}$$

$$q_f(\bar{x}_j) = \frac{\frac{\Delta Q_i(x_j)}{\Delta x_j} \, [1 - T(\bar{x}_j)]}{1 - \sum_{x_{min}}^{x_{max}} T(\bar{x}_j) \cdot \Delta Q_i(x_j)} \, . \tag{49b}$$

Numeric example for calculating total collection efficiency ϕ from a known flue gas particle size population $Q_i(x)$ and known grade efficiency $T(x)$.

$x_j/\mu m$	$\bar{x}_j/\mu m$	$Q_i(x_j)$	$\Delta Q_i(x_j)$	$T(\bar{x}_j)$	$T(x_j) \cdot \Delta Q_i(x_j)$
0		0			
	1		0.03	0.02	0.0006
2		0.03			
	3		0.29	0.20	0.058
4		0.32			
	5		0.32	0.72	0.231
6		0.64			
	7		0.19	0.90	0.171
8		0.83			
	9		0.09	0.95	0.085
10		0.92			
	11		0.04	0.98	0.039
12		0.96			
	13		0.02	0.99	0.02
14		0.98			
	16		0.015	1.0	0.015
18		0.995			
	20		0.005	1.0	0.005
22		1.0			

$\phi = 0.624$

1.3 Separation in Fibre Assemblies

Current theories of separation deal primarily with those processes within deep bed filters that are responsible for separation in the interior of the filter medium. They explain the interaction between particle and single fibre on a micro scale. Theoretical treatment together with detailed experimental investigations have done much to clarify the underlying physical processes. They have brought to light the significance of various factors; the most recent published works show good quantitative agreement between theoretical prediction and measurement in some areas. Such concordance was not possible in general, because of the many approximations used in the theoretical calculations, viz. the complex structure of real fibre surfaces and the consequent unpredictability of the interaction between fibre and air flow. These have hitherto not been elucidated sufficiently.

A similar approach can be transfered to cleanable filters for the initial collection phase when separation is initiated within the interior of the fibre assembly. This condition is peculiar to neddlefelts and similar media, in their virgin state; it applies also, with some exceptions, to the beginning of each filtering cycle after cleaning. For this reason a review is presented of the theoretical basis for particle separation onto fibres in an air stream. As will be shown subsequently, other phenomena become important during the sequence of operation of a cleaned fitration system.

1.3.1 Collection Efficiency in Fibre Assemblies

The grade efficiency of a fibre assembly is derived from a mass balance. It is assumed in this case that the air stream is perpendicular to the fibre axis, and that the particles are uniformly distributed within the available space. Also, that particles are uniformly dispersed in the gas phase and across the direction of flow. Thus, the concentration of particles of size x in the flow direction is (fig. 1.11) —

$$-\frac{dc(x)}{c(x)} = \frac{4}{\pi} \left(\frac{1-\epsilon}{\epsilon}\right) \frac{1}{D_F} \varphi(x) \, dz \, . \tag{50}$$

ϵ is the porosity of the fibre bed, $\varphi(x)$ is the single fibre collection efficiency and expresses the separation process within a single fibre element within the layer. Integrating eq. (50) yields:

$$c_f(x) = c_i(x) \exp\left[-\frac{4}{\pi} \left(\frac{1-\epsilon}{\epsilon}\right) \frac{Z}{D_F} \varphi(x)\right] . \tag{51}$$

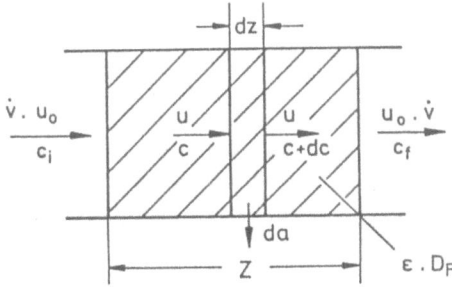

Fig. 1.11
Mass balance of a deep bed filter

Thus, the particle concentration across the fibre layer decreases exponentially, so that the collection efficiency becomes —

$$T(x) = 1 - \exp[-f\varphi(x)] . \qquad (52)$$

f characterizes the filter geometry and can be determined for monodisperse cylindrical fibres as follows:

$$f = \frac{4}{\pi} \left(\frac{1-\epsilon}{\epsilon}\right) \frac{Z}{D_F} \qquad (53a)$$

or

$$f = \frac{4}{\pi} \frac{1}{\epsilon} \frac{W}{D_F \cdot \rho_F} \qquad (53b)$$

where W is the mass per unit area, f is the ratio of the projected fibre cross section (perpendicular to the flow direction) to the filter surface area. For coarse filters typical values for f are 5 to 10, while 100 to 300 are more typical for high efficiency filters.

A theoretical determination of the separation function $T(x)$ requires a value for the single fibre collection efficiency $\varphi(x)$. Possible methods of determining this are set out below. Davies [11] expands on the various theories of separation.

1.3.2 Collection Efficiency of a Single Fibre

What enables fibres to trap particles? The mean spacings between fibres are significantly larger than the particles to be attracted. The sieving effect, i.e. mechanical entrainment makes little contribution to depth filtration within fibre assemblies (fig. 1.12).

To make particle separation possible —
1. they must be transported to the fibre surface.
2. they must be retained there.

Thus the separation efficiency φ can be described by —

$$\varphi = \eta h . \qquad (54)$$

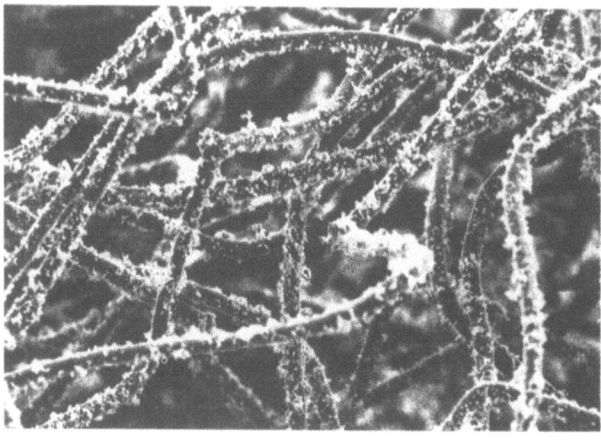

Fig. 1.12

Deep bed filter $D_f \approx 40 \ \mu m$ with deposited quartz particles

(Scanning electron microscope, Institute for MVM, University Karlsruhe)

1.3 Separation in Fibre Assemblies

η is the collision efficiency and expresses the transport mechanism, while h is that portion that is retained on the fibre from the impinging particles. The objective now is to describe η and h in terms of their function on the various limiting values.

1.3.2.1 Transport Mechanisms (fig. 1.13)

Particles can reach the fibre surface by:

a) Stochastic motion of particles around their mean trajectory. This effect arises from thermal motion of the gas molecules (Brownian motion). This can be calculated with the aid of Fick's diffusion equation.

b) Deterministic motion of particles towards fibre surfaces by inertial, gravitational or electrostatic forces. These effects can be expressed by a general equation of motion.

The collision rate is defined as the quantity of particles per unit length of fibre striking the fibre surface to the total quantity of particles passing through the projected fibre cross section.

Fuchs and co-workers [13] made a close study, from both a mathematical and empirical viewpoint, of the diffusion effect in the region of very small Reynold's numbers (Re ≤ 1). Lee and Liu [14] recently published new investigations in this field; they have calculated the collision rate due to diffusion using the flow field equations as developed by Kuwabara: —

$$\eta_D = 2.6 \left(\frac{1-\alpha}{k}\right)^{1/3} \text{Pe}^{-2/3} \tag{55}$$

by using a hydrodynamic factor k:

$$k = -\frac{1}{2} \ln \alpha - 0.75 + \alpha - 0.25 \alpha^2 \tag{56a}$$

where α is the fibre volume fraction, $\alpha = 1 - \epsilon$

$$P_e = \frac{U_0 D_F}{D} \quad \text{Peclet No.} \tag{56b}$$

$$D = \frac{k^* T Cu}{3 \pi \mu x} \quad \text{Diffusion coefficient (Diffusivity of particles)} \tag{56c}$$

U_0 = face velocity
μ = gas viscosity
T = absolute temperature
k^* = Boltzmann constant
Cu = Cunningham correction factor

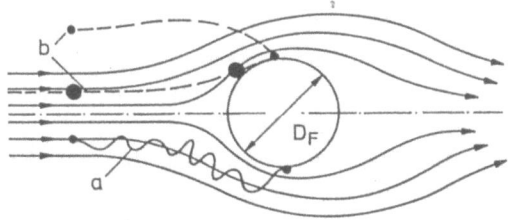

Fig. 1.13
Transport mechanisms for particle deposition on fibres

1 Fundamental Principles of Particle Separation in Fabric Filters

Table 1.3 Diffusion coefficients and Cunningham correction factors

x/μm	0.01	0.1	0.5	1	2	5
$D \cdot 10^{12} \frac{m^2}{s}$	52600	684	62.6	27.4	12.8	4.8
Cu (eq. (7))	22.31	2.899	1.328	1.162	1.082	1.032

Table 1.3 sets out some values for the diffusion coefficient and Cunningham correction factor as determined by Millikan [27].

Since the diffusion coefficient $D \sim \frac{1}{x}$, $\eta_D \sim (U_0 \cdot D_F \cdot x)^{-2/3}$, i.e. the collision rate decreases with increasing velocity and/or increasing particle size.

Separation by diffusion within a technically useful range occurs with particles of less than 0.1 μm to 0.5 μm, at velocities below 10 cm/s; this is within the range of high efficiency filters. Inertial forces do not come into play until these limits are exceeded. If electrostatic effects are disregarded, the separation function transverses a minimum between diffusion determined and inertia forces determined separation. Particle interception appears to play some part here. Particles that are carried in the gas stream without deflection are trapped by their surfaces contacting that of the fibre. In this case there is no special transport mechanism towards the fibre, it is simply a case of interception.

For calculation purposes that gas stream is considered only which moves past the fibre at a distance of up to one particle radius from the fibre surface. This results in an collision efficiency due to interception [14] for Re < 1: –

$$\eta_R = \left(\frac{1-\alpha}{k}\right) \frac{R^2}{1+R}, \tag{57}$$

where $R = x/D_F$, and k is a hydrodynamic component derived from eq. (56) η_R increases with increasing R. The effect of particle dimensions during collection will be mentioned again in connection with force effects.

Lee and Liu have estimated the contribution of diffusion and interception by postulating – incorrect in the strict sense – an additive superimposition of eqs. (55) and (57), since one effect only is operative at any one time. By using correction factors the authors obtained good experimental verification of their theoretical predictions.

$$\eta_{DR} = 1.6 \left(\frac{1-\alpha}{k}\right)^{1/3} \cdot Pe^{-2/3} + 0.6 \left(\frac{1-\alpha}{k}\right) \frac{R^2}{1+R}. \tag{58}$$

Fig. 1.14 compares calculated values (eq. (58)) with measured results. The curve shows a characteristic shape with a minimum in the range 0.4 to 0.7 μm.

To arrive at a formula for the deterministic particle transport mechanism towards the fibre surface by different forces a model was developed where the fibre is represented by a circular cylinder perpendicular to the direction of gas flow (fig. 1.15).

The theoretical problem is to calculate the particle trajectory and to find that path that allows particle and fibre surface to just touch. These are termed the limiting particle

1.3 Separation in Fibre Assemblies

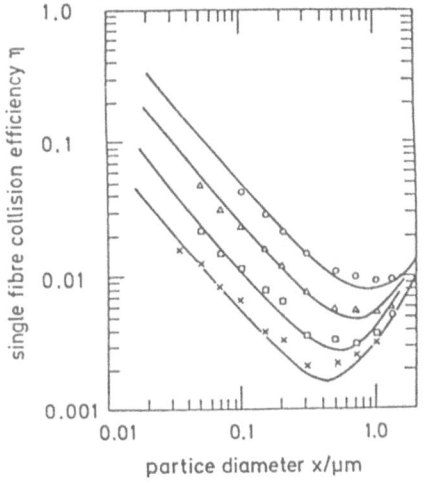

Fig. 1.14 Single fibre collision efficiency [14]

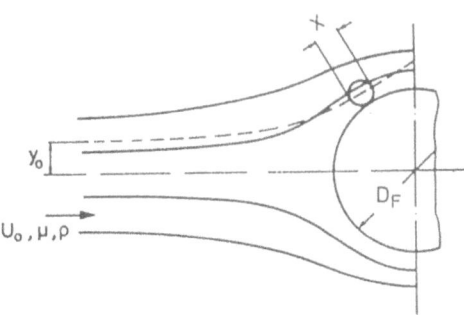

Fig. 1.15 Model for calculating deposition on a fibre

trajectories, that are y_0 distant from the axis of symmetry. Collision efficiency is then defined as: —

$$\eta = \frac{2 y_0}{D_F} .\tag{59}$$

It is therefore assumed that all arriving particles of similar size within y_0 strike the fibre, while those further distant move on.

Particle paths are obtained from the equation of motion assuming: Sum of all forces = zero. *All* acting forces must be included in this expression; to generalize —

$$F_i + F_d + F_s + F_{el} = 0$$

F_i = inertia forces
F_d = drag forces
F_s = gravitational forces
F_{el} = electrostatic forces

The collision efficiency obtained in this manner is denoted as: η_{TES} (T = inertia, E = electrostatic, S = gravity). A dimensional analysis shows that collision efficiency η_{TES}, in the most general case, is derived from five parameters:

$$\eta_{TES} = f(\psi, Re, S, Fr_p, N_{el}) \tag{60}$$

where

$$\psi = \frac{\rho_p \, x^2 \, U_0}{18 \, \mu \, D_F}$$ Inertia parameter, describes particle dynamic.

$$Re = \frac{U_0 \, D_F \, \rho}{\mu}$$ Reynolds No., expresses the fluid dynamic.

$$S = \frac{\rho_p}{\rho}$$ Density ratio.

$$Fr_p = \frac{U_0^2}{xg} \cdot \frac{\rho}{\rho_p}$$ Particle Froude No., takes gravity into account.

$$N_{el} = \frac{F_e(r^*)}{3 \, \pi \, \mu \, x \, U_0}$$ Charge parameter, takes electrostatic forces into account.

$F_e(r^*)$ signifies an electrostatic force with distance r^* from the centre of the fibre. Several modifications of the electrostatic interactions are possible depending on which element carries the charge and whether external electrical fields polarise fibre or particle. Such situations are very complex and not yet sufficiently defined. Important fundamental considerations have been documented by Zebel [16].

A comparative study showed [17] that of all the variables, the Coulomb effect, i.e. charged fibres and particles, is the most significant. Here experimental evidence supports theoretical predictions. The charge parameter for Coulomb forces is:

$$N_{el} = N_{Qq} = \frac{Q \cdot q}{3 \, \pi^2 \, \epsilon_0 \, \mu \, x \, D_F \, U_0} \tag{61}$$

Q = charge on fibre per unit length
q = particle charge
ϵ_0 = Permittivity = $8.859 \cdot 10^{-12}$ As/Vm.

Equations of motion can be analytically solved for simplified conditions only. For a more general condition, the equations have to be solved numerically. This limits their immediate application to practical situations. One approach is to develop approximate solutions for typical sets of parameters which are easy to evaluate. However, this is at the price of generality and precision of such solutions.

A further problem is that for numerical solution, it is necessary to know the flow field of the air stream around the fibre. Finite solutions have so far only been developed for the region Re < 1 (Lamb, Kuwabara) and for Re > 100 (potential flow). For technically significant levels 0.1 < Re < 50 only few numerical solutions for discrete Reynolds No are available (e.g. Suneja and Lee). Furthermore, the calculated flow fields are valid only for a flow vertical to the cylindrical fibres. The real life situation in technical filters is far more complex and cannot be calculated precisely. Thus the deviation from this idealized model for single fibre separation must be obtained empirically by very carefully designed experiments. Care must be taken to isolate the effects of filter structure from other influences that are specific to the material. It should also be noted that in any experimental studies the collision efficiency is not measured directly, but the collection efficiency, i.e. collision efficiency times adhesion efficiency on the fibre.

1.3 Separation in Fibre Assemblies

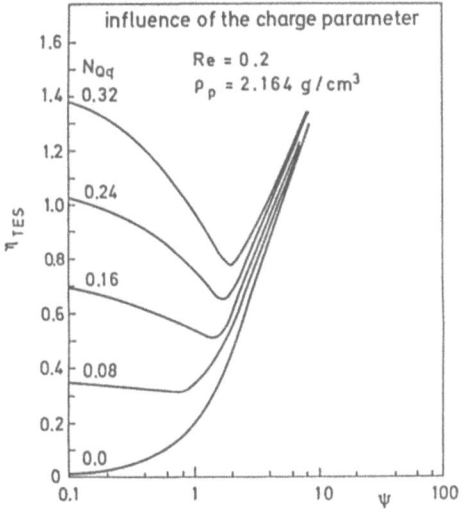

Fig. 1.16
Theoretical collision efficiencies η_{TES} (Muhr [18])

To illustrate the influence of Coulomb forces, fig. 1.16 shows the relationship between calculated collision efficiencies η_{TES} and the inertia parameter ψ for different charge parameters N_{Qq} (Muhr [18]).

The curve $N_{Qq} = 0$ is for the condition of zero charge, i.e. when inertia and gravity solely determine transport. As ψ increases, e.g. as particle size increases at constant velocity, η increases steadily. The shape of this curve is typical for calculated solutions and valid for all Reynolds No. By definition it is possible to have η greater than 1. The limiting condition for pure inertial effects on particles moving in a straight line is reached when $\eta_T = 1 + x/D_F$.

For $N_{Qq} > 0$, collision efficiencies rise again with increasing ψ below the minimum at $\psi \approx 1$. Above $\psi \approx 1$ these curves asymptotically approach the limiting condition of $N_{Qq} = 0$. One may conclude from this that electrostatic forces become significant in the region $\psi < 1$, i.e. in fig. 1.16 at the conditions below about 3 μm and below $\simeq 25$ cm/s. Coulomb forces dominate in this region.

The shape of these curves further makes it evident that all acting forces must simultaneously be included in an equation of motion. If partial solutions for all the acting forces are superimposed, as is often reported in the literature, although a simple approach, it leads to quite erroneous conclusions. Evidently, both inertia and gravity effects counteract electrostatic forces in the region of this minimum. By disregarding inertia effects, the curves would become horizontal, independent of ψ.

Muhr was able to substantiate these theoretical predictions in experiments with filter models of parallel fibres and with particles in the region of 1 to 10 μm.

Similarly, Bergmann [19] was able to show with his investigations that electrostatic forces could significantly improve separation in the transition zone between diffusion and inertia effects.

Approximations for the collision efficiency η

Listed below are simplified equations to generate numeric solutions for single fibre collision efficiencies – in many instances these have been verified experimentally. The grouping is according to Reynolds No:

1) Re < 1

According to Hiller [20] the collision efficiency due to inertia η_T can be estimated as follows:

$$\eta_T = 1.03 + (0.6 \, \text{Re} - 1.5) \, 0.85^{(\psi + 0.5)} \, . \tag{62a}$$

Valid for the Range $\quad 1 < \psi < 10$
$\qquad\qquad\qquad\qquad 0.01 < \text{Re} < 1$
$\qquad\qquad\qquad\qquad \eta_T > 0.1$
Maximum error $\quad \pm 0.05$

Gravitational effects have been neglected, which can be done without significant error for particle Froude Nos. above 0.1, i.e. particle size below $10 \, \mu m$, at velocities above 15 cm/s.

For precise calculations the gravity effects for a vertically downwards flow can be read off the diagrams given in [20].

Muhr [18] took account of the electrostatic forces by taking the difference of the equation with and without electrostatic effects, i.e. $\eta_E = \eta_{TES} - \eta_{TS}$

$$\eta_E = \frac{1.22 \, (2 - \ln \text{Re})}{1 + \left(\frac{\psi}{2 \, \text{Re}}\right)^{1.5}} N_{Qq} \, ; \quad \text{(for } N_{Qq} \text{ ref. to eq. (61))} \tag{62b}$$

Equation (62b) thus contains a component to take account of the inertial forces on the electrostatic influence; this can be seen in fig. 1.16. η_E increases as ψ and Re decrease. For $\psi \to 0$, η_E approaches a limiting value:

$$\eta_E = 1.22 \, (2 - \ln \text{Re}) \, N_{Qq} \, . \tag{62c}$$

2) 0.5 < Re < 50

For this technically significant region the measurements obtained by Suneja and Lee [21] can be approximated using a modified form developed by Muhr:

$$\eta_{TS} = \{[1 + (1.53 - 0.23 \ln \text{Re} + 0.0167 \ln^2 \text{Re})/(2 \cdot \psi)]^{-2}$$

$$+ \left(\frac{2\rho}{\text{Re} \, \rho_p \, \psi}\right)^{1/2}\} \left[1 + 3 \left(\frac{2\rho \psi}{\text{Re} \, \rho_p}\right)^{1/2}\right] \tag{62d}$$

Muhr confirmed this equation in his experiments using particles with good adhesive properties.

3) 50 < Re < 500

For this range it is useful to divide η_T in to two parts [22]:

$$\eta_T = \eta_1 + \eta_2 \tag{62e}$$

1.3 Separation in Fibre Assemblies

where

$$\eta_1 = \frac{\psi^3}{\psi^3 + f_1(Re)\,\psi^2 + f_2(Re)\,\psi + f_3(Re)} \tag{62f}$$

$$\eta_2 = \frac{x}{D_F} = 3\left(\frac{2\rho\psi}{Re\,\rho_p}\right)^{1/2} \tag{62g}$$

with

$f_1(Re) = -0.0133 \ln Re + 0.931$
$f_2(Re) = +0.0353 \ln Re - 0.36$
$f_3(Re) = -0.0537 \ln Re + 0.398$
validity: $1\,g/cm^3 \leq \rho_p \leq 6\,g/cm^3$, $\quad \eta_T \geq 0.1$
maximum error: ± 0.01

Caution should be excercised when making any comparisons using collision efficiencies derived from eq. (62), since only the transport mechanisms are considered. Nothing has been said so far about adhesion of particles to the fibre surface.

1.3.2.2 Adhesion Efficiency

When the particles reach the fibre they must adhere to the surface in order to be collected. The particles must neither bounce on impact, nor be released after deposition. To clarify the conditions for particle release numerous measurements of the adhesion forces have been made using centrifugal techniques, and the velocity of blow-off as a measure [23]. In order to carry out these measurements particles were deposited on the fibre surface by filtration techniques.

The centrifugal technique yielded a wide distribution of adhesion forces for mono sized particles. This can be represented well on a logarithmic normal distribution plot. The variance for these distributions is almost constant for differing experimental conditions. Thus the median F_{50} can be used to compare the variances for adhesion values.

Fig. 1.17 shows median values for quartz particles on polyamide fibres. These are plotted against the face velocity U_0 for various particle sizes, and at 50% relative humidity. The adhesion force increases with increasing particle size. Notable is the large effect of face velocity, particularly with particles above 5 μm. Presumably this is due to selective deposition. This is based on an observation that for particles $> 5\,\mu$m the number deposited fell as the face velocity increased. Evidently there is a selective impact depending on the surface adhesion conditions. As the face velocity increases weakly bonded particles become detached rapidly. The number of bonded particles decreases, while F_{50} increases.

When attempting blow-off it was found that larger particles are detached at lower velocities despite higher bonding forces. Though this occurred only at fairly high blow-off velocities. Incipient detachment occurred at above 1 m/s; to blow off around 50% of deposited particles air speeds had to be raised to 10 to 20 m/s. This is illustrated in fig. 1.18 where the velocity to remove 50% of particles U_0 is plotted against particle size. Different curves show the filtration velocity U_0 at which particles were first deposited on to the fibres.

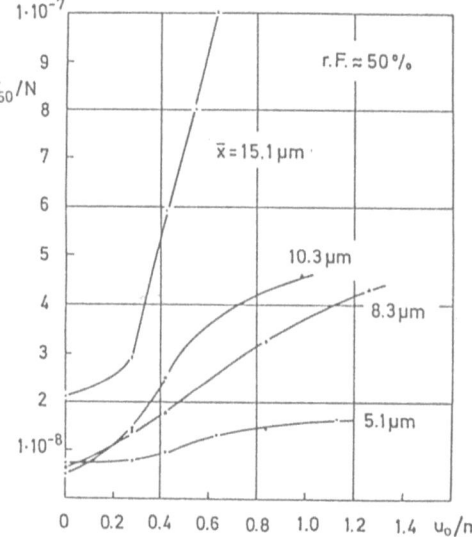

Fig. 1.17

Mean values of adhesion force distributions for quartz particles on polyamide fibres

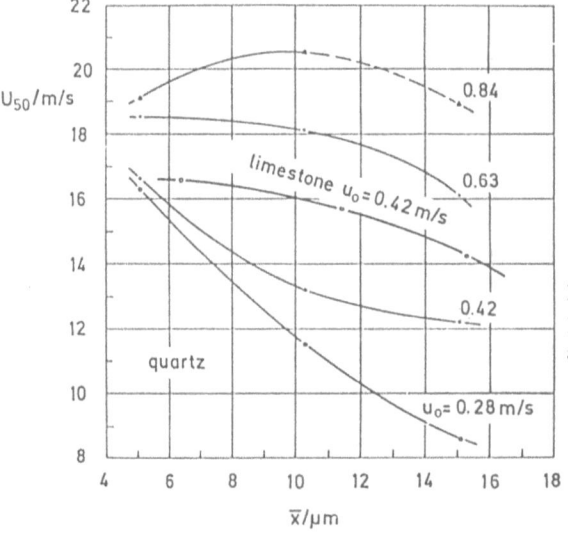

Fig. 1.18

Blow-off velocity U50 for quartz and limestone particles

It was possible to remove larger particles at lower air velocities, presumably because they project further into the air stream surrounding the fibre, thus creating higher separation forces. The higher the filtration velocity, the higher also the required blow-off velocity. Furthermore, a size fractionation becomes evident that is also found in centrifugal measurements.

These measurements of adhesion forces and blow-off velocities have shown that the probability of particles becoming detached once deposited is rather remote within the

1.3 Separation in Fibre Assemblies

filtration conditions found in practice. The fate is sealed at the time of initial particle/fibre contact.

Direct optical observation has confirmed that particles may bounce from fibre surfaces after impact. The criteria for particle retention is that the adhesion energy is greater than the elastically stored deformation energy on impact. Based on this assumption Hiller [20] developed an equation for the critical velocity v_c, above which the probability of particle bounce increases. The equation assumes that adhesion is due to van der Waal interaction.

$$v_c = \frac{1}{x} \frac{(1-k_{pl}^2)^{1/2}}{k_{pl}^2} \cdot \frac{A}{\pi z_0^2 (6 \, p_{pl} \, \rho_p)^{1/2}} \cdot \qquad (63)$$

Where

A = Hamaker constant ($\approx 10^{-19} - 10^{-18}$ J)
z_0 = minimum spacing ($\approx 4 \cdot 10^{-10}$ m)
p_{pl} = microscopic yield pressure ($\approx 5 \cdot 10^8 - 5 \cdot 10^9$ Pa)
k_{pl} = the "plastic" coefficient of restitution, i.e. that part of the impact energy lost through irreversible deformation at the point of contact.

The three parameters A, p_{pl}, k_{pl} are specific to the material under consideration and connot be predicted theoretically. They must be empirically determined. To date very few determinations for the impact index k_{pl} have been made.

The results collected so far can be expressed thus — that for glass spheres k_{pl} lies between 0.6 and 0.9; for quartz and limestone values lie scattered in the range 0.4 to 0.6. Surface contamination and roughness have a significant effect. It would seem useful to categorize particle types according to their behaviour on impact into 3 to 4 groups.

By calculating the local impact velocity from the equation of motion those zones can be defined in which the impact velocity lies below the critical velocity v_c. If this zone is referenced to the zone determined by the limiting particle trajectory, it is possible to obtain the adhesion efficiency h.

Fig. 1.19 shows calculated adhesion efficiencies and collision efficiencies for various coefficients of restitution k_{pl} (for $A = 5 \cdot 10^{-19}$ J) and for various Hamaker constants A (for $k_{pl} = 0.8$). In contrast to the collision efficiency, the adhesion efficiency is not only goverened by ψ, but also very significantly by the material specific values of A and k_{pl}. The calculation of separation ratio $\varphi = \eta h$ thus requires unequivocally the inclusion of empirically derived material properties.

A general representation of the interrelationship between all variables has not been possible so far. For "average" materials where $A = 5 \cdot 10^{-19}$ J, $p_{pl} = 5 \cdot 10^9$ Pa and $k = 0.7$ Hiller calculated an approximate h according to: —

$$h = 1.37 \cdot \psi^{-1.09} \, Re^{-0.37} \qquad (64)$$

for a validity range of $1 < \psi < 20$
$0.01 < Re < 1$
$h > 0.1$

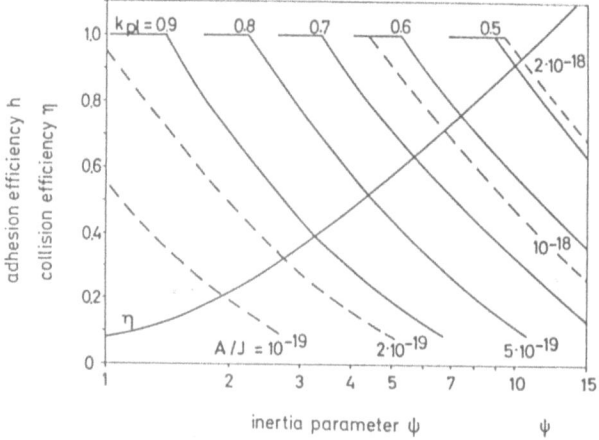

Fig. 1.19
Relationship between adhesion efficiency h and inertia parameter ψ. Hamaker constant A and coefficient of restitution k_{pl} for Re = 0.2

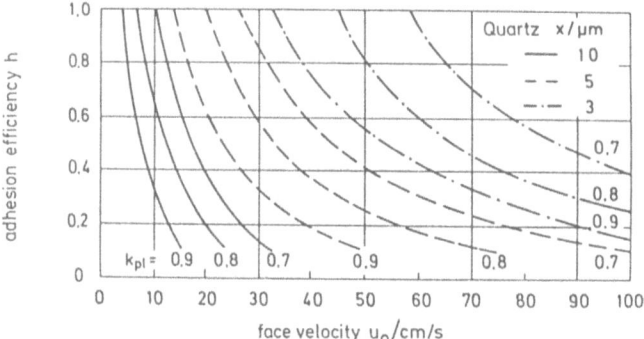

Fig. 1.20
Calculated distribution of adhesion efficiency h

Fig. 1.20 is a scaled diagram of the adhesion efficiency for particle sizes 3 μm, 5 μm and 10 μm, and for k_{pl} = 0.7, 0.8 and 0.9, respectively, as a function of face velocity. It is evident that bouncing may occur at 5–50 cm/s and that the adhesion efficiency decreases with increasing velocity, increasing particle size and increasing coefficient of restitution.

Such calculations are limited by several assumptions and simplifications. Adhesion efficiencies were therefore determined directly from high-speed cinematography [20]. Fig. 1.21 is an example of such measurements. These experiments gave good confirmation not only of trends and interrelationships, but also confirmed the order of magnitude.

This applies especially to the onset of bouncing. The increased adhesion of quartz particles compared with glass spheres corresponds with the significantly lower k_{pl} values for quartz; this is caused by shape and surface effects. Similarly, the higher adhesion efficiency on fibres with larger diameter agrees with theoretical predictions. As expected also, treating the fibre surface to improve adhesion increases the adhesion efficiency substantially.

1.3 Separation in Fibre Assemblies

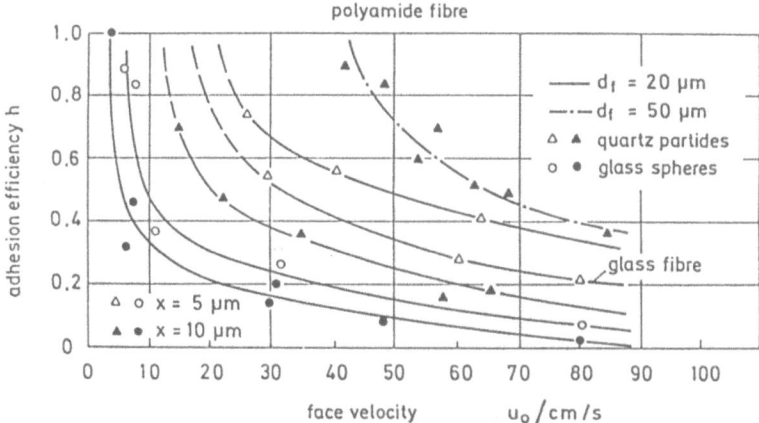

Fig. 1.21 Experimentally determined distribution of adhesion efficiency h

1.3.2.3 Summarising Particle Separation on Single Fibres

If all the mechanisms of separation outlined above are assembled into a single diagram, we obtain a qualitative relationship as shown in fig. 1.22. Below 0.5 μm separation is accomplished either by diffusion or by electrostatic forces. Above 0.5 μm inertial effects take over, while electrostatic forces may still operate in the region up to 3 to 5 μm. From 1 to 3 μm onwards the adhesion phenomena become even more significant, and ultimately become the dominant factor above 5 μm at velocities of 20 to 30 cm/s. Particle bounce has been recorded not only on single fibres, but also on fibre assemblies (deep bed filters); at least where the fibre volume proportion lies below approx. 5 %.

Useful equations are available today for calculating the transport mechanism and adhesion efficiency in deep bed filter, substantiated both by experiment and on simplified model

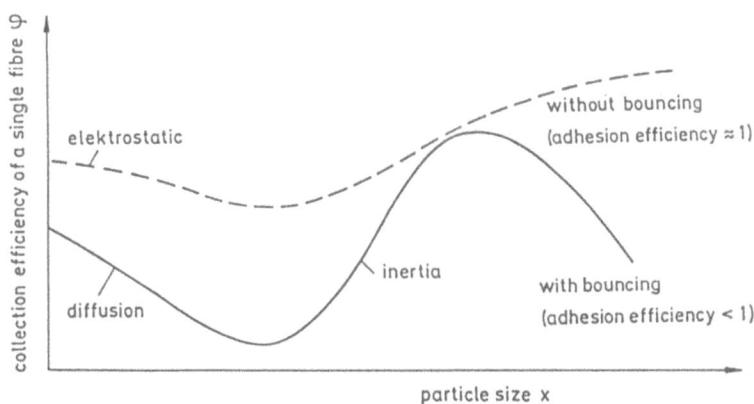

Fig. 1.22 Deposition at a single fibre

filters. They provide useful information on general trends at least, and on the order of magnitudes. The inclusion of the electrostatic effect during particle transport on the one hand, and the intrinsic material parameters — which cannot be calculated — on the other, has brought significant advances.

Discrepancies between theory and measurements on actual filters are due to presumably the differences between real filter structures and the models on which current mathematical calculations are based (angled paths of flow, interaction between fibres). Even in idealized situations questions arise due to the nature of the various flow fields.

A further important restriction should be kept in mind. The mathematical modelling quoted above is based on virgin filters, i.e. for the initial clean filter condition. It is well known that filtration behaviour changes as particles are accumulated — it usually improves —; however, the mathematical treatment of this behaviour is still in its infancy. Tien [24] and Payatakes [25] have published papers that deal with the behaviour of dendritic particle depositions on fibres in simplified filters.

Approximations of empirically determined values are also useful. E.g. Davies [26] describes the behaviour of the permeability ratio $P = 1 - \phi$ of deep bed filters as a function of time —

$$P_t = P_0 \exp[-\alpha t] \tag{65}$$

where P_0 is the permeability of the unused filter and α is an empirical correction factor. This equation is valid only for the permeability of known systems, i.e. known P_0 and α. Elucidation of the time dependent behaviour of depth filters still requires much research.

1.3.3 Dust Separation in Bag Filters

Bag filters are used preferentially for high dust load situations (around 100 g/m^3). Because the differential pressure rises continuously, they require periodic cleaning (ref. chap. 3); they are therefore often referred to as cleaning filters.

Today nonwoven materials such as needlefelts (ref. chap. 2) are used increasingly in place of woven fabrics. Needlefelts and nonwovens have a similar structure to deep bed filters, though their fibre volume ratio is considerably higher. Thus during the initial dust deposition phase, while in the clean state, a similar mechanism operates on such filters as described previously for deep bed filters under section 1.3.2.

Fig. 1.23 is an example in point, where the grade efficiency is plotted for a 600 g/m² PE felt — in the uncontaminated state; dust concentration was 73 mg/m³ with a 5 min. recording time. The dust particles (limestone) were electrostatically neutral so that only inertia effects on particles above 0.5 μm were present. The shape of the curve confirms this; it is characteristic for separation by inertia in deep bed filters.

As dust deposition increases the retention of particles of all sizes (fig. 1.24) increases. The curve is plotted for several pressure drops Δp; these in turn depend on the dust load. Even a small increase in Δp, i.e. as the result of more dust depositing, produces a significantly higher separation efficiency. At Δp = 109 Pa the dust retained was 33 g/m², while at 174 Pa it rose to 80 g/m². Separation efficiency was above 95 % for all particle sizes at 109 Pa and above. Photomicrographs from a scanning electron microscope [28] show that the 33 g/m² dust is deposited primarily at the surface of the fibre assembly,

1.3 Separation in Fibre Assemblies

and that a filter cake has formed (fig. 1.25a and 1.25b). As more dust is deposited it is evident that separation proceeds mainly within this dust layer. This separation takes place by geometrical interception, i.e. a sieving effect. Rothwell [29] has already drawn attention to this. This explains why in commercial installations the dust emission remains almost independent of the raw gas dust content. This observation together with fig. 1.24 also explains the high separation efficiency found in commercial filters, even for particle sizes below 1 μm.

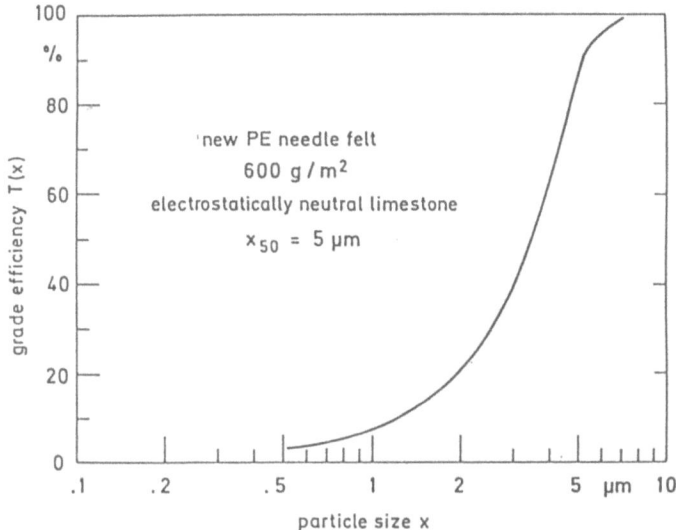

Fig. 1.23 Grade efficiency of a new PE needlefelt 600 g/m² with v = 180 m³/m² h

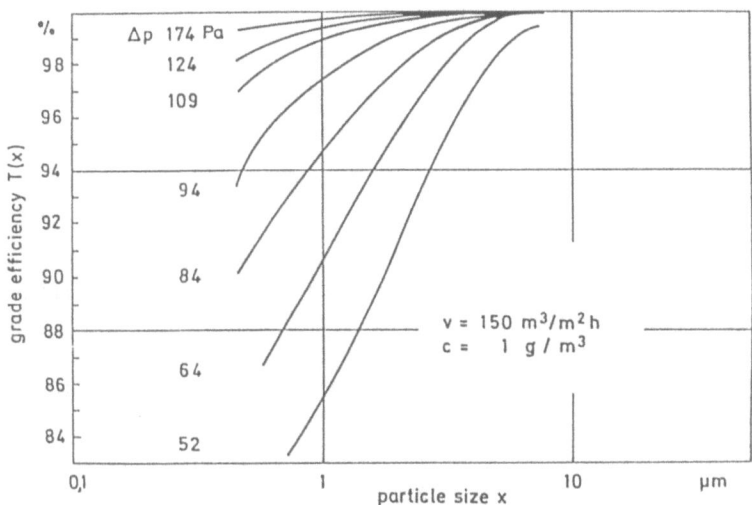

Fig. 1.24 Grade efficiencies of a PE needlefelt in relation to dust hold-up Klingel [28]

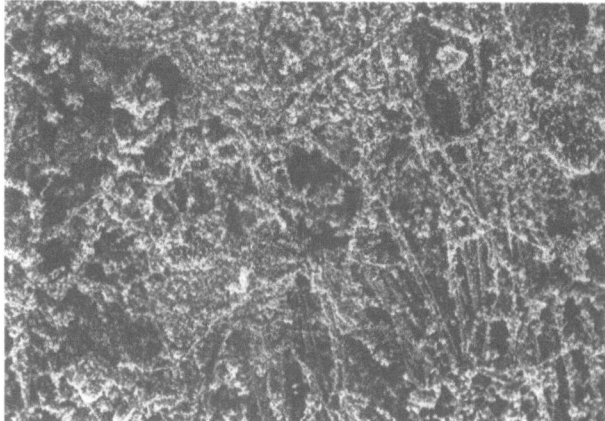

Fig. 1.25a
Raw gas side of PE needlefelt
(REM) with filter cake
(W = 32 g/m²)
(Δp = 110 Pa)
200 μm

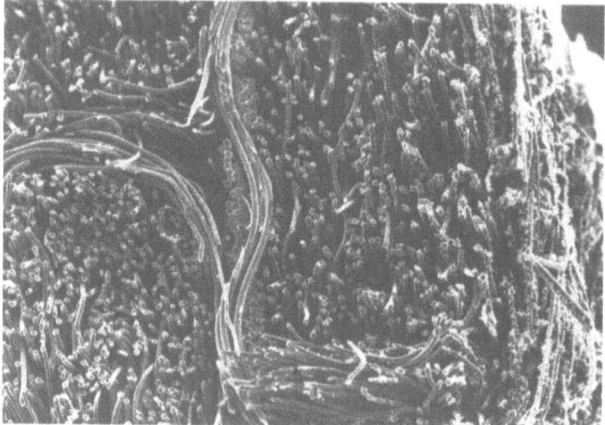

Fig. 1.25b
Cross sectional view (REM)
of a PE needlefelt with
entrained dust
(W = 33 g/m²)
(Δp = 110 Pa)
200 μm

In some of the published literature attempts have been recorded to plot the dust penetration as a function of time using empirical equations. These authors use the equation developed by Valentin [30]:

$$P = \exp[-k\, W_0^n] \tag{66}$$

here W_0 is a multiple of the dust layer referenced to 1 g/m². k and n are experimentally determined parameters and depend on many other variables. Some authors [31, 32] have tried to use k to determine the effect of gas velocity. In the opinion of the current author these equations are still heavily dependent on experimental conditions and equipment, it is not possible to arrive at generalisations. Therefore this approach will not be expanded.

Leith and First [33] divide dust transmission in bag filters into 3 types:

1. "Straight through" penetration, especially at the beginning of the filtering phase. This depends on dust concentration;

1.3 Separation in Fibre Assemblies

2. Dust leakage through "pin holes" in the dust cake. Prevalent in filter fabrics and at structural defect points in needlefelts. This effect depends on dust type and concentration;
3. "Seepage" of particles through the filter medium. Most likely during cleaning.

Several authors [34] have reported the "pinhole effect" associated with a simultaneous change in differential pressure.

Individual operating conditions markedly affect the seepage during cleaning. As will be shown later, this phenomenon may set the level of the emission.

1.3.4 Cleaning and its Effect on Dust Emission

Dust concentration in the cleaned gas as well as pressure drop are heavily dependent on the filter dust hold-up, i.e. on the filtration time (fig. 1.26). There is a prominent concentration peak when the filter is new, but this decreases rapidly. When the pressure drop reaches a preset level, cleaning commences. This lowers pressure drop, but is accompanied by an increase in emission. As filtration proceeds the peaks recur at regular intervals; their size and frequency depend on dust level, type, filter medium, gas composition, operation conditions and cleaning mode.

Luetzke and Wilkes [35] have examined different industrial installations equipped with either pulse jet cleaning or reverse air flow. They recorded substantial concentration peaks parallel to the cleaning cycles. This penetration is very significant in that for a particular type of filter medium a doubling of cleaning frequency results in a doubling of emission.

The cleaning intensity plays an important role here. On the one hand a base line intensity is necessary to achieve any cleaning at all, on the other hand excessive cleaning is to the detriment of emmision level. Klingel [28] has made a systematic study of this problem. With the aid of a light scattering particle counter capable of a 30 second measuring time it was possible to measure particle distribution and concentration on the clean gas side of a pulse jet cleaned filter bag.

The instrument was synchronised with the cleaning pulse. Curves A to D in fig. 1.27 illustrate the first 4 intervals of 30 seconds each of particle counts as a function of

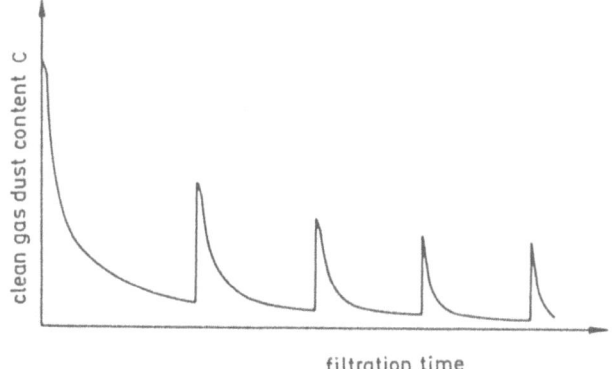

Fig. 1.26
Variation with time of clean gas dust content

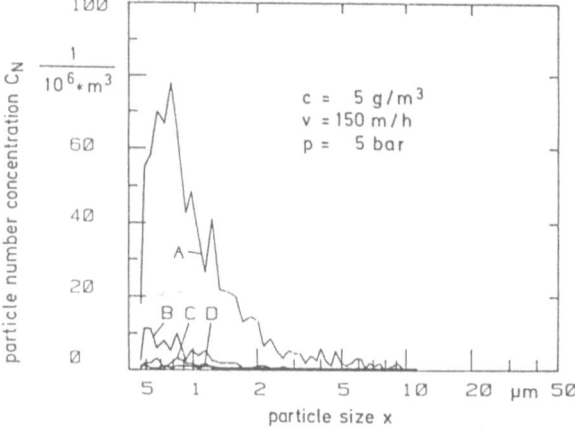

Fig. 1.27
Particle number concentration in the cleaned gas during the cleaning phase [28]

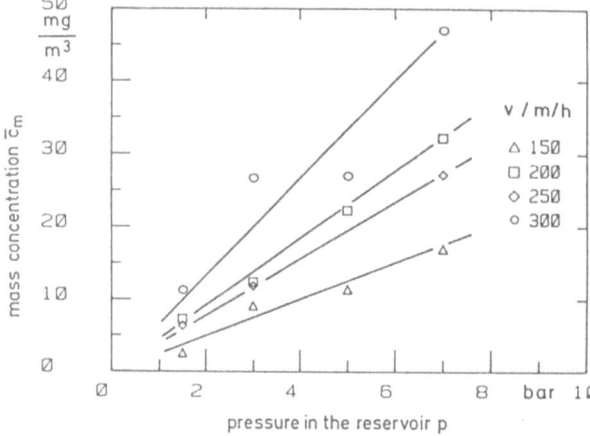

Fig. 1.28
Mean mass concentration as a function of pressure in the reservoir and face velocity (recording time 30 s [28]

particle size. Face velocity for the trial was 150 m/h, with a flue gas dust concentration of 5 g/m³ and 5 bar pressure in the compressed air reservoir.

The initial 30 second recording (curve A) produced an elevated concentration across all particle sizes. This would appear to be as a result of the filter bag deforming during the jet pulse, with subsequent recoil of the bag onto the support cage. For a mean 0.3 s pulse duration this rebound takes place within the first few seconds, once the pulse is triggered. Its magnitude depends on the pressure inside the bag. Within the sampling period from 30 to 60 seconds (curve B) the penetrating particle concentration has already decreased substantially, while it has almost disappeared after 120 s (Curve D).

It is possible to derive mean mass concentrations from the particle counts. Fig. 1.28 shows the mass concentration c_m as a function of compressed air pressure p in the reservoir for various face velocities v [28]. As the pressure increases, i.e. with increasing cleaning intensity, the initial 30 second concentration increases. The rate of rise increases also with higher face velocities. The more energetic rebounding of the filter bag may explain this. At higher face velocities the permeability of the loaded filter is lower which leads to a higher pressure build-up inside the bag during the cleaning pulse.

Klingel [28] found that up to 80% of the total emission for a given filtration period can originate during these first 30 seconds. At higher face velocities particulate emissions increases even further than the values set out in fig. 1.28, because cleaning frequency must be increased to maintain a preset differential pressure.

These investigations show how important the cleaning system is to the separation behaviour; it emphasises also the importance of selecting filter materials and cleaning conditions to achieve optimum results. As cleaning intensity increases more cleaning air is required, and the clean gas dust concentration increases. Careful study is required to select filtering speeds that will keep emission levels within acceptable limits. The answer to these questions must be determined specifically for each particular set of circumstances (dust behaviour, filter medium, cleaning mode). Our present state of knowledge does not provide universal solutions; more research is required.

1.4 Pressure Loss in Cleanable Fabric Filters

The pressure drop in bag or envelope filters affects the filtration process more than the dust separation itself. If cleaning is inadequate, problems arise during operation, because of a rapid rise in filter drag. For industrial installations a judicious choice of filter medium, face velocity, cleaning frequency and intensity must be arrived at by experiment and experience.

Various attempts have been made to represent the behaviour of pressure drop over time; however, all require variables that must be quantified empirically. This has brought together valuable data on the relative importance of variables and their order of magnitude. In fig. 1.29 Δp is plotted against time. Δp_0 is the initial pressure drop of an unused filter, Δp rises rapidly as dust accumulates. This rise is based on the assumption that gas velocity v remains constant; for simplicity the change in pressure drop is represented here as a straight line function. It is further assumed that the initial separation phase within the interior of the filter medium is negligible compared to that phase in which the dust cake builds up on the surface.

Cleaning commences when Δp_{max} is reached; Δp drops to Δp_1, where Δp_1 is higher than Δp_0, because some dust remains within the fibre assembly. From the point of view of separation efficiency this residual dust is desireable. The behaviour of Δp_1 with time will determine whether stable operating conditions have been reached, i.e. whether the operating conditions chosen were appropriate, or whether the filter medium will blind gradually and will have to be replaced prematurely.

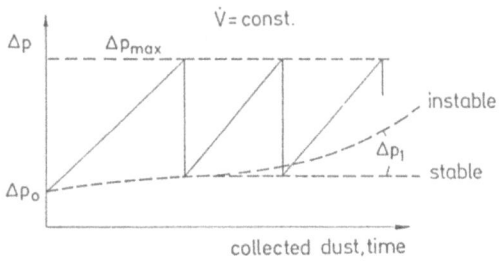

Fig. 1.29
Change in the pressure loss as a function of time. $\dot{V}$ = constant

1.4.1 Approximations derived from Experimental Measurements

There is a group of publications that deals with Δp and the attempts to fit the experimental data to a curve by regression analysis. Billings [36] is a case in point: —

$$\Delta p = a_1 + a_2 c^{a_3} \cdot p^{-a_4} \cdot N^{-a_5} . \tag{67}$$

Similarly Leith and First [37]

$$\Delta p = b_1 \Delta W^{b_2} \cdot p^{-b_3} \cdot v^{b_4} \tag{68}$$

c = dust concentration
p = compressed air pressure in the reservoir
N = number of cleaning pulses
v = face velocity
W = weight per unit area of deposited dust
a_i, b_i = regression coefficients

Both equations were developed for pulse jet filters and describe the mean pressure drop for a filter installation. They contain factors unique to that installation referred to by the authors, as well as types of dust and parameter limits. The coefficients a_i and b_i are to be redetermined for each set of conditions. It is surprising that the face velocity has no part in eq. (67).

1.4.2 Attempts at Filter Modelling

The second group of publications takes another approach and assumes that the pressure drop, Δp_1, within the filter medium after cleaning, and the differential pressure, Δp_2, within the dust cake, is additive; the basic assumption being that the resistances contributed by the dust cake and by the fabric, respectively, are in series. —

$$\Delta p = \Delta p_1 + \Delta p_2 . \tag{69}$$

Since the flow lies in the low range of Reynolds numbers (Re < 1), Darcy's law can be applied to calculate Δp_1 and Δp_2 i.e.:

$$\Delta p = \frac{1}{B_1} \mu L_1 v + \frac{1}{B_2} \mu L_2 v \tag{70}$$

where

μ = gas viscosity
v = face velocity
L_1 = thickness of filter medium
L_2 = thickness of dust cake
B_1 = permeability of fabric plus residual dust after cleaning
B_2 = permeability of dust layer

L_2 is time dependent and can be calculated from a mass balance:

$$L_2 = \frac{c v t \phi}{\rho_s (1 - \epsilon)} = \frac{W}{\rho_s (1 - \epsilon)} \tag{71}$$

1.4 Pressure Loss in Cleanable Fabric Filters

where

c = dust concentration
t = filtration time
ϕ = total separation efficiency
ρ_s = density of particles
ϵ = porosity of dust cake
W = mass per unit area of dust cake

by combining eq. (71) with eq. (70): —

$$\Delta p = \frac{\mu}{B_1} L_1 v + \frac{\mu}{B_2} \frac{c t \phi}{\rho_s (1 - \epsilon)} v^2 \qquad (72a)$$

or

$$\Delta p = \frac{\mu}{B_1} L_1 v + \frac{\mu}{B_2} \frac{W}{\rho_s (1 - \epsilon)} v \qquad (72b)$$

which is frequently expressed as

$$\Delta p = K_1 \mu v + K_2 \mu W v \qquad (73)$$

K_1 = filter material resistance after cleaning, viz. residual resistance, e.g. in 1/m;
   $K_1 = L_1/B_1$
K_2 = specific resistance of the dust cake, e.g. in m/kg.

$$K_2 = \frac{1}{B_2 \rho_s (1 - \epsilon)}$$

This is the filter equation used by many authors which forms the basis for cleaning filter layouts. Eq. (72) contains all the important variables. It becomes clear that as face velocity increases the differential pressure rises — all other conditions being equal —, because the filter cake thickness L_2, viz. the mass per unit area of the filter cake, increases proportionally. This implies that for a constant Δp and for a higher velocity the filtration interval ($T \sim 1/v^2$) must be reduced.

Permeabilities B_1 and B_2, i.e. resistances K_1 and K_2 are the significant parameters for eqs. (72) and (73). These are not constants, as is often assumed, but complicated functions which in themselves are made up of a host of other variables.

The residual resistance K_1 depends on the properties of the filter materials and dust, as well as the face velocity and cleaning intensity. The specific filter cake resistance K_2 depends on the particle size distribution, the structure and porosity of the filter cake, and thus is also affected by all the filtration conditions. Fig. 1.30 illustrates a method for determining K_1 and K_2. Thus —

$$K_1' = K_1 \mu = \frac{\Delta p_R - \Delta p_s}{v}$$

$$K_2' = K_2 \mu = \frac{\Delta p_E - \Delta p_H}{W_E v} \qquad (74)$$

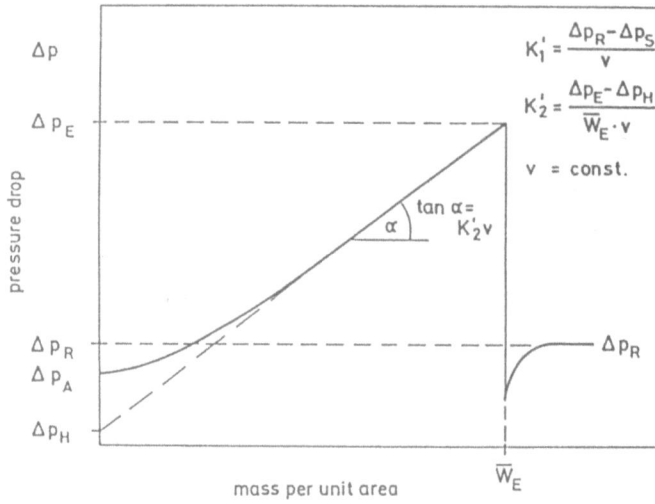

Fig. 1.30 Determination of K1 and K2

where

Δp_E = pressure drop before cleaning
$\overline{W}_E$ = mean mass per unit area of the collected dust at Δp_E
Δp_R = residual pressure loss after cleaning
Δp_S = differential pressure arising from the mechanical filter components
Δp_H = an auxiliary value

In addition to the pressure drop it is necessary to determine the mass per unit area of the dust cake. Klingel [28] has developed a radiometric method that allows bags to be measured in place. Fig. 1.31 ($K_1' = K_1 \mu$) and fig. 1.32 ($K_2' = K_2 \mu$), respectively, give some examples which were determined with limestone dust with a mass median size $x_{50.3} = 5\,\mu m$ and a polyester needlefelt with pulse jet cleaning.
The residual resistance K_1' can be approximated with –

$$K_1' = 17.45\, p^{-0.96} \qquad (76)$$

if K_1' is expressed in Pah/m and p in bar. Eq. (76) is, of course, valid only for one particular set of conditions. However, other authours [37, 39] have reported similar relationships between K_1' and the reservoir pressure p.

There is an inherent difficulty in any experimental determination of K_1 in that this variable can only be defined when the system has reached equilibrium. K_1 increases continuously during the transition phase. Equilibrium conditions – depending on the behaviour of all other operating variables – may take several hundred filtering cycles to reach. An additional complication is that K_1 depends also on the amount of dust that is redeposited immediately after the cleaning pulse ceases. The redeposited dust in turn is a function of face velocity and spacial arrangements within the filter. Thus, measurements of K_1' obtained under laboratory conditions may not be representative of values found in practice.

1.4 Pressure Loss in Cleanable Fabric Filters

K_2' rises with face velocity, and indicates that the filter cake increases in density. This effect exacerbates the rise in pressure drop, as discussed in reference to eq. (72). The increase in density leads to a further rise in cleaning frequency; more compressed air is required, dust penetration increases and the filter bags suffer more wear and tear.

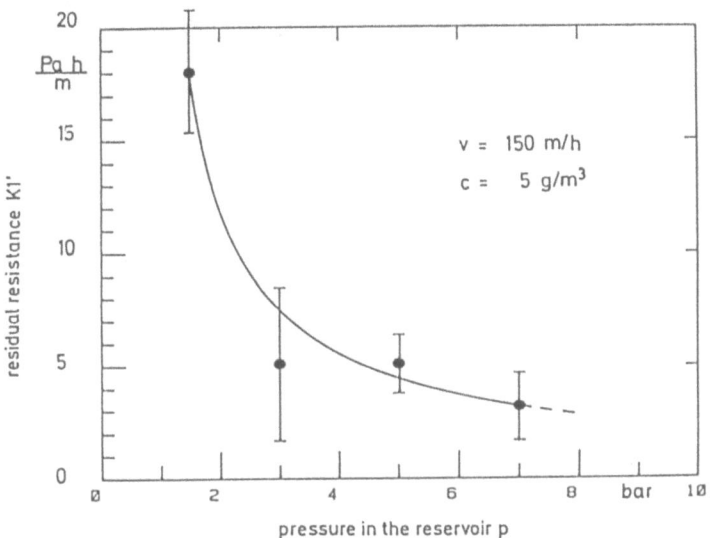

Fig. 1.31 Residual resistance K_1' as a function of cleaning pressure

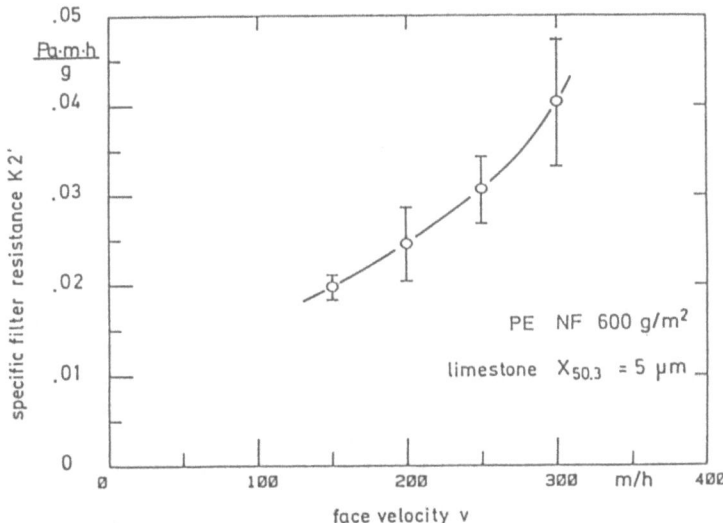

Fig. 1.32 Specific filter cake resistance K_2' as a function of face velocity

Published values of K_2 are to be treated with caution, and need to to be referenced carefully to the experimental conditions under which they were obtained. Essential information is often missing, indicating that the true significance of circumstances under which K_2 was determined is not always fully appreciated.

Table 1.4 surveys suggested methods for determining K_2 and for some cases K_1. The table also indicates some problems in calculating values of K_2. Some approaches are based on assumptions derived from physical models, others are based solely on empirical approximations. Probably none of these take all variables into account, because not all are fully understood. It cannot be emphasized enough that all measurements of filter resistance should be made under conditions as close as possible to actual in-use conditions.

Table 1.4 Summary of Resistance Values

K_1	K_2	limiting conditions	authors
–	$k_{CK} S_V^2 \dfrac{(1-\epsilon)^2}{\epsilon^3}$	$k_{CK} = f(\epsilon)$	Carman, Kozeny [45]
–	$5.6\, k\, \epsilon^{-5.5}\, \bar{x}_2^{-2}$	$0.35 < \epsilon < 0.7$ $k \simeq 1$	Rumpf, Gupte [46]
–	$k S_V^2 \dfrac{1-\epsilon}{\epsilon^3} \dfrac{1}{\rho_s} \dfrac{1}{\rho_f}$	$k = f$ (shape, surface)	Williams et al. [47]
$K_1' \sim (p_3 - p_1)^{-0.8}$ p_3, p_2 s. fig. 1.34	$1.25 \cdot \dfrac{1}{x_{50.3}}$	$v = 90 - 162$ m/h $x_{50.3} = 0.3 - 50\ \mu m$	Strangert [38]
–	$116.75\, x_{50.3}^{-0.878}$	$v = 72$ m/h, $c = 11.4$ g/^{3} $x_{50.3} = 3.9 - 15.3\ \mu m$	Davis, Kurzyske [48]
–	$(K_2)_2 = (K_2)_1 \left(\dfrac{v_2}{v_1}\right)^n$	fly ash $0.2 \leqslant n \leqslant 1.5$	Dennis, Klemm [49]
$18500\, p^{-0.66}$	–	p storage pressure k Pa	Dennis, Klemm [39]
$1600\, (d\,\Delta p/dt)^{-1.13} \cdot v^{-1}$	–	$d\,\Delta p/dt$/k Pa/s, v/m/min	Dennis et al. [50]
–	$a_1\, v^{a_2}\, W^{-a_3}$	a_{1-3} regression values	Atzger [32]

1.4.3 The Effect of Cleaning Intensity on Residual Pressure Drop

The resistance K_1 of the filter cake immediately after cleaning and the resultant pressure drop Δp_R depend on the filter medium, type of dust, filter condition, and above all on the cleaning intensity. The question arises whether the success of this cleaning can be adequately described by measuring the residual pressure drop, e.g. as suggested by Ritscher and Frenzel [51]. Very few investigations have been documented on this problem. Klingel [28] has made a systematic study of the processes that occur on a filter bag during pulse jet cleaning (length 2.5 m, diameter 11.4 cm). Pressure transducers monitored

1.4 Pressure Loss in Cleanable Fabric Filters

the pressure rise at several places along the bag as the pulse travels through the bag. Using radiometric techniques he was able to record the mass per unit area along the length of the bag as well.

Fig. 1.33 shows the effect of different cleaning pressures on the residual mass per unit area W_R. $W_R + W_E$ is plotted as a function of v. The shaded area is bounded above by curve W_E — the terminal area mass before cleaning commences — and below by W_R — the retained and redeposited dust before recommencement of filtering. If the pressure drop at the start of cleaning is kept constant ($\Delta p_E = 2500$ Pa) both curves approach one another as face velocity increases, until the difference between them becomes very small. The graph shows clearly that residual dust rises with v.

At a compressed air reservoir pressure of 1.5 bar cleaning is ineffectual. The plotted values are scattered around the W_E curve, so that in this case $W_R = W_E$. On the other hand it is surprising that as reservoir pressure increases (3 to 7 bar) the cleaning effect improves little. Thus cleaning is initiated only at 1.5 bar and beyond; however, above 3 bar no further reduction in W_R is achieved. This effect becomes even more evident, if one examines the pressure generated within the bag during cleaning. Fig. 1.34 shows a pressure profile inside the bag.

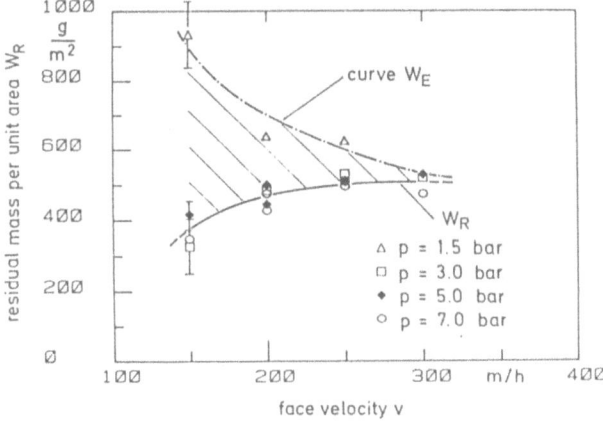

Fig. 1.33
Residual mass per unit area as a function of face velocity and storage air pressure

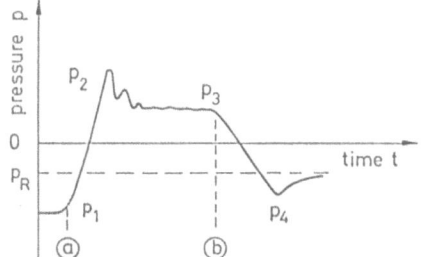

Fig. 1.34
Profile of a pressure pulse

The various symbols represent: —

$p_1 = \Delta p_E - \Delta p_S$ final — system pressure loss
p_2 pressure peak
p_3 quasi stationary pressure
p_4 recovery phase
p_R residual pressure
ⓐ opening of diaphragm valve
ⓑ closing of diaphragm valve

The diaphragm valve opens at point ⓐ. p_1 at this point is equal to the final differential pressure in the bag. Once the valve is fully open the pressure maximum p_2 is reached. Cleaning between these points is achieved by throwing off the filter cake. As the valve closes, and the pressure in the bag falls into a negative region and eventually recovers to the residual differential pressure p_R.

The integral of the pressure pulse P_D is introduced here to describe the cleaning.

$$P_D = \int_0^T p(t)\,dt. \tag{77}$$

The pressure pulse integral is the area between the positive part of the p(t) curve and the zero line (ref. fig. 1.34). It also takes into account the total blow time T.

Fig. 1.35 illustrates the relationship between the dust mass remaining after cleaning W_R and the pulse p_D, for face velocities 150 to 300 m/h. As the pulse pressure increases (due to higher pressure in the compressed air reservoir W_R decreases initially. However, the curves for different face velocities soon merge.

It appears that cleaning is at a maximum when $P_D \cong 50$ Pas. This corresponds to an internal pressure of ≈ 250 Pas inside the bag when the blow period T = 200 s. Even if P_D is increased substantially, there is no further reduction in the residual dust mass.

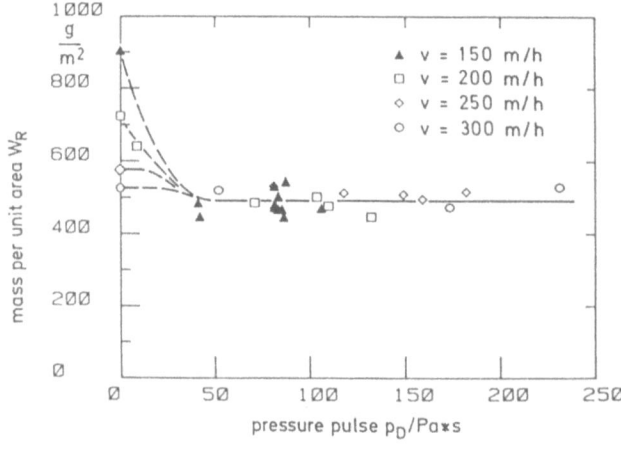

Fig. 1.35

Relationship between residual mass per unit area W_R and pressure pulse P_D

1.4 Pressure Loss in Cleanable Fabric Filters

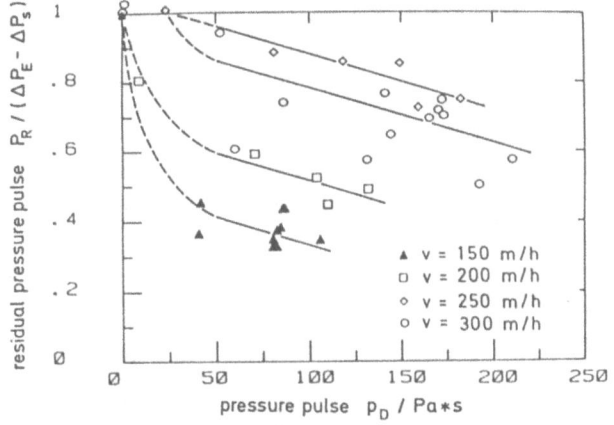

Fig. 1.36
Relationship between residual pressure and pressure pulse

To enable a direct comparison the effect of pressure pulse on residual pressure drop is plotted in fig. 1.36. The pressure loss of the total system Δp_{syst} becomes more significant with increasing face velocity, so that the residual pressure drop is related to the effective differential drop before cleaning ($\Delta p_E - \Delta p_S = p_1$). When the filter cake is thrown off, there is an initial large decrease in pressure drop; no limiting condition could be established within the range of conditions examined.

An increase in the cleaning intensity can reduce the residual pressure drop; however, not the residual area mass. Fig. 1.34 shows that no further dust is removed. Two factors may account for this: The structure is opened and/or channels are formed within it.

Once filtration recommences the opened pores and channels close rapidly with a consequent rapid rise of filter resistance. It follows then that nothing is gained by a more intensive cleaning, once the filter cake has been dislodged. Any increase in cleaning intensity raises running costs (compressed air consumption); it is also detrimental to particle separation as shown above.

It is also clear that the residual pressure loss by itself is not an adequate indicator to judge the cleaning efficiency. The curves in fig. 1.36 could tempt the use of unnecessarily high reservoir air pressures, i.e. to work with high compressed air consumption. An instantaneous record of residual dust is a much clearer indicator for proper cleaning. (The $P_D \approx 50$ Pas limit at 3 bar reservoir pressure shown in fig. 1.35 is of course derived for one particular system and cannot be generalized.) Thus, to arrive at optimal cleaning conditions it is useful to measure directly the amount of dust remaining on the bag, viz. the detached portion. It is possible to contemplate systems other than radiometric techniques [28] to accomplish this.

1.4.4 The Effect of Electrostatic Forces on Pressure Drop

Section 1.3.2.1 covered the effect of electrostatic forces on the separation of particles in deep bed filters. Attempts have been made recently to improve dust separation in bag filters by electrostatic techniques. The aim is to reduce filter resistance, i.e. to reduce the flow resistance K_2 of the filter cake.

Two avenues have been followed. One is to charge the particles before they enter the filter bag [40]. The other is to polarize both fibre and particles by external electric fields which are externally mounted [41]. Both approaches show promise on a laboratory scale. Pilot studies [42] with system 2 (external fields) have yielded significant reductions in pressure drop as well as in residual filter resistance. Current explanation for this is that the particles no longer penetrate into the filter medium, but settle out entirely on the surface to form a filter cake that is more porous than if collected non-polarized. Further studies need to be conducted on these techniques.

1.5 Design of Multichamber Fabric Filters

The comments made in section 1.4 apply to single bags, i.e. bags or envelopes working in groups and in parallel within a single cell, under conditions of identical differential pressure and flow rate. Filter installations often consist of multiple cells that are connected in series. Each cell operates under differing conditions.

For such systems an average velocity $\bar{v}$, viz. a mean pressure drop Δp_m applies. Solbach [43] and Winter [44] have reported calculations based on an analogy to the single cell. Those of Winter are briefly outlined here.

The pressure drop across the entire installation is assumed as constant. Depending on dust deposition differing gas velocities will prevail in each cell, so hat the filter eq. (73) can be rewritten as –

$$\Delta p = K_1' v + K_2' W v . \tag{77}$$

The quantity of dust collected per unit filter area is given by the integration over the operating period T:

$$W = \int_0^T c \phi v(t) \, dt . \tag{78}$$

Assuming a constant raw gas dust concentration c and collection efficiency $\phi \approx 1$, eq. (77) can be rewritten as eq. (78):

$$\frac{\Delta p}{v(t)} = K_1' + K_2' c \int_0^T v(t) \, dt . \tag{79}$$

Winter assumes that K_1' und K_2' is independent of time or velocity, i.e. constant. Undoubtedly this is a gross simplification.

Eq. (79) can be differentiated: –

$$\frac{1}{v(t)^3} \frac{dv}{dt} = - \frac{K_2' c}{\Delta p} \tag{80}$$

with initial conditions

$$t = 0 \rightarrow v_0 = \frac{\Delta p}{K_1'} .$$

1.5 Design of Multichamber Fabric Filters

It follows from eq. (80):

$$v(t) = \frac{\Delta p}{(2 K_2' c \Delta p t + K_1'^2)^{1/2}} . \tag{81}$$

Equation (81) describes the velocity change with time in a cell assuming Δp = constant.

For a collection plant with m cells and filter surface area f per cell, the total filter surface area is $F = m \cdot f$. For this case Winter calculated a mean velocity $\bar{v} = \dot{V}/F$ with the assumption that the mean value $\bar{v}$ will apply to a multiple of cells, and that it is equal to the time averaged mean for one cell over the filtration period:

$$\bar{v} = \frac{1}{T} \int_0^T v(t) \, dt . \tag{82}$$

When eq. (82) is encorporated into eq. (81), the mean velocity is expressed by —

$$\bar{v} = \frac{(2 K_2' T c \Delta p + K_1'^2)^{1/2} - K_1'}{K_2' T c} . \tag{83}$$

From eq. (83) the mean differential pressure Δp_m can be derived —

$$\Delta p_m = K_1' \bar{v} + \frac{1}{2} K_2' T c \bar{v}^2 . \tag{84}$$

Eq. (84) corresponds to eq. (72), whereby for eq. (72) v = constant and $\Delta p = \Delta p(t)$. Winter terms eq. (84) "the fundamental multiple cell filter equation". K_1' and K_2' filter resistances must be determined experimentally to take account of the limitations outlined under sect. 1.4.

Once these constants are known for a set of conditions (dust type, filter medium, cleaning mode) eqs. (83) and (84) can be used to calculate a multiple cell filtration system.

Example 1:

Given: Filter installation with fibreglass bags
Raw gas dust concentration $c = 1.5 \text{ g/m}^3$

$K_1' = 3 \frac{\text{Pa h}}{\text{m}}$; $\quad \frac{1}{2} K_2' T = 0.42 \frac{\text{Pa m h}^2}{\text{g}}$ (experimentally determined)

To be determined:

a) Mean pressure difference Δp_m?
b) What will be the new mean gas velocity $\bar{v}$ at differential pressure calculated under a) when raw gas dust concentration rises to 3 g/m^3?

Solution:

a) substituting into eq. (84) —

$$\Delta p_m = 39 \frac{\text{Pa h}}{\text{m}} \cdot 40 \frac{\text{m}}{\text{h}} + 0.42 \frac{\text{Pa m h}^2}{\text{g}} \cdot 1.5 \frac{\text{g}}{\text{m}^3} \cdot 40^2 \frac{\text{m}^2}{\text{K}^2}$$

$$\underline{\Delta p_m = 2568 \text{ Pa}}$$

b) substituting into eq. (83) —

$$\bar{v} = \frac{\left(4 \cdot 0.42 \, \frac{Pa\,m\,h^2}{g} \cdot 2568 \, Pa \cdot 3 \frac{g}{m^3} + 39^2 \, \frac{Pa^2 \, h^2}{m^2}\right)^{1/2} - 39 \, \frac{Pa\,h}{m}}{2 \cdot 0.42 \, \frac{Pa\,m\,h^2}{g} \cdot 3 \frac{g}{m^3}}$$

$$\underline{\underline{\bar{v} = 32.2 \, \frac{m}{h}}}$$

i.e. when the raw gas concentration rises from 1.5 to 3 g/m³ the mean gas velocity will decrease by 20 %.

Example 2:

Given: $\dot{V} = 520{,}000 \, m^3/h$, $c = 1.5 \, g/m^3$
filter surface area per cell $f = 1210 \, m^2$

$$K_1' = 39 \, \frac{Pa\,h}{m} \,; \quad \frac{1}{2} K_2' \, T = 0.42 \, \frac{Pa\,m\,h^2}{g}$$

$$\Delta p_m \leqslant 2500 \, Pa$$

Determine: No. of cells required, m = ?

Solution:
from eq. (84) it follows

$$\frac{\Delta p}{\bar{v}} = K_1' + \frac{1}{2} K_2' \, T \, c \, \bar{v}$$

with $\bar{v} = \dot{V}/m \cdot f$ it follows:

$$\frac{\Delta p\,f}{\dot{V}} m^2 - K_1' \, m - \frac{1}{2} K_2' \, T \, c \, \frac{\dot{V}}{f} = 0$$

$$m = \frac{K_1' \, (\overset{+}{-}) \, (K_1'^2 + 2 K_2' \, T \, c \, \Delta p)^{1/2}}{2 \, \frac{\Delta p}{\dot{V}} \, f}$$

substituting the values —

$$m = \frac{39 + (39^2 + 4 \cdot 0.42 \cdot 1.5 \cdot 2500)^{1/2}}{2 \cdot \frac{2500}{520.000} \cdot 1210}$$

$$\underline{\underline{m = 10.9, \text{ i.e. } m = 11 \text{ cells}}}$$

Literature

[1] *Löffler, F.:* Die Abscheidung von Partikeln aus Gasen in Faserfiltern. Chemie-Ing.-Technik **52** (1980), 312/323

[2] DIN 66141: „Darstellung von Korngrößenverteilungen" 1974.

[3] DIN 66111: „Sedimentationsanalyse im Schwerefeld" 1973.

[4] *Rumpf, H.; Ebert, K. F.:* Darstellung von Kornverteilungen und Berechnung der spezifischen Oberfläche. Chemie-Ing.-Technik **36** (1964), 523/537

[5] *Leschonski, K.; Alex, W.; Koglin, B.:* Teilchengrößenanalyse. Chemie-Ing.-Technik **46** (1974).

[6] *Allen, T.:* Particle Size Measurement. 2nd Ed. 1975, Chapman and Hall, London

[7] *Pahl, M. H.; Schädel, G., Rumpf, H.:* Zusammenstellung von Teilchenformbeschreibungsmethoden. Aufbereitungs-Technik **14** (1973). 1. Teil 5, 257/264, 2. Teil 10, 672/683, 3. Teil 11, 759/764.

[8] *Köster, E.:* Granulometrische und morphometrische Meßmethoden. Verlag Ferd. Enke, Stuttgart, 1964.

[9] *Huller, D.; Weichert, R.:* Quantitative Shape Analysis of Particle Contours, in N. G. Stanley-Wood, T. Allen, Ed.: Particle Size Analysis 1981. Wiley Heyden Ltd. 1982.

[10] DIN 66142: „Darstellung und Kennzeichnung von Trennungen disperser Güter" 1981.

[11] *Davies, C. N.:* Air Filtration. Academic Press London, 1973.

[12] *Löffler, F.:* Zur Theorie und Praxis der Partikelabscheidung in Faserfiltern. Chemische Rundschau **29** (1976), 12, 9/11.

[13] *Kirsch, A. A.; Fuchs, N. A.:* Studies on Fibrous Filters. Ann. Occup. Hyg. **11** (1968), 299/304.

[14] *Lee, K. W.; Liu, B. Y. H.:* Theoretical Study of Aerosol Filtration by Fibrous Filters. Aerosol Sci. and Technol. **1** (1982), 147–161

[15] *Löffler, F.:* Problems and recent advances in aerosol filtration. Separation Science and Technology **15** (1980), 297/315.

[16] *Zebel, G.:* Deposition of aerosol flowing past a cylindrical fiber in a uniform electric field. J. Colloid Sci. **20** (1965), 522/543.

[17] *Muhr, W.; Löffler, F.:* Abscheideverhalten von Faserfiltern bei elektrostatischer Aufladung von Staub und Faser. Maschinenmarkt **82** (1976), 669/672.

[18] *Muhr, W.:* Theoretische und experimentelle Untersuchung der Partikelabscheidung in Faserfiltern durch Feld- und Trägheitskräfte. Dissertation Universität Karlsruhe 1976.

[19] *Bergmann, W.; Hebard, H. D.; Lum, B. Y.:* Electrostatic Filters Generated by Electric fields. Proc. 2nd World Filtration Congr. London, 1979.

[20] *Hiller, R.:* Der Einfluß von Partikelstoß und Partikelhaftung auf die Abscheidung in Faserfiltern. Dissertation Universität Karlsruhe 1980. Fortschr.-Berichte VDI-Ztg., Reihe 3, Nr. 61, 1981

[21] *Suneja, S. K.; Lee, C. H.:* Aerosol filtration by Fibrous Filters at intermediate Reynolds numbers. Atmos. Environm. **8** (1974), 1081/1094.

[22] *Löffler, F.; Muhr, W.:* Die Abscheidung von Feststoffteilchen und Tropfen an Kreiszylindern infolge von Trägheitskräften. Chemie-Ing.-Technik **44** (1972), 510/514; VDI-Bericht Nr. **187** (1972), 212/216.

[23] *Löffler, F.:* Über die Haftung von Staubteilchen an Faser- und Teilchenoberflächen. Staub – Reinh. der Luft **28** (1968), 456/462.

[24] *Emi, H.; Wang Chin-Sen; Tien, C.:* Transient Behavior of Aerosol Filtration in Model Filters. AICHE-J. **28** (1982), 397/405.

[25] *Payatakes, A. C.; Gardon, L.:* Dendritic deposition of aerosol particles in fibrous media by inertial impaction and interception. Chem. Eng. Sci. **35** (1980), 1083/1096.

[26] *Davies, C. N.:* The clogging of fibrous aerosol filters. Aerosol Sci. **1** (1970), 35/39.

[27] *Millikan, R.:* On the velocity of fall of a small spherical body through a gas and its bearing upon the nature of molecular reflection from surfaces. Phys. Rev. **22** (1923), 1.

[28] *Klingel, R.:* Untersuchung der Partikelabscheidung aus Gasen an einem Schlauchfilter mit Druckstoßabreinigung. Dissertation Universität Karlsruhe 1982. Fortschr.-Berichte VDI-Ztg., Reihe 3, Nr. 76, 1983.

[29] *Rothwell, R.:* Concepts of Fabric Dust Filtration. Filtration & Separation 13 (1976), 477/486.

[30] *Valentin, F. H. H.:* Cloth Filtration of fine Aerosols. Brit. Chemical Engineering 7 (1962), 268/271.

[31] *Wiemann, H.-J.:* Geschwindigkeitsabhängigkeit des Durchlaßgrades, ein Beitrag zu den Grundlagen der Gewebestaubabscheidung. Dissertation T. H. für Chemie, Leuna-Merseburg 1969.

[32] *Atzger, J.:* Zusammenfassende Untersuchungen der Staubabscheidung in filternden Abscheidern unter besonderer Berücksichtigung des instationären Filterbetriebs. Dissertation Universität Essen 1976.

[33] *Leith, D.; First, M. W.:* Performance of a Pulse-Jet Filter at High Filtration Velocity. I. Particle Collection. Journal of the Air Pollution Control Association 27 (1977), 534/539.

[34] *Holland, C. R.; Rothwell, E.:* Model Studies of Fabric Dust Filtration. 2. A Study of the Phenomenon of Cake Collapse. Filtration and Separation 14 (1977), 224/231.

[35] *Lützke, K.; Wilkes, R.:* Untersuchung des Einflusses verschiedener Aerosole, Filtermaterialien und Abreinigungsverfahren auf den Abscheidegrad filternder Abscheider. Umweltforschungsplan des BMI, Forschungsbericht 10403 180, September 1981.

[36] *Billings, C. E.:* Aerosol Filtration Technology for Source Emission Control. AIChE Symposium Series 70 (1974) 137, 341/350.

[37] *Leith, D.; First, M. W.:* Pressure Drop in a Pulse-Jet Fabric Filter. Filtration & Separation 14 (1977), 473/474.

[38] *Strangert, S.:* Predicting Performance of Bag Filters. Filtration & Separation 15 (1978), 42/55.

[39] *Dennis, R.; Klemm, H. A.:* Modeling Concepts for Pulse-Jet-Filtration. Journal of the Air Pollution Control Association 30 (1980), 38/43.

[40] *Helfritch, D. J.:* Performance of an Electrostatically Aided Fabric Filter. Chem. Eng. Prog. 73 (1977), 54/57.

[41] *Lamb, G. E. R.; Constanza, P. A.:* A Low-Energy Electrified Filter System. Filtration and Separation 17 (1980), 319/322.

[42] *Greiner, G. P. et al:* Electrostatic Stimulation of Fabric Filtration. J. Air Poll. Contr. Assoc. 31 (1981), 1125/1130.

[43] *Solbach, W.:* Ableitung eines Berechnungsverfahrens für Mehrkammer-Gewebefilter aufgrund von Versuchsergebnissen. Staub – Reinh. der Luft 29 (1969), 24/28.

[44] *Winter, K.:* Zur Kinetik der Entstaubung mit Mehrkammerfiltern. Staub – Reinh. der Luft 37 (1977), 390/392.

[45] *Dickey, G. D.:* Filtration. Reinhold Publishing Corp., New York. Chapmann & Hall, LTD., London 1961, insbes. S. 24/33.

[46] *Gupte, A. R.; Rumpf, H.:* Einflüsse der Porosität und Korngrößenverteilung im Widerstandsgesetz der Porenströmung. Chemie-Ing.-Technik 43 (1971), 367/375.

[47] *Williams, C. E.; Hatch, T.; Greenburg, L.:* Determination of Cloth Area for Industrial Air Filters. Heating Piping and Air Conditioning 12 (1940), 259/263.

[48] *Davies, W. T.; Kurzyske, F. R.:* The Effect of Cyclonic Precleaners on the Pressure Drop of Fabric Filters. Filtration & Separation 16 (1979), 451/454.

[49] *Dennis, R.; Klemm, H. A.:* Modeling Coal Fly Ash Filtration with Glass Fabrics. EPA – 600/7-78-087. NTIS Springfield, Va., 1978.

[50] *Dennis, R.; Wilder, J. E.; Harmon, D. L.:* Predicting Pressure Loss for Pulse-Jet Filters. Journal of the Air Pollution Control Association 31 (1981), 987/992.

[51] *Ritscher, G.; Frenzel, W.-P.:* Filternde Abscheider mit Druckluftabreinigung. Luft- und Kältetechnik (1982) 187/189.

2 Filter Media

Hans Dietrich

2.1 The Nucleus

Without vanity or exaggeration we may surely say that the filter medium is the most important member of the dust collector plant. Without doubt it is the nucleus of the installation, because it alone fulfills the purpose for which the plant has been built: to separate solid matter from gas. It is therefore the focus of the extractor. The filter medium is like a lung, equipped with millions of pores, like the bronchia in the breathing apparatus of living organisms.

Just like the natural lung cannot exist without its body, so the filter medium needs a protective housing, which channels or regulates the gas feed and exhaust; it must perform reliably and fit harmoniously into the overall engineering concept of the entire installation. Filter media discriminate, they are made from high grade natural or synthetic fibres. They constitute the textile suit of the metal constructed dust collector. In service they form an alliance with the carrier gas and dust particles to enable a separation of the two under such manifold conditions as – high or low temperatures, dry or humid conditions, inert or chemically corrosive gases, laminar or turbulent flow, fine or coarse, soft or hard particles, harmless or abrasive dusts. The dust particles may flow singly or in agglomerates, with high or low velocity. The dust load may be concentrated or dilute. Frequently the surface of the filter medium is the "battle field", where gas and particles react, where adsorption or absorption takes place and where gases that are potentially reactive form their ultimate molecular structure while passing through the catalytic layer of dust. Exothermic reactions are not rare. This imaginative profusion of possible physical chemical reactions is exceeded only in actual practice. Much scientific and empirical experience exists, but each new application even the experienced filtration expert adds to his know-how.

Now, if the filter medium is the "lung", the "core", the "made to measure suit", no matter how inappropriate this analogy may sound, it is worth stressing that the medium must be suitably equipped for its difficult and exacting task. To this end we must have knowledge about its properties, methods of assembly and performance, and most importantly, we must know how to assess all the relevant processing parameters: the skill and ability to construct the "suit" in such a way that it is balanced both physically and climatically. It is the aim of the following sections to expound the state of technology in the field of filter media as it relates to its mechanical and chemical application.

2.2 Fibres as Building Blocks

In 1886 the German Patents Office granted Patent No 38396 to the grinding mill engineer Wilhelm Friedrich Ludwig Beth titled "Suction Hose Filter with Automatic Cleaning Action". If an exposé were to be written on the fibres available at that time for filter media, its length would be shorter than the patent specification itself. Wool and cotton were indeed the only two fibres our grandfathers had to make filters; though by no means the worst choice for effective filter media [1].

2.2.1 Cotton

Cotton is a natural fibre. It is chemically a cellulose with a density of 1.5 g/cm³. It is the fruit of the bushy or tree-like cotton plant (fig. 2.1) which grows in Egypt, Africa, North- and South-America, China, India and the USSR. Cotton fibres are single celled and have a convoluted ribbonlike structure. These convolutions change in direction and frequency (fig. 2.2). The fibre diameter lies between 12–22 µm with an average length of 25–30 mm. Cross sections (fig. 2.3) are oval, kidney or bean shaped [2]. As for all celluloses cotton has low resistance to acids. In contrast it withstands alkali very well. In dust collectors it can withstand a continuous service temperature of 70 °C. The fibre has a low static charge potential, but also a high water retention capacity of 45–50 % compared to polyester 3–5 %. Under standard room conditions (21 °C and 65 % relative humidity) it holds 7–11 % water. Cotton is subject to biological attack and lower strength than the synthetic fibres.

Because of its short staple length only woven or knitted structures can be made: needle-felts cannot be made from cotton. The fibre has no felting ability. Cotton fabrics have increasingly been replaced with synthetic fabrics in filtration systems during the

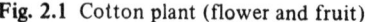

Fig. 2.1 Cotton plant (flower and fruit)

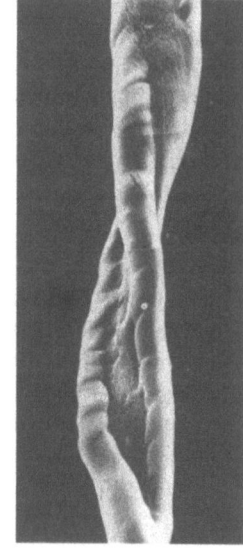

Fig. 2.2 Cotton fibre magnified 1 : 1500

2.2 Fibres as Building Blocks

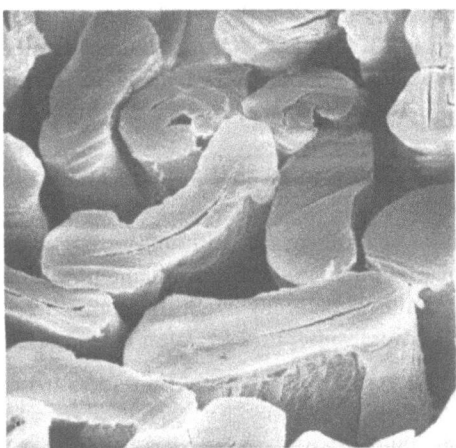

Fig. 2.3 Cross sections of cotton fibres, magnified 1 : 2400

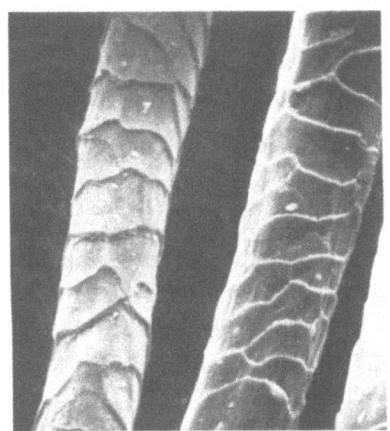

Fig. 2.4 Scales on wool fibres [3]

last 25 years. In turn, the last 10 years have seen the introduction of needlefelts from synthetic fibres in place of fabrics. However, cotton fabrics are found as filter media even today in simple applications, e.g. the timber and textile industry where low cost is essential.

2.2.2 Wool

In contrast to cotton, wool is an animal fibre. It belongs chemically to the keratins or proteins. It is shorn from sheep. This wool mantle, called fleece, is graded according to length, fineness and cleanliness. Quality differentiations are made between shoulder (especially valuable), back (weather exposed), or belly wool. Before spinning the wool (greasy wool) is washed, possibly combed — to separate the more valuable long fibres — or carbonized, i.e. vegetable contamination is removed through the action of weak sulphuric acid and heat. The density of wool is 1.32 g/cm^3. Depending on the length of the growing season, wool has a length of up to 100 mm. This makes spinning especially easy. The quality of wool is determined by the type of sheep and the breeding (e.g. climate and feed conditions in the origin country). Specially fine and high quality wools come from South Africa and from Australia. New Zealand wools are coarser and considered stronger.

Sheep wools and animal hairs have relatively good acid resistance, but low alkali resistance. In dust filtration long term temperature exposure should not exceed 70 °C with short peaks of 100 °C. Wool is biologically attacked by moths and bacteria (fungi and rotting). For this reason it must be chemically protected. The moisture retention capacity of wool lies between 40–55 %. Under normal room conditions it absorbs 15–17 % moisture. In consequence the electrostatic chargeability is low and moisture dependent.

Under the microscope, wool fibres show a scale-like surface structure (fig. 2.4). Humidity, heat and chemicals swell the fibre. As the scales expand they lock to each other. Heat and humidity, pressure and movement intensify this effect — the wool starts to felt. This effect is used industrially in the wool and hairfelt industry. Fabrics and needlefelts of wool are very filter active. They have low biological, thermal and chemical resistance which limits their application [3].

2.2.3 Chemical and Mineral Fibres

2.2.3.1 Introduction

First attempts by man to produce chemical fibres go back several hundred years. Their model was the silkworm, the first producer of a two-component monofilament with a length of more than 3000 metres. The caterpillar builds a cocoon when in the pupa phase. The silkworm builds an oval protection cocoon in several superimposed layers (fig. 2.5).

In 1883 the English scientist Swan was able to manufacture for the first time fibres from cellulose nitrate.

1931 to 1938 saw the development of polyamide (Nylon and Perlon), Polyvinylchlorides (PVC) followed in 1939 and 1941 Polyesters, 1942 the Polyacrylonitrile fibres. These are all known as "synthetic fibres" or "artificial fibres". In English speaking countries the term "man made fibres" is used.

Man made fibres are formed according to wet, dry or melting techniques [4]. The spinning substance is pressed through fine jets — spinnerettes — (fig. 2.6).

Fig. 2.5 Cocoon of a silkworm

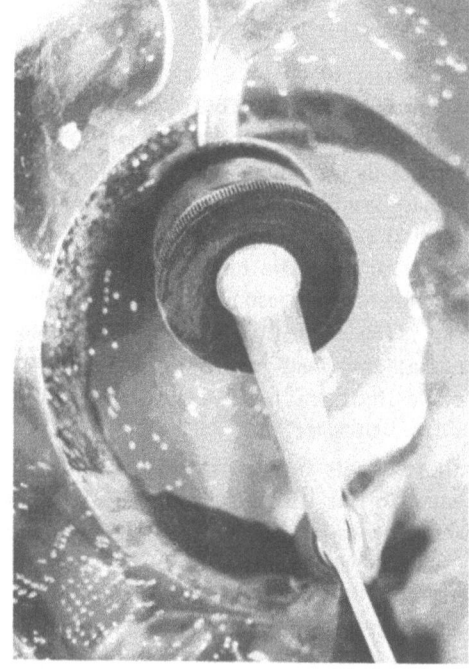

Fig. 2.6 Birth of a thread through a spinneret [4]

2.2 Fibres as Building Blocks

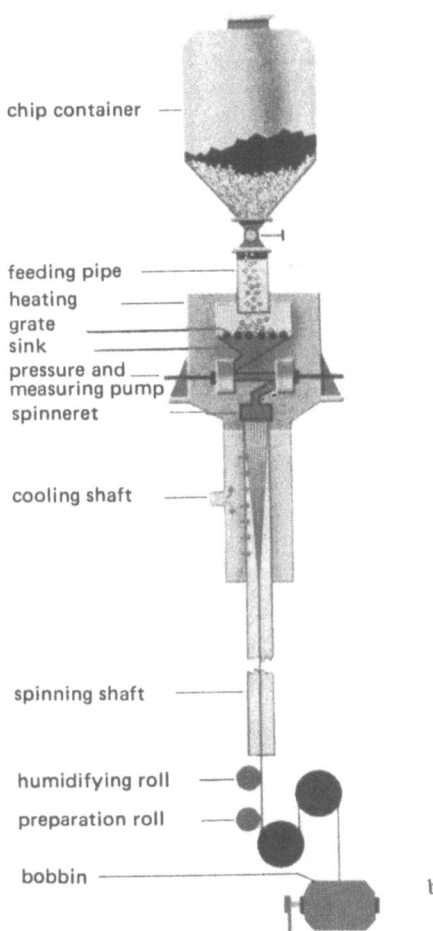

Fig. 2.7
Manufacture and cross sectional example of synthetic fibres
a) Star-shaped Perlon 1 : 6000
b) Schematic spinning head for perlon [4]

In wet spun fibres the spinning solution is precipitated from the spinning bath. Dry spun fibres are made by evaporating the solvent after extruding the spinning substance through the spinneret. Melt spun fibres (fig. 2.7a) are thermoplastic; they reach their strength during cooling. The fibres are drawn after spinning which further increases their strength Aftertreatments may consist of washing, bleaching, desulphuring, drying, annealing or heat setting. The spinneret orifice may be round, oval, dumbell shaped, lobal, triangular or other configurations (fig. 2.7b). The spinning mass may contain additives such as titanium dioxide to achieve dullness or other surface characteristics of the fibre. Shape of the cross section, friction characteristics but also fineness have a strong influence on the filtration ability of woven or felted materials (fig. 2.10 and table 2.1).

In earlier days, fibre fineness was expressed in denier (latin − denarius). This unit is still used in the USA. If this unit of measure is used for silk, then the silk thread with a titre

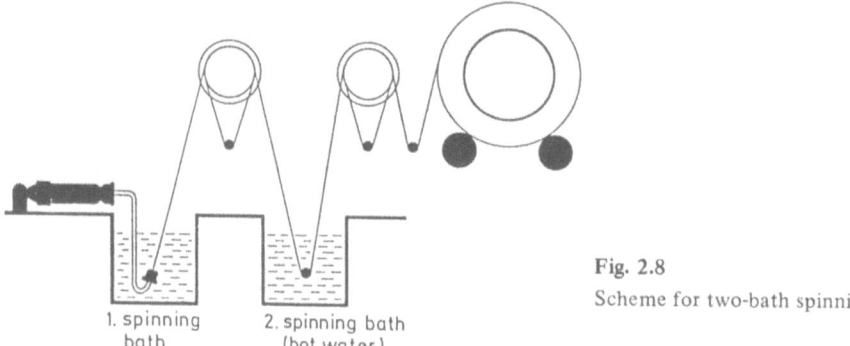

Fig. 2.8 Scheme for two-bath spinning

1. spinning bath
2. spinning bath (hot water)

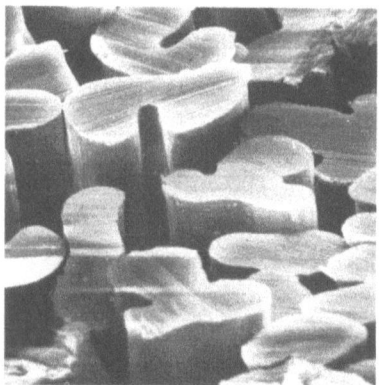

Fig. 2.9 Cross section of arcylic fibres 1 : 600

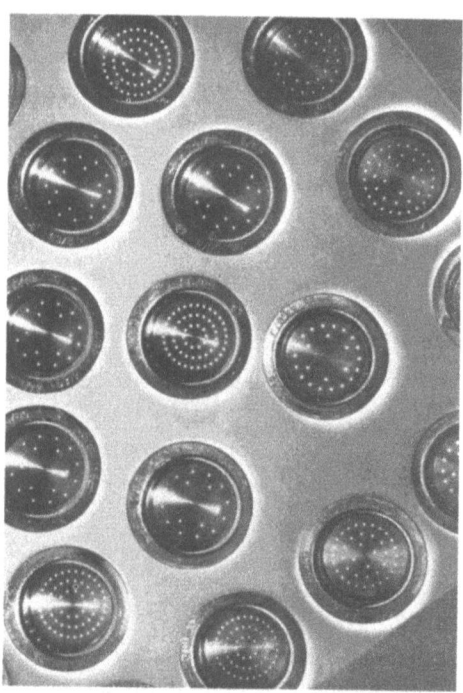

Fig. 2.10 Assorted spinnerettes to give different fibre deniers [4]

Table 2.1 Fibre denier

Coarse	5.5 dtex and above
Medium	3.7 dtex
Fine	1.7–2.7 dtex
Superfine	0.05–1.7 dtex

2.2 Fibres as Building Blocks

of 1 denier is equivalent to 450 m weighing 0.05 g. This unit of measure was transferred to man made fibres, 1 denier equates to 9000 m of filament weighing 1 g. In order to unify fineness measurement the TEX System was introduced:

$$1 \text{ tex} = \frac{1 \text{ g}}{1000 \text{ m}}.$$

2.2.3.2 Percentage Share by Fibre Types

Polyamides and Acrylics had a major share in filter fabrics during the 1950's and 60's. Since the development of needlefelts the trend has moved markedly to polyesters. A cautious estimate of fibre consumption for filter media in the German Federal Republic for 1981 is around 3700 tons per annum. This is equivalent to 9.5 million m^2 of filtering surface (needlefelts and fabrics).

Table 2.2 estimates the contribution of different fibre types by weight (15–25 μm).

Table 2.2 Percentage fibre contribution in filter media [5, 6]

Natural Fibres (Cotton, Wool)	12 %
Polyester	45 %
Acrylics	15 %
Polypropylene	12 %
Fibreglass	3 %
Others (PTFE, Metal, Mineral, PVC, etc)	1 %

2.2.3.3 Polyester (PES)

Table 2.2 shows polyester being the most frequent fibre worked into filter media. Chemically (fig. 2.11a) we are dealing with an ester derived from an alcohol (glycol) and an acid (terephthalic acid). During polymerisation polyethylene glycol terephthalate is formed [4]. This fibre is stable to 150 °C and has high mechanical strength (fig. 2.11b and 2.11c). The critical temperature for a polyester lies between 218–222 °C, i.e. even a brief excursion beyond this temperature leads to serious fibre damage. Polyester felts and fabrics should be thermofixed at 190–200 °C. This compacts the molecular chain of the polymer and achieves an absolute dimensional stability. Polyester hydrolises when exposed to "moist heat" or "hot humidity" above 70 °C. Hydrolysis is accelerated by alkali. This process is covered in more detail below. Acid resistance of polyester is good.

2.2.3.4 Polyacrylonitrile (PAN)

Among these fibres (fig. 2.12a und 2.12b) we differentiate between copolymers (Dralon®, Orlon®) with at least 85 % polyacrylonitrile content, and a polymer made of pure polyacrylonitrile) (Dralon® T). Copolymers have a service temperature of 115 °C, and 125 °C for pure polymers, respectively. During process drying or when running close to an acid dewpoint a difference of 10 °C will decide which type to use. In general acrylics are chosen, if the resistance to hydrolysis of polyester is insufficient. Like all members of the acrylate family (e.g. Plexiglas®) these fibres are resistant to ageing. However, they are thermoplastic and deform easily, if the above temperatures are exceeded. They have

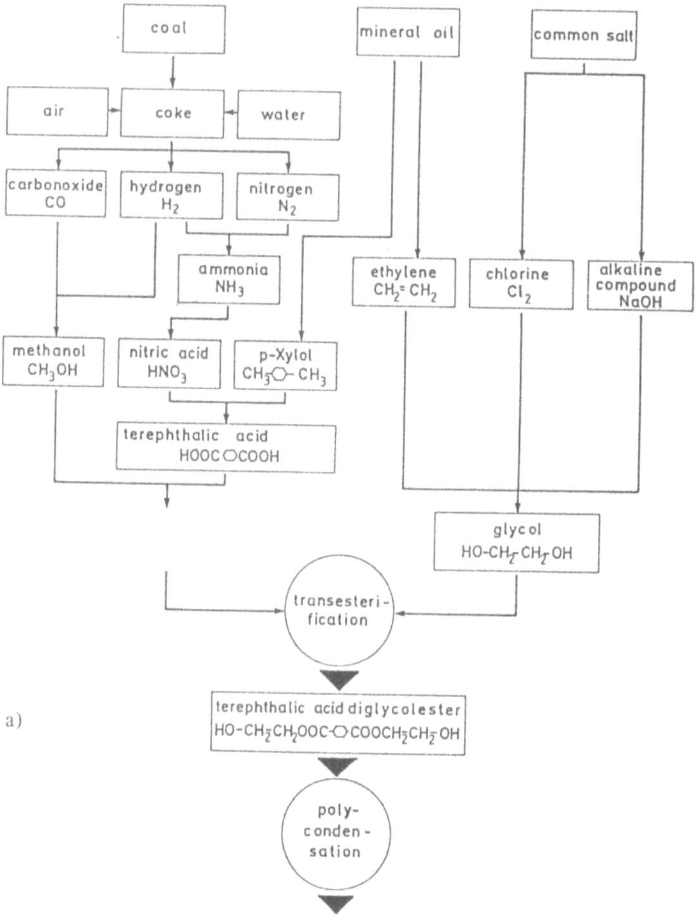

a)

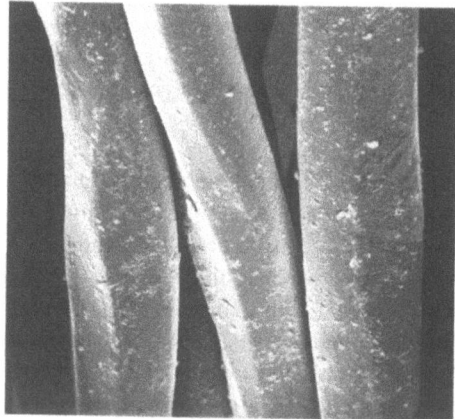

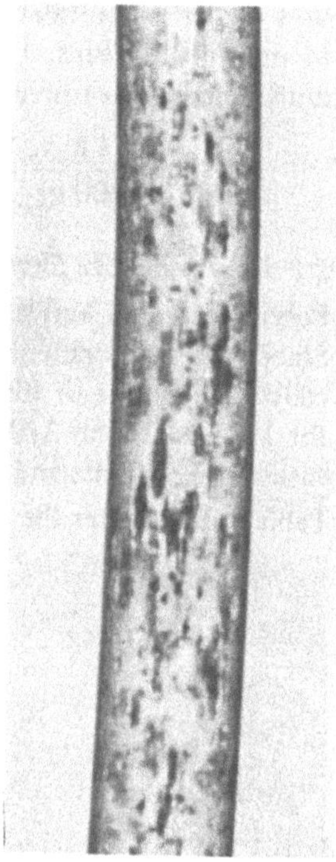

b)

c)

Fig. 2.11
Manufacture of polyester with differing characteristics

a) Polymerisation flowchart
b) Matte polyester fibre
c) Textured polyester

2.2 Fibres as Building Blocks

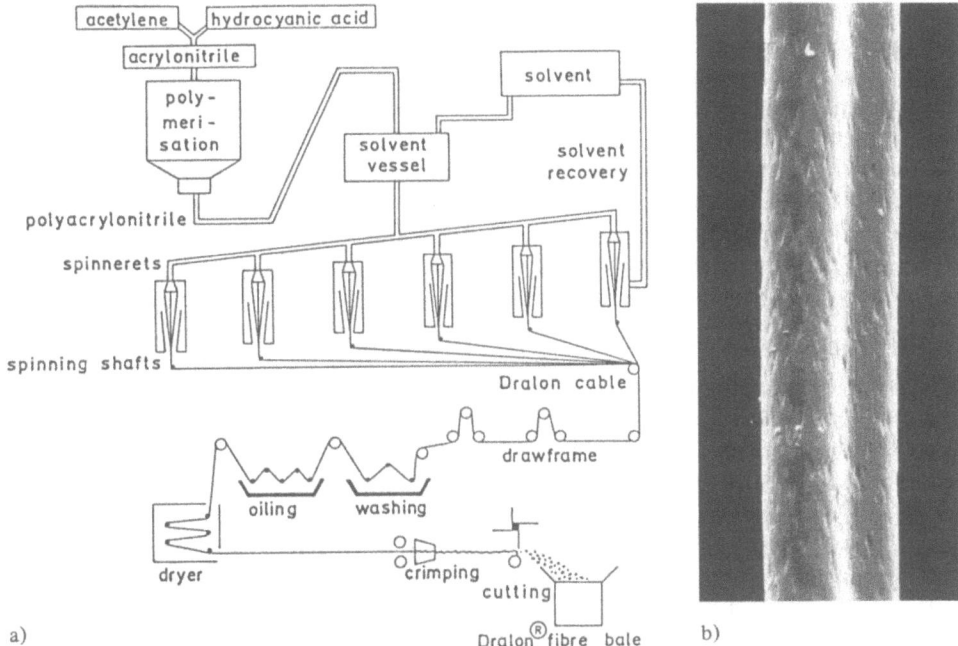

Fig. 2.12 Manufacture of polyacrylonitrile and typical cross section
a) Schematic flow diagram, Bayer [2]
b) Dumbell shaped PAN

adequate mechanical strength, even if somewhat less than polyester or polyamide fibres. On the other hand fabrics and needlefelt of this chemical type excel in their resistance to wet steam, alkali and acid. Caution should be excersised when mixing polyesters and acrylics. This applies to fabrics as well as needlefelts, because glycol is released during hydrolytic decomposition of polyester. This causes the acrylic to swell and to open the molecular structure. Thus hydrolysis attack gains access to the fibre interior. Certain metal salts such as zinc chloride or barium sulphate act similarly.

2.2.3.5 Aliphatic Polyamides (PA)

Aliphatic polyamides (Nylon®, Perlon®) have achieved great importance in dust separation, and in the filtration of liquids where they are used in fabric form. As needlefelts for filter media they have not enjoyed the same success. Limited temperature rating (110 °C); more difficult compaction during needling, because of its high resilience; insufficient resistance to acid and probably also commercial aspects are the reasons for this. Polyamides excel in high mechanical strength. They should be treated with caution in applications likely to meet high temperatures in order to preserve their dimensional stability.

2.2.3.6 Aromatic Polyamides (Aramides)

The increasing demand for hot gas cleaning has led to a corresponding increase in the demand for heat resistant filter media. Aramides (e.g. NOMEX®, Conex®) have gained in popularity. These fibres can withstand service temperatures of 180 °C in a neutral environment and may even survive short peaks close to 220 °C (fig. 2.13). Aramides do not melt, but merely char. The fibre has moderate acid and limited alkali resistance. It is sensitive to humidity, especially at elevated temperatures. Hydrolysis protective treatments can be applied which retard chemical attack sufficiently to make the costly filter material commercially viable. Aramides have high strength: their fineness, crimping and friction characteristics give finished products a high filtration efficiency.

2.2.3.7 Polypropylene (PP)

Polypropylene is a polyolefine fibre and originates as a byproduct of crude oil refining. In jest it is sometimes called "the TEFLON® of the poor man", because polypropylene has excellent resistance to acid, alkali and solvents. It absorbs negligible amounts of moisture and does not retain it. It is therefore exceptionally non-adhesive, dirt repellent and easy to clean. Polypropylene rarely builds up high differential pressures even with compacting dusts. However, its service temperature is only 100 °C. It has a high electrostatic chargeability (specific resistance 10^{18} $\Omega \cdot$ cm) so that antistatic precautions are recommended when dealing with dusts that have very low conductivity. Mechanical strength of polypropylene lies below those of polyester, but is sufficient for many dust conditions to yield a viable service life. Under strongly oxydative conditions or in the presence of UV

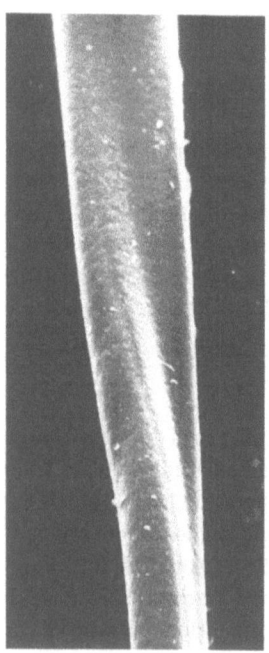

Fig. 2.13 NOMEX®

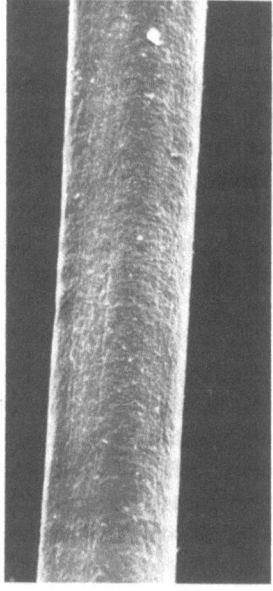

Fig. 2.14 TEFLON®

2.2 Fibres as Building Blocks

radiation polymer damage will occur, if the fibre is not sufficiently protected. The thermoplastic nature of the fibre allows surface treatments on the filter medium to smooth the surface. Caution is advised when treating needlefelts to prevent pore closure leading to lower porosity and possible blinding.

The above mentioned fibres are those most frequently encountered in filter media. Cellulose, PVC or copolymer fibres are only mentioned for completeness sake.

2.2.3.8 Polytetrafluoroethylene (PTFE)

The heat resistant fibres tell another story. They have made significant progress and have opened completely new areas in dust collection which have up to now not been accessible for temperature reasons. First in line is the PTFE fibre (Polytetrafluoroethylene). Polymer chemistry "magic" has made it possible to introduce fluorine into the molecule and to achieve hitherto unobtainable chemical resistance (ref chemical structure). This fibre is marketed under the trade name of TEFLON® (fig. 2.14), Toyoflon® and Hostaflon®. It withstands service temperatures of 260 °C with short peaks of 280 °C. At 290 °C the evaporation rate of fluorine is 0.0002 % per h: at 430 °C it is 4.5 % per h. This fibre has almost unlimited acid, alkali and solvent resistance and will only be attacked by a mixture of one part concentrated nitric acid and 3 part hydrochloric acid. Some perfluoro organic combinations can act as solvents.

$$\left(\begin{array}{cc} F & F \\ | & | \\ -C & -C- \\ | & | \\ F & F \end{array} \right)_n$$

chemical structure of PTFE

At 2.1 g/cm^3 the density of TEFLON® is nearly twice that of polyester (1.38 g/cm^3). In consequence a servicable TEFLON® felt has a density of 0.6 g/cm^3 against 0.3 g/cm^3 for polyester. To ensure high dimensional stability and to avoid sagging under its own weight requires a setting temperature just below 300 °C. Also, high gas temperature applications necessitate such high setting temperatures, because shrinking filter elements can no longer be cleaned nor loosened from their cages. The physical properties (strength, abrasion resistance and extensibility) are lower for PTFE fibres. Proper manufacturing of needlefelts (dense needling, full heat sealing of the fibre which can shrink up to 40 %) results in a fully servicable filter medium with a service life of 3 to 5 years.

PTFE is slippery, smooth and non-sticky. The fibre diameter is coarser than those of other fibres after finishing. Processing is difficult, because the fibre has little friction and is prone to electrostatic charging. This makes a homogeneous fibre distribution difficult. These limitations work against good separation efficiencies and a careful blend of production and finishing treatments are necessary to compensate with higher linear weight for these defficiencies.

Further improved, mechanical and thermal properties have been announced, e.g. the American fibres Ryton® and PBI®. It is still too early to speculate on their application and cost effectiveness.

2.2.3.9 Mineral Fibres (Natural and Man Made)

Organic fibres with high temperature resistance are expensive and have a limiting service temperature of 280 °C. Inorganic fibres or fibres from inorganic materials have always had a great attraction. Glassfibre fabrics (area weight 200–600 g/m^2) (fig. 2.15) and sometimes up to 900 g/m^2 with antislip elastomeric coatings such as graphite, silicone or PTFE have found application for many years in low pressure reverse cleaned baghouses. A highly refined E Glass as used in the electronic industry with low alkali content with a fibre diameter of 3–13 μm is the base. Fibreglass does not burn: it softens at 500–700 °C and depending on glass type melts between 900 °C and 1300 °C. Initial strength loss occurs above 230 °C. Low alkali fibreglass has relatively good chemical resistance; however, it is rapidly attacked by fluoric acid (fig. 2.16).

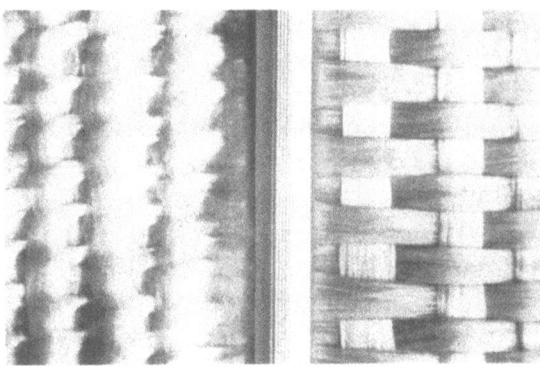

Fig. 2.15 Types of fibreglass weaves

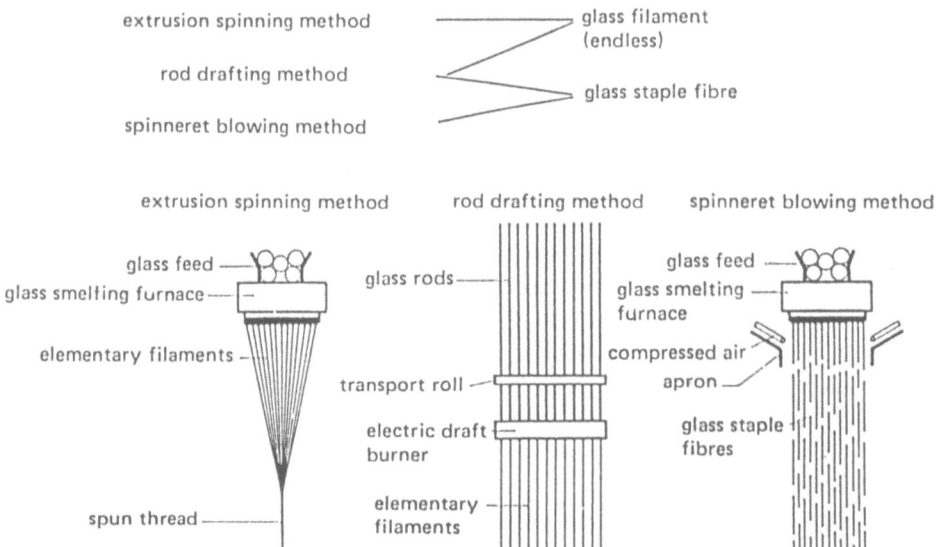

Fig. 2.16 Various methods of making fibreglass [7, 8]

2.2 Fibres as Building Blocks

Besides the long established fibreglass fabrics, there have appeared on the market needle-felts with a woven support scrim. For the fibreglass web (fig. 2.17) E glass is used with a fineness of around 3 μm. The weight of such felts, which are impregnated with up to 10 % of organic binders, lies between 800 and 1000 g/m². Temperature is limited by the binder and ranges between 220–250 °C.

Webs and needlefelts of metal fibres (e.g. stainless steel X 5 CrNi 18 8 and X 2 CrNiMc 18 10) with a fibre diameter of 8–12 μm are suitable for a hot gas temperature of 400–600 °C. The development of such fibres has reached a plateau. It remains to justify the cost effectiveness of such high-priced fibres. By reason of their potential face velocities and ease of cleaning this appears feasible (fig. 2.18).

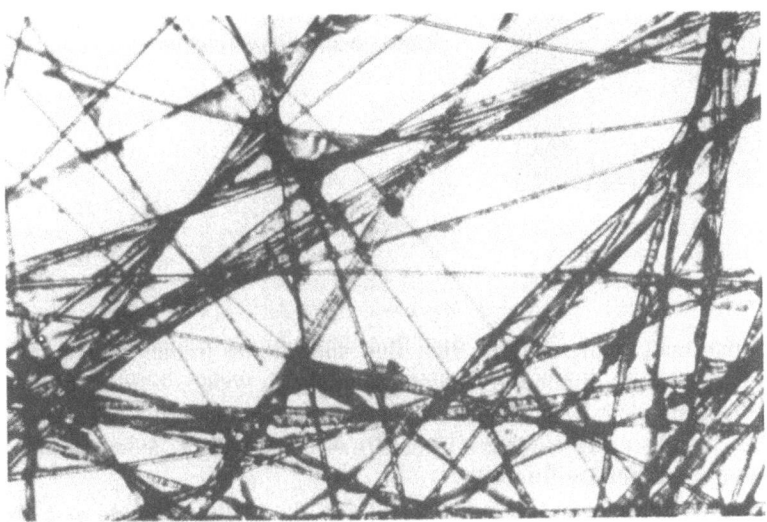

Fig. 2.17 Structure of fibreglass web

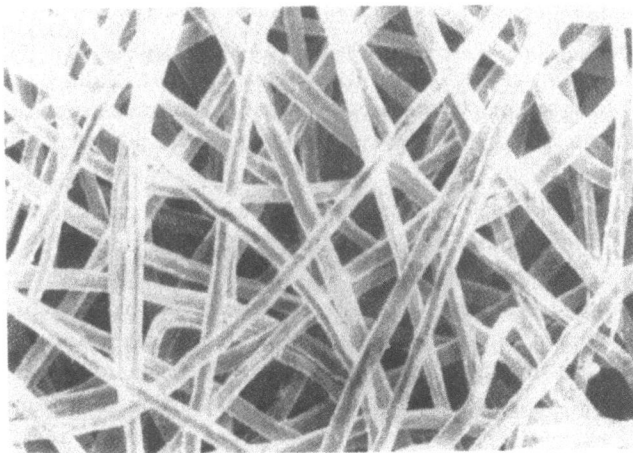

Fig. 2.18
Structure of metal fibre web

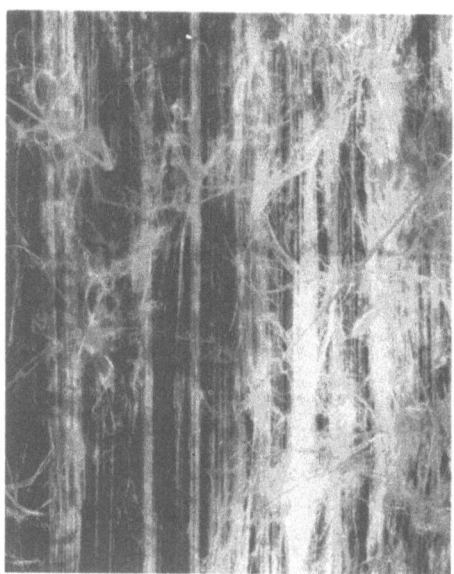

Fig. 2.19
White asbestos, fibre structure

A further class of man made mineral fibres that find applications in high temperature situations are quartz, silicium dioxide, rockwool, aluminium oxide, borosilicates or modified glasses. For some years white asbestos played a significant part. Its ability to split into fibrils gives it high surface activity (fig. 2.19). Asbestos is now recognized as a carcinogen and no longer used in dry filtration.

Woven fabrics only can be made from the above mentioned fibres with high area weights. They have a high service temperature of 600–800 °C which has been pushed to 1400 °C. The production of needlefelts has so far not been possible, because the fibres are brittle and break during needling. Also temperature resistant flexible binders to hold the textile structure together are not available to achieve commercially tolerable service lives.

When assessing temperature stability one must differentiate between behaviour of the bulk of the material and that of the fibres made from it. The much greater temperature sensitivity of the fibrous structure compared to the mother material is often disappointing. The origin lies in the huge surface area that is presented to the gas or the aggressive dust.

At the end of this chapter on fibres [9] we wish to point out why it is necessary to consider such a wide variety of fibres as building elements for textile filter media. This is contrary to economic thinking. The reason is simple: The chemical industry owes us a fibre which is resistant to hydrolysis, resistant to all other chemicals, can be used cold or hot, that separates effectively and can be easily cleaned, that can be easily processed, is durable and cheap. To date such a fibre has not come forward. We are therefore left to optimise to the best of our ability and to overcome the many processing limitations and economic barriers.

2.2 Fibres as Building Blocks

Table 2.3 Tradenames and Manufacturers of Synthetic Fibres [2]

Trade Name	Raw Material acc. to DIN 60001	Manufacturer, Country
A		
Acetat	CA	Rhodia, FR Germany
Acrilan	PAC	Monsanto, USA
Acrilan 45	PAC (bicomp.)	Monsanto, USA
Acrilan 57	PAC (bicomp.)	Monsanto, USA
Acrilan 71	PAC (bicomp.)	Monsanto, USA
Acrilan 94	PAC	Monsanto, USA
Acrilan 96	PAC	Monsanto, USA
Acrilan-Bi-Loft	PAC (bicomp.)	Monsanto, USA
Acrylast	PST	Dawbarn, USA
Airon PL	CM (CP)	Montedison, Italy
Airon TK	CM (HWM)	Montedison, Italy
Algoflon	PTFE	Montecatini, Italy
Alon	CT	Toho Rayon, Japan
Alphaquartz	KE	Alpha, USA
Aluthen	MT (foil)	Wolff, FR Germany
Anilana	PAC	LZWS, Poland
Anso	PA 6 (hollow fibre)	Allied Chemical, USA
Antron	PA 6.6	DuPont, USA
Antron III HF	PA 6.6 (antist.)	DuPont, USA
Antron T 838	PA 6.6 (hollow fibre)	DuPont, USA
Arenka	PA (arom.)	Enka, FR Germany
Arnel	CT	Celanese, USA
Artilana	CV	Svenska Rayon, Sweden
Asota	PP	Chemie Linz, Austria
Astroquartz	KE	Stevens, USA
A-Tell	PES (PEE)	A-Tell, Japan
ATF 1017 (s. Dunova)	PAC (absorptive)	Bayer, FR Germany
Atlas Wire	PA 6	Bayer, FR Germany
Austrophan	CV (foil)	Lenzing, Austria
Avceram	KE	American Viscose, USA
Avceram S	KE	American Viscose, USA
Avceram CS	KE	American Viscose, USA
Avisco PFR	CV (flame retardent)	American Viscose, USA
Avisco Vinyon	PVC/PVAA	American Viscose, USA
Avril	CM (HWM)	American Viscose, USA
Avril	CV (flame retardent)	American Viscose, USA
B		
Beamette	PP	Dawbarn, USA
Bedor	MT	Benedict & Dannheisser, FR Germany
Bemberg	CC	Bemberg, several countries
Bemberg-Zellglas	CC (foil)	Bemberg Folien, FR Germany
Bemberg-Folie	PE (foil)	Bemberg Folien, FR Germany
Beslon	PAC	Toho Rayon, Japan
Beta	GL	O.C. Fiberglas, USA and Belgium
Bidim	PES (spun Bond)	Rhône-Poulenc, France
Bekinox	MT	Bekaert, Belgium
Biolan	PVAA	USSR

Trade Name	Raw Material acc. to DIN 60001	Manufacturer, Country
Bluebell	PP	Belfast Rope, Northern Ireland
Bri-Nylon	PA 6.6	ICI, England
Bristex	PVC/PVA	Polymers, USA
Bristrand	PVC/PVA	Polymers, USA
Brulon	PA 6.10	ICI, England
Brunsmet	MT	Brunswick, USA
BX-Saran	PVD	Bakalite Xylonite, England
C		
Cadon	PA 6.6	Monsanto, USA
Cambrelle	PA 6/PA 6.6 (bicomp.)	ICI, England
Cantrece	PA 6.6/PA 10 (bicomp.)	DuPont, USA
Carbolon	CAR from PAC	Nippon Carbon, Japan
Casein	PR	–, Belgium
Cashmilon	PAC	Asahi Chemical, Japan
Caslen	PR	Rubberset, USA
Cellometall	MT (cellulose-glass)	Kalle, FR Germany
Cellophan	CV (foil)	Kalle, FR Germany
Cellophan PE	PE (foil)	Kalle, FR Germany
Cellophan WEKA	CV (foil)	Kalle, FR Germany
Celon	PA 6	Courtaulds, England
		Svensky Rayon, Sweden
Celon antistat	PA 6 (antistat)	Courtaulds, England
Chinon	PAC/PR	Toyobo, Japan
Chlorin	PVCC	–, USSR
Claron	PST	Rosenhirsch, USA
Clevyl F	PVC	Rhône Poulenc, France
Clevyl T	PVC	Rhône Poulenc, France
Clorène	PVD	Rhône Poulenc, France
Colback	PES/PA 6 (spun bond)	Enka, Netherlands
Colbon	PES (spun bond)	Enka, FR Germany
Conex	PA arom.	–, Japan
Cordacel	CV (cord)	Celanese, Mexico
Cordenka 700	CV	Enka, FR Germany
Cordenka EHM	CV (cord)	Enka, FR Germany
Cordron	CV (cord)	Courtaulds, Australia
Cordyl	CV (cord)	Rhône Poulenc, France
Corovin	PP (spun bond)	Benecke, FR Germany
Courtelle	PAC	Courtaulds, several countries
Courtelle LC	PAC (bicomp.)	Courtaulds, England
Cremona	PVAA	Kuraray, Japan
Creslan	PAC	American Cyanamid, USA
Creslan 68 CS	PAC (bicomp.)	American Cyanamid, USA
Crinovyl	PVCC	Rhône Poulenc, France
Crinvil	PVC	Polifiber, Italy
Crylor	PAC	Rhône Poulenc, several countries
Cuprothen	PE (foil)	Bemberg, FR Germany
Cuprammonium Rayon Hollow Fiber	CC (hollow fibre)	Asahi; Gosen, Japan
Cupro-Hohlfaser	CC (hollow fibre)	Bemberg; FR Germany

2.2 Fibres as Building Blocks

Trade Name	Raw Material acc. to DIN 60001	Manufacturer, Country
D		
Dacron	PES	DuPont, USA
Daiflon	PTFE	Osakakinzoku, Japan
Danamid	PA 6	−, Hungary
Danufil	CV	Süddeutsche Chemiefaser, FR Germany
Danufil K	CV	Südd. Chemiefaser/Hoechst, FR Germany
Danufil W	CV	Südd. Chemiefaser/Hoechst, FR Germany
Danuflor	CV	Südd. Chemiefaser/Hoechst, FR Germany
Darelle	CV	Courtaulds, England
Dawbarn	PST	Dawbarn, USA
Dawbarn-Saran	PVD	Dawbarn, USA
Dayan	PA 6	Perlofil, Spain
Dederon	PA 6	several plants, DR Germany
Dederon brillant	PA 6	several plants, DR Germany
Dederon flirret	PA 6	several plants, DR Germany
Delebion	PP	−, Italy
Delfion	PA 6	Snia, Italy
Depron	PST (foam foil)	Kalle, FR Germany
Diolen	PES	Enka, FR Germany
Diolen 23	PES	Enka, FR Germany
Diolen 51	PES	Enka, FR Germany
Diolen BC	PES	Enka, FR Germany
Diolen SV	PES/CV	Enka, FR Germany
Diolen Ultra	PES/PA (bicomp.)	Enka, FR Germany
Dolan	PAC	Hoechst, FR Germany
Dolan 81	PAC	Hoechst, FR Germany
Dolan 88	PAC (flame retardent)	Hoechst, FR Germany
Dorcolor	PA 6, PAC (spun dyed)	Bayer, FR Germany
Dorix	PA 6	Bayer, FR Germany
Dorlastan	PUE	Bayer, FR Germany
Dorosuisse	PA 6.6	Viscosuisse, Switzerland
Dorvivan	PA 6 (light stabilized)	Bayer, FR Germany
Draka-Saran	PVD	Draadenkabelfabriek, Netherlands
Dralon	PAC	Bayer, FR Germany
Dralon C	PAC (flame retardent)	Bayer, FR Germany
Dralon T	PAC (fil.)	Bayer, FR Germany
Dralon K	PAC (bicomp.)	Bayer, FR Germany
Drylon	PST	Plasticisers, England
Dunova	PAC (absorptive)	Bayer, FR Germany
E		
Eftrelon	PA mod.	−, DR Germany
Elana	PES	TZTS, Poland
Elaston	PUE	Chemiefaserwerk J. G., Poland
Elura	MOD	Monsanto, USA
Encron	PES	American Enka, USA
Enka N 20	PA arom.	Enka, FR Germany
Enka N 40	PA arom.	Enka, FR Germany
Enka comfort antistatic	PA (antistatic)	Enka, FR Germany
Enkafort	GL	Enka, Netherlands
Enkalon	PA 6	Enka, several countries
Enkalure	PA 6	Enka, several countries

Trade Name	Raw Material acc. to DIN 60001	Manufacturer, Country
Enka-Nylon	PA 6.6	Enka, FR Germany
Enkasa	PR	Enka, Netherlands
Enka-stat	PA (antistat.)	Enka, FR Germany
Enkaswing	PUE	Enka, Spain
Enkatherm (Versuchsfaser)	PTO	Enka, FR Germany
Enkatron	PA 6/PES (bicomp.)	Enka, Netherlands
Enkrome	CV	American Enka, USA
Envilon	PVC	Toyo Chemical, Japan
Epitropic-Nylon	PA (antistat.)	ICI, England
Espa	PUE	Toyobo, Japan
Euroacril	PAC	ANIC, Italy
Evlan	CV	Courtaulds, England
Evluxe	CV	Courtaulds, England
Exlan	PAC	Exlan, Japan
Exlan	(bicomp.)	Exlan, Japan
F		
Ferenka	MT	Enka, FR Germany
Fiber 700	CM (HWM)	American Enka, USA
Fiber HM	CM (HWM)	Rhône Poulenc, France
Fiberfrax	GL, KE	Carborundum, USA
Fiberglas	GL	Fiberglas Canada, Canada
Fiberglas	GL	Owens Corning Fiberglas, USA
Fibrafinn	CV	S.O., Finland
Fibravyl	PVC	Rhône Poulenc, France
Fibrolane	PR	Courtaulds, England
Filtrona	PP	Filtrona, England
Finilon	PA arom.	–, USSR
Fluon	PTFE	ICI, England
Forlion	PA 6	Orsio Mangelli, Italy
Fortanese	CA	Celanese, England
Forte	CM (CP)	Nitto Spinning, Japan
Fortisan	CA	Celanese, England
Fortrel	PES	Celanese, England
Fosfol	PVAA	–, USSR
Ftorlon	PTFE	–, USSR
Fujibo-Spandex	PUE	Fuji Spinning, Japan
Fulon	PTFE	Toray, Japan
Furlon	PVD	Nippon Geon, Japan
G		
Garan	GL	Johns-Manville, USA
Garanmat	GL	Johns-Manville, USA
Genolon	PVC (foil)	Kalle, FR Germany
Genopak	PVC (foil)	Kalle, FR Germany
Genotherm	PVC (foil)	Kalle, FR Germany
Gerrix	GL	Gevetex Textilglas, FR Germany
Gevetex	GL	Gevetex Textilglas, FR Germany
Glassion	GL	Asahi Fiber Glass, Japan
Glospan	PUE	Globe, USA and England
Grafil	CAR from PAC	Courtaulds, England
Grilene	PES	Grilon, Switzerland

2.2 Fibres as Building Blocks

Trade Name	Raw Material acc. to DIN 60001	Manufacturer, Country
Grilene	PES (PEE)	Grilon, Switzerland
Grilene SAP	PES	Grilon, Switzerland
Grilon	PA 6	Grilon, Switzerland
Grilon-Hohlfaser	PA 6 (hollow fibre)	Grilon, Switzerland
Grilon CS	PA 6	Grilon, Switzerland
Grilon K 115	PA 12	Grilon, Switzerland
Grilon K 140	PA 12	Grilon, Switzerland
Grisuten	PES	several plants, DR Germany
Guttagena	PVC (foil)	Kalle, FR Germany
Gymlene	PP	–, England

H

Helion	PA 6	Montefibre, Italy
Herculon	PP	Hercules, USA
Hochmodul 333	CM (HWM)	Lenzing, Austria
Hostaphan	PES (foil)	Kalle, FR Germany
Hostaphan PE	PES/PE (foil)	Kalle, FR Germany
Hyfil	CAR from PAC	Hyfil, England

I

Ibetherm	PES/PE (foil)	Bemberg, FR Germany
Islon	PA 6	Islon, Turkey
Isover-TR	ST	Grünzweig u. Hartmann, FR Germany
Isover-Tel	GL	Grünzweig u. Hartmann, FR Germany
Istrona	PP	–, CSSR

J

Junion	CM (CP)	Fuji Spinning, Japan

K

Kanebo Nylon 22	PA 6 (bicomp.)	Kenebo, Japan
Kanekalon	MOD	Kanegafuchi, Japan
Kermel	PA-Imid	Rhône Poulenc, France
Kevlar	PA arom.	DuPont, USA
Khlorin	PVCC	–, USSR
Klingerflon	PTFE	Klinger, England
Kodel	PES (PCHT)	Eastman, USA
Koplon	CM (CP)	Snia, Italy
Krehalon	PVD	Kureha, Japan
Krehalon S	PVC	Kureha, Japan
Kuralon	PVAA	Kuraray, Japan
Kuraray	PES	Kuraray, Japan
Kureha-Pitch	CAR	Taiyo Kakon, Japan
Kynol		Carborundum, USA

L

Lavsan	PES	C.K. Kursk, USSR
Leacril	PAC	Montefibre, Italy
Leavil	PVC	Montefibre, Italy
Lenasal	CV	Snia, Italy
Lenzing acryl	PAC	Lenzing, Austria
Letilan	PVAA	–, USSR

Trade Name	Raw Material acc. to DIN 60001	Manufacturer, Country
Letyn	PVAA	–, USSR
Lilion	PA 6	Snia, Italy
Lilion antistatic	PA (antistat.)	Snia, Italy
Lufnen	PAC (flame retardent)	–, Japan
Lurex	MT (foil)	Lurex, several countries
Lutrasil	PP (spun bond)	Freudenberg-Gruppe, FR Germany
Lycra	PUE	DuPont, USA
M		
Malon	PAC	Ohis-Organsko, Jugoslavia
Manryo	PVAA	Kuraray, Japan
Marglass	GL	Marglass, England
Mawol	PVAA	–, USSR
Melana	PAC	Uzina, Rumania
Melkwol	PR	Snia, Italy
Meraklon	PP	Montefibre, Italy
Merinova	PR	Snia, Italy
Merolon	PVAA	Toyobo, Japan
Metalltransparit	MT/cellulose glass	Wolff, FR Germany
Metlon	MT	Metlon Co., USA
Mewlon	PVAA	Unitika, Japan
Microlith	GL	Schuller, FR Germany
Mixel AN	CA/PA 6.6 (heteroyarn)	Teijin, Japan
Mixel AT	CA/PES (heteroyarn)	Teijin, Japan
Mobilon	PUE	Nisshin Spinning, Japan
Modmor	CAR from PAC	Morganite Modmor, England
Monsanto SEF Modacrylic	PAC (flame retardent)	Monsanto, USA
Monvelle	PA 6.6/PUE (bicomp.)	Monsanto, USA
Morganite	CAR from PAC	Morganite Modmor, England
Movil	PVC	Montefibre, Italy
Movil F	PVC	Montefibre, Italy
Movil N	PVC	Montefibre, Italy
Movil T	PVC	Montefibre, Italy
MP-Faser	PVC/PVAC	Wacker Chemie, FR Germany
Mtilon	PAC/CV	–, USSR
Mylar	PES (foil)	DuPont, USA
N		
Nailon	PA 6.6	Montefibre, Italy
Nalophan	PES (foil)	Kalle, FR Germany
Neva-Viscon	CV/PA 6 (heteroyarn)	Enka, FR Germany
Nichiray	PES	Unitika, Japan
Nipla	PVC	Nichici Chemical, Japan
Niplon	PVC	Nichici Chemical, Japan
Nitlon	PAC	Nitto Boseki, Japan
Nitron	PAC	–, USSR
Nomex	PA arom.	DuPont, USA
Novotherm	PVC (foil)	Bemberg, FR Germany
Numa	PUE	American Cyanamid, USA
Nurel	PA 6	Fibras Esso, Spain

2.2 Fibres as Building Blocks

Trade Name	Raw Material acc. to DIN 60001	Manufacturer, Country
Nylfrance	PA 6.6	Rhône Poulenc, France
Nylfrance nostatic	PA (antistatic)	Rhône Poulenc, France
Nylsuisse	PA 6.6	Viscosuisse, Switzerland
O		
Omni-Saran	PVD	Omni de Mexico, Mexico
Opelon	PUE	Toyobo, Japan
Oplon	PP	Jeil Synthetic Fibers, South Korea
Orlon (Sayelle)	PAC	DuPont, USA
Orlon-T 21	PAC (bicomp.)	DuPont, USA
Orlon-T 34	PAC (bicomp.)	DuPont, USA
Orlon-T 75	PAC (bicomp.)	DuPont, USA
P		
Panacryl	PAC	VEB Budapest, Hungary
Perlonfaser EX 301	PA (antistatic)	Bayer, FR Germany
Perlon-Glitzer	PA 6	several manufacturers, FR Germany
Permene	PST	Modglin, USA
Pewlon	PAC	Asahi Chemical, Japan
Piviacid (früher PeCe)	PVC	VEB Wolfen, DR Germany
Platon	PA 6	Plate, FR Germany
Platon	PA 6.10	Plate, FR Germany
Pliana	PP	
Polifen	PTFE	–, USSR
Polital	PP (split fibre)	Adolff, FR Germany
Polsilon	GL	–, Poland
Polyfiber	PST	Dow Chemical, USA
Polyflon	PTFE	several manufacturers, Japan
Polyfluff	PST	Polymers, USA
Polyno	CM (CP)	Daiwa Spinning, Japan
Polynosic BX	CM (CP)	Rhône Poulence, France
Polynosica	CM (CP)	Safa, Spain
Polypro	PP	Syn-Pro, USA
Polysteen PP	PP	Steen FR Germany
Prima	CM (HWM)	Snia Viscosa, Italy
Prolene	PP	several manufacturers
Proplon	PP	Lardanais, France
Protel	PP	Chemcell, Canada
Pylen	PP	Mitsubishi, Japan
Q		
Qiana	PA alicycl.	DuPont, USA
R		
Redon	PAC	ehemals Phrix, FR Germany
Reemay	PES	DuPont, USA
Refrasil		Thompson Fiber Glass, USA
		Chemical Insulting, England
Reilen	PP	Reinhold KG, FR Germany
Retractyl	PVC	Rhône Poulenc, France
Rhonel	CT	Rhône Poulenc, France
Rhovylon	PVC/PA 66	Rhône Poulenc, France

Trade Name	Raw Material acc. to DIN 60001	Manufacturer, Country
Rhovyl	PVC	Rhône Poulenc, France
Rigilor	CAR from PAC	Rhône Poulenc, France
Rilsan	PA 11	France, Italy, Brazil
Rovan	CV	Rhône Poulenc, France
Rovana	PVD	Dow Badische, USA
R-P-Acrylfaser Typ 910	PAC	Rhône Poulenc, France
Ryton	PPS	Phillips Fibers, USA
S		
Safacril	PAC	SAFA, Spain
Salfil	KE	ICI, England
Sanderit	PA 6	Sander, FR Germany
Saran	PVD	Fegafil, FR Germany
Sarlon	PVD	Universal Products, Australia
Scaldyl	CV	Fabelta, Belgium
Seflon	PA 6	Seflon, Turkey
Shalon	PST	Bakalite Xylonite, England
Shalon	PST	Shawinigan, Canada
Shalon	PST	Polymers, USA
Sideria	PA 6/PES (bicomp.)	Kanebo, Japan
Sigrafil	CAR from PAC	Sigri Elektrographit, FR Germany
Silenka	GL	Silenka, Netherlands
Silfas	PA 6	–, Turkey
Sillione	GL	Verre Textile, France
Sillook	PES	Toray, Japan
Silon	PA 6	Silon, CSSR
Silpalon	PAC	–, Japan
Sinitex	PVD	Industrias Sinimbue, Brazil
Snia Super	CV-Cord	Snia, Italy
Solvron	PVA	Nitivy, Japan
Source	PA 6/PES (bicomp.)	Allied Chemical, USA
Soviden	PVD	–, USSR
Spanzelle	PUE	–, England
Ssaniw	PVD	–, USSR
Stex	PST	Vectra, USA
Stilon	PA 6	–, Poland
Stratifil	GL	Isoverbel, Belgium
Styroflex	PST	Norddeutsche Seekabel, FR Germany
Styron	PST	Asahi-Dow, Japan
Sulfon T	PA arom.	–, USSR
Super Rayflex	CV (cord)	American Viscose, USA
Super-Supercord	CV (cord)	–, Rumania
Supralan	CV	Borregard, Norway
Supralon	PA 6	Progress, Jugoslavia
Suprenka	CV (cord)	American Enka, USA
Supronyl	PA 6.6 (foil)	Kalle/Hoechst, FR Germany
Supronyl-PE	PA 6/PE (foil)	Kalle/Hoechst, FR Germany
Suprotherm	PVC (foil)	Kalle/Hoechst, FR Germany
Survon	PA 6.10	ICI, England
Syphibers	KE	Aluminium Co, USA

2.2 Fibres as Building Blocks

Trade Name	Raw Material acc. to DIN 60001	Manufacturer, Country
T		
Tapiflor	CV	Lenzing, Austria
Tapilon	PA 6/PA 66 (bicomp.)	Toray, Japan
Teflon	PTFE	DuPont, USA
Teijin-Cord	CV (cord)	Teijin, Japan
Teklan	PVD	Courtaulds, England
Teklan.	MOD	Courtaulds, England
Tenasco	CV (cord)	Courtaulds, England
Tenax		Enka, Netherlands
Terel	PES	UFSS, Rumania
Tergal	PES	Rhône Poulenc, several countries
Tergal X 403	PES (bicomp.)	Rhône Poulenc, France
Teriber	PES	SAFA, Spain
Terital	PES	Montefibre, Italy
Terlenka	PES	Enka, Netherlands
Terlenka 620	PES (hollow fibre)	Enka, Netherlands
Terlenka 623	PES (hollow fibre)	Enka, Netherlands
Terlenka 2000	PES	Enka, Netherlands
Ternel	PVC	Polymer Industrie, Italy
Tersuisse	PES	Viscosuisse, Switzerland
Terylene	PES	ICI, several countries
Tesil	PES	Silon, CSSR
Tetoron	PES	Teijin, Japan
Teviron	PVC	Teijin, Japan
Texover	GL	Vidrireria, Argentina
Texvil	PVC	SAIR, Italy
Thermovyl	PVC	Rhône Poulenc, France
Thornel P	CAR	Union Carbide, USA
Thornel 50	CAR	Union Carbide, USA
Titanol	PVAA	–, USSR
Tohalon	CT	Toho Rayon, Japan
Tolon	PVC	Toyo Chemical, Japan
Torayca	CAR	Toray Ind., Japan
Toraylon	PAC	Toray, Japan
Toraylon	PAC (bicomp.)	Toray, Japan
Torlen	PES	–, Poland
Toyobo	PES (flame retardent)	Toyobo, Japan
Toyoflon	PTFE	Toray, Japan
Transparit	CV (foil)	Wolff, FR Germany
Trespaphan	PP (foil)	Kalle/Hoechst, FR Germany
Trevira	PES	Hoechst, FR Germany
Trevira brillant	PES	Hoechst, FR Germany
Trevira VF	PES	Hoechst, FR Germany
Trevira 350	PES	Hoechst, FR Germany
Triaceta	CT	Daicel, Japan
Triacetat-Spinnf.	CT	Celanese, several countries
Tricel	CT	Celanese, England
Tricel Duracol	CT	Celanese, England
Tricella	CT	Celanese, England
Tricelon	CT/PA 6	Courtaulds, England
Trilan	CT	Celanese, Canada

Trade Name	Raw Material acc. to DIN 60001	Manufacturer, Country
Trinese	CT	Celanese, Mexico
Trinyl	PA 6	Inquitex, Spain
Triton-Kaowool	KE	several manufacturers, USA, Belgium
Trofil Typ P	PP	Dynamit Nobel, FR Germany
Typar	PP (spun bond)	DuPont, USA
Tygan	PVD	Fothergill Harvey, England
U		
Ultron	PA (antist.)	Monsanto, England
V		
Vairin	PUE	Pirelli, Italy
Vairen	PVC	Teijin, Japan
Varlen	PVC	Teijin, Japan
Vectra	PP	Enjay, USA
Vectra-Saran	PVD	Enjay, USA
Vegon	PP	Faserwerk Bottrop, FR Germany
Velana	PES	Silon, CSSR
Velicren	PAC (bicomp.)	Snia, Italy
Velicren FR	PAC (flame retardent)	Snia, Italy
Velon	PVD	Firestone Plastics, USA
Verel	MOD	Eastman, USA
Verel Modacrylic	PAC (flame retardent)	Eastman, USA
Vestan 21	PES	Bayer, FR Germany
Vestan W	PES (PCHT)	Bayer, FR Germany
Vetrolon	GL	Gevetex Textilglas, FR Germany
Vetrotex	GL	Vetreria Balzareffi, Italy
Viloft	CV	Courtaulds, England
Vilon	PVAA	Nitivy, Japan
Vincel 28	CM (HWM)	Courtaulds, England
Vinylon	PVAA	several manufacturers, Japan
Visca-Flock	CV (flock)	Enka, FR Germany
Viscolan	CV	Lenzing, Austria
Viscolen	CV	Lenzing, Austria
Viscothen	CV (foil)	Wolff, FR Germany
Viscothen	CV/PE (foil)	Wolff, FR Germany
Viskose FR	CV (flame retardent)	Lenzing, Austria
Vitro-Flex	GL	Johns Manville, USA
Vitron	GL	Johns Manville, USA
Vitrotex	GL	Johns Manville, USA
Vivalon	PA 6	Bayer, FR Germany
Vonnel	PAC	Mitsubishi, Japan
Vonnel V-57	PAC (bicomp.)	Mitsubishi, Japan
Voplex	PVC/PVD	Vogt, USA
Vulflon	PTFE	Nippon Valqua, Japan
Vyrene	PUE	USA, England, Italy
Vytacord	PES	Goodyear, USA
W		
Walomid	PA 6 (foil)	Wolff, FR Germany
Waloplast	PE (foil)	Wolff, FR Germany
Walothen	PP (foil)	Wolff, FR Germany

2.2 Fibres as Building Blocks

Trade Name	Raw Material acc. to DIN 60001	Manufacturer, Country
Walothen-Combi	PP (foil)	Wolff, FR Germany
Walotherm	PE (foil)	Wolff, FR Germany
Wetrelon	PA mod.	–, DR Germany
Wipolan	PR	Lodzkie, Poland
Wistel	PES	Snia, Italy
Wolpryla	PAC	–, DR Germany
Wolpryla-Se	PAC (flame retardent)	–, DR Germany
Woolon	PVAA	Japan Synthetic Fiber, Japan
Y		
Yulon	PA 6	KTM, Jugoslavia
Z		
Zantrel	CM (CP)	American Enka, USA
Zefran	PAC	Dow Badische, USA

Table 2.4 Physical properties of fibres[1]

Fibre		Cotton	Wool	Polyester	Amide	Acrylic	Metal[2]	Fibreglass	Arom. PA	PTFE
Trade name				Diolen® Trevira®	Perlon (P)® Nylon (N)®	Dralon T® Orlon 81®	BEKINOX®		Nomex®	Teflon®
Density	g/cm³	1.5–1.54	1.32	1.2–1.4	1.14	1.14–1.18	7.7–7.9	2.45–2.6	1.38	2.1
Tensile strength	daN mm²	35–70	15–25	35–90 Hi tenacity 95–130	45–70 Hi tenacity 70–100	23–40 Hi tenacity 40–55	500–1000	175–300	75	unbleached 16–28
Regain	21 °C 65 % RH	7–11	15–17	0.2–0.5	3.5–4.5	1.0–1.5	–	–	4.5–5	–
Moisture retention	%	40–50	40–50	3–5	10–15	5–12	zero	NA	NA	NA
Softening point	°C	400 decomposes	240	230–265	P: 180–200 N: 220–235	190–330[3]	1400–1450 Melting pt.	850–950	375 decomposes	327
Specific resistance	Ω cm	10⁶–10⁷	5·10⁸	NA	P: 4.9·10⁹ N: 4.1·10⁹	5·10⁸	conducts	10¹⁵	NA	NA
Static charging	1) 21 °C 25 % RH 2) 24 °C 24 % RH	low low	low medium	medium high	medium high	medium very high	none	NA	NA	NA

1) Extract "Reutlinger Fibre Table 1974"
2) VDI Standard 3677
3) Extremes found in literature

2.2 Fibres as Building Blocks

Table 2.5 Resistance of fibres to chemicals in vapour or gas form [10]

x = resistant
o = moderately resistant
– = unsuitable

	Cotton	Wool	Fibreglass	Metal	Polyacrylonitrile (Dralon T)®	Polyacrylonitrile (Orlon)®	Aliphatic Aramides (Nylon G®/66 Perlon®)	Aromatic Polyamides (Nomex)®	Polyester (Trevira®, Diolen®)	Polyethylene	Polypropylene (Meraklon)®	Polytetrafluoroethylene (Teflon)®	Polyvinylchloride (Thovyl®, Thermovyl®)
Acids													
Chromic acid	–	–	x		x	x	–	–	x	x	o	x	x
Hydrochloric acid	–	–	x		x	x	–	–	x	x	x	x	x
Fluoric acid	–	o	–		x	x	–	–	o	x	x	x	x
Nitric acid	–	o	x		x	x	–	o	x	o	o	x	x
Phosphoric acid	o	o	x		x	o	o	x	x	x	x	x	x
Sulphuric acid	–	o	x		o	o	–	–	o	x	x	x	x
Organic acids													
Acetic acid	x	o	x		x	x	o	o	x	x	x	x	o
Benzoic acid	x	o	x		x	x	–	o	x	x	x	x	
Phenolic acid	x	o	x		x	x	–	–	o	o		x	–
Formic acid	x	o	x		x	x	o	o	x	x	x	x	x
Lactic acid	x	o	x		x	x	o	o	x	x	x	x	x
Oxalic acid	x	o	x		x	x	x	–	x	x	x	x	x
Salicylic acid	–	–	x		x	x	o	o	x	x	x	x	
Bases													
Ammonium hydroxide	o		o		o	o	o	o	–	x	x	x	x
Calcium hydroxide	x	x			x	x	x	x	x	x	x	x	x
Potassium hydroxide	x	–			o	–	o	o	o	x	o	x	x
Potassium carbonate	x	–			o	o	x	x	o	x	x	x	x
Sodium hydroxide	x	–			o	o	o	o	o	x	o	x	x
Sodium carbonate	x	–			x	x	x	x	x	x	x	x	x
Salts													
Calcium chloride	x	o			x	x	–	o	x	x	x	x	x
Ferric chloride	–	–			x	x	o	o	x	x	x	x	x
Sodium acetate	x	o			x	x	x	o	x	x	x	x	x
Sodium bisulphite	x	x			–	x	o	o	x	x	x	x	
Sodium bromide	x	–			x	x	x	x	x	x	x	x	
Sodium chloride	x	–			x	x	x	x	x	x	x	x	x
Sodium cyanide	x	o			x	x	o	o	x	x	x	x	
Sodium nitrate	x	o			x – x		o	o	x	x	x	x	x
Sodium sulphate	x	o			x	x	x	x	x	x	x	x	x
Sodium sulphide	x	o			x	x	x	x	x	x	x	x	x
Zinc chloride	o	–			–	o	–	o	–	x	x	x	x
Oxidising Agents													
Bromine	o	x			o	o	–		o	–	x	x	–
Calcium hypochlorite	o	x			x	x	o		x	x	x	x	x
Chlorine	o	x			o	o	–		o	–	x	x	o
Fluorine	–	–			o	o	–		o	o	x	x	–
Hydrogenperoxide	x	–	x		x	x	o		o	–	x	x	
Iodine	x	–	x		x	x	–		x	–	x	x	
Ozone		–	x		x	x			x	o	o	x	
Potassium chlorite	o	–	x		x	x	o	o	x	x	x	x	x
Sodium hypochlorite	o	–	x		o	x	o		o	o	x	x	x
Sodium chlorate	x	–	x			x				x	o	x	x

2.3 Textile Structures as Dust Filters

2.3.1 Fabrics

Fabrics are assemblies of threads which interlace at right angles (fig. 2.20). The warp threads run in the longitudinal direction while the weft threads run across. The threads or yarns may be made of continuous filaments (fig. 2.21) or short fibres (synthetic fibres are cut 30–80 mm long) which are spun and twisted. Staple fibres (fig. 2.22) are converted into staple fibre yarns. When several threads are assembled by twisting together they are known as folded yarns. Both, continuous filament and stable fibre yarns can be arranged in groups. In monofilament fabrics (fig. 2.23) each thread is a solid "wire" extruded from a spinneret. Monofilaments are rarely used in dry filtration, but appear more frequently in wet filtration.

Fig. 2.20
Structure of a fabric

Fig. 2.21
Cross section of a fibre bundle

2.3 Textile Structures as Dust Filters

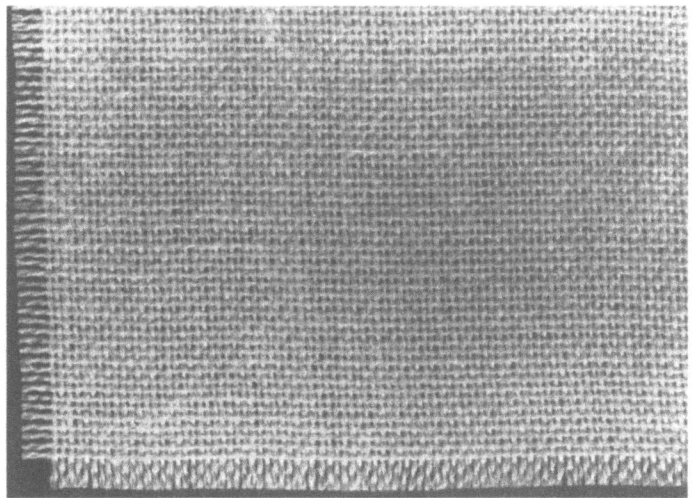

Fig. 2.22
Staple fibre weave

Fig. 2.23
Monofilament weave

The quality of a woven fabric depends on the following characteristics:
- count (number of threads per cm in warp and weft)
- thickness of yarn (yarn number, denier, dtex)
- weave
    a) plain weave (fig. 2.24)
    b) twill weave (fig. 2.25)
    c) satin weave (fig. 2.26)
- type of fibre (raw material)

Plain Weave

applicable to canvas
damask
calico
muslin
taffeta

The plain weave structure is the simplest and most durable assembly. It has similarities with darning. Each thread interlaces alternately in both directions. The draft pattern looks like a chess board; its pattern repeat is made up of 2 warps and weft threads, respectively. The face and back of a plain weave looks alike.

Typical plain weave fabric types:
damask, calico, batiste, linen, poplin

Variations: panama weave
repp

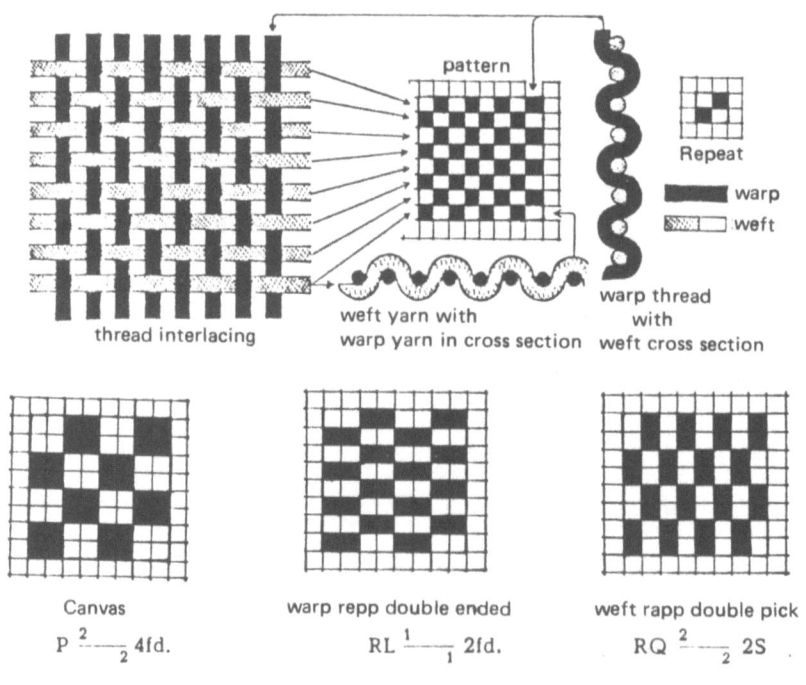

Canvas
$P \frac{2}{2} 4fd.$

warp repp double ended
$RL \frac{1}{1} 2fd.$

weft rapp double pick
$RQ \frac{2}{2} 2S$

Fig. 2.24 Plain weave interlacing

2.3 Textile Structures as Dust Filters

Twill Weave

The twill structure is not as tight and not as durable as the plain weave. It is woven on a minimum of 3 shafts up to 20; in rare cases even more. This construction enables the yarn that is more expensive to be shown on the surface. The twill is characterised by diagonal lines that run to the upper right or left — twill lines. Each float is bound into the fabric at its extremety.

A minimum repeat is 3 (3 × 1 Twill). Depending on whether warp of weft threads predominate on the surface of the fabric it is called a warp or weft twill. If the floats on the face and back are of equal length, it is termed a double sided twill.

Twills are typical in gabardines and serges.

Variations: double twill
herringbone twill
cross twill
multiple twill

By increasing the weft density the twill angle can be flattened, conversely a denser warp surface steepend the twill line. Some gabardines are derived fron satin constructions.

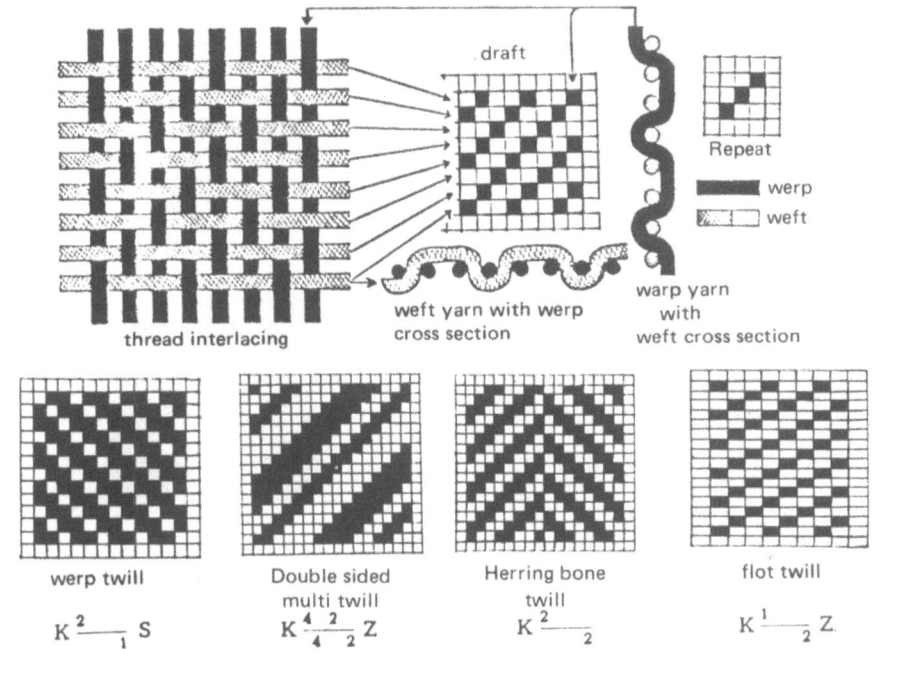

Fig. 2.25 Twill weave interlacing

Satin Weave

In the satin weave the regular interlacings are hidden. This produces a smooth surface which can be given a lustre or gloss when appropriate threads are chosen. Satins are repeats of 5 or up to 12, five being the minimum. Depending on whether warp or weft threads predominate on the surface one speaks of a (warp) satin or (weft) sateen. This structure is most appropriate to bring the decorative or expensive component to the surface.

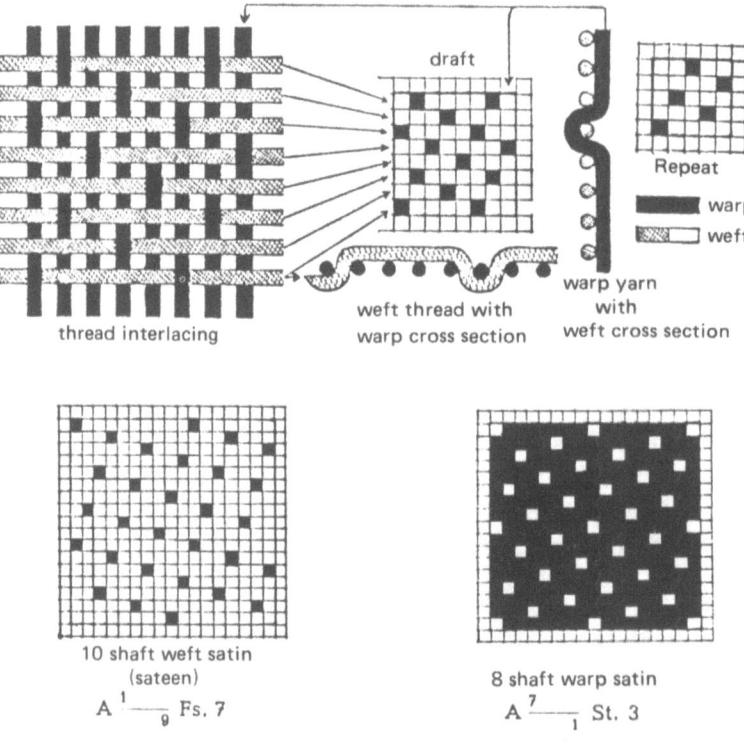

Fig. 2.26 Satin weave interlacing

2.3 Textile Structures as Dust Filters

For dust separation fabrics range in weight from 200–450 g/m² with an air permeability of 300–100 1/dm² min at 196.2 Pa (20 mm water gaupe) as specified in DIN 53 887.

Fabrics are "two dimensional" filter media. Their filter action is on the surface. To improve their filter efficiency they require an initial dust cake, the build-up of which can be encouraged by brushing the surface. In the virgin condition fabrics behave according to the law of the sieving effect, viz. all particles which are larger than the pores between warp and weft are retained [11].

The basic element of a fabric is the thread or yarn. It is therefore important to differentiate between yarn pores (fig. 2.27a) and weave pores (fig. 2.27b). In hard twisted staple or multifilament yarns the yarn pores play a subordinate role. They are completely absent when dealing with monofilaments and the weave pores alone make up the spaces between the warp and the weft. These pores then determine the filtration properties. These pores contribute only a small part of the total surface area of the fabric. On average a filter fabric has a free sieving surface of around 40 % which is available for filtration. The remaining 60 % is taken up by the yarn and considered "dead filter surface" with sole function to give strength and stability. This results in large filtration surfaces which in turn require very large baghouses.

a)

b)

Fig. 2.27
Thread and fabric

Due to their lighter weight filter fabrics tend to be cheaper than needlefelts. This advantage is negated by a larger housing to accommodate the larger filter area.

Good separation efficiencies can be achieved with fabrics, if they are densely constructed. This results in higher filter resistance and lower face velocities. Air to cloth ratios for dust collectors with woven fabrics lie between 0.6 and 1.0 m^3/m^2 min. When a conventional woven filter is replaced with a needlefelt, experience shows that the air permeability can be doubled compared to woven fabric without adversely effecting filter efficiency. The result is a 30–50% higher air to cloth ratio with the same or lower differential pressure.

Demand for lower emission values, increasing use of jet cleaning and the economic advantages outlined above have resulted, at least in Europe, in filter fabrics losing their dominance to the needlefelt. Most weaving mills take a flexible approach and have followed the trend to produce support scrims for needlefelts or to engage in making needlefelts themselves.

Fibreglass fabrics for the hot gas filtration up to 200 °C or more have not been affected by this development so far. On the contrary, their significance has increased, because in this temperature region there is no price competitive alternative for a baghouse with low pressure reverse air cleaning [12].

2.3.2 Needlefelts

Primitive needlefelts from natural fibres were first manufactured towards the end of the 19th century. This porous, fibrous structure came to a climax after the 1st World War when efficient machinery and needles were developed to allow more advanced manufacturing techniques. The Standards Institutions gave this young offspring of textile art a great deal of attention; needlefelts can take pride to be mentioned in 3 sections of the German Industry Standards. Little wonder that specialists today argue which of these standards is the right one [13].

DIN 61205, No. 1.2 describes needlefelts as "surface and body structures made from assorted fibres. They consist of nonwoven webs which are consolidated through alternate insertions and extractions of a bank of specially designed needles". This process is called "Needling". The same DIN has an alternative ready under Sec. 2.2.2, "The Needled Nonwoven Felts; Textile structures made from a base material and single or multiple layers of fibre webs. These are consolidated by the alternative insertion and extraction of rows of special needles so as to combine them with the base fabric". DIN 61210 states half-heartedly: "Mechanically consolidated nonwoven are a textile structure manufactured from fibre webs that are consolidated solely by means of a mechanical process, e.g. needling". In an annotation it says: "Nonwovens which are reinforced with threads or textile structures (e.g. yarn sheets, fabrics, knitted fabrics, knits or foils) are called reinforced nonwovens".

For the industrial engineer who wishes to solve a filtration problem with a needlefelt (fig. 2.28): the product definition is far simpler. For him the needlefelt is a three dimensional medium (fig. 2.29) whose filtration properties originate from the surface as well as from the depth of the structure. The basic building block is the single fibre, not the thread or yarn. The fribres lie adjacent, on top of or under one another, paralle, crossed or random, horizontal, vertical or at any angle. Each fibre is simultaneously substance

2.3 Textile Structures as Dust Filters

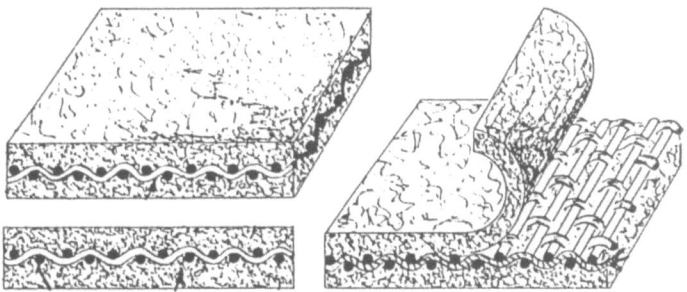

Fig. 2.28 Schematic diagram of needlefelt with a scrim reinforcement

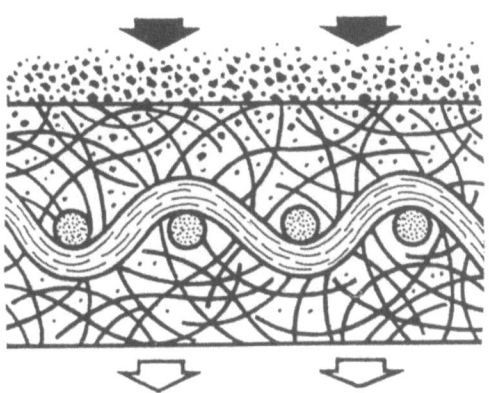

Fig. 2.29
The threedimensional concept

and binding agent. The intersticies are the pores. These differ in size and are proportional to the degree of consolidation of the felt and the fineness of the fibre. Depending on the compaction the pore volume is around 80%. Needlefelts may carry a support weave or scrim in their interior for reinforcement. For filtration purposes this results in higher tensile strength and lower extension. As raw material synthetics and mineral fibres are preferred with a minimum staple length of 40 mm [11].

To manufacture needlefelts the fibres are firstly blended and carded (fig. 2.30). Antistatic agents or spinning lubricants may be added at that point. When fibres have been sufficiently opened they are fed to the laying machine [14]. One may use either a conventional card (fig. 2.31) or a suction drum. During carding a uniform thin, opened, and parallelized fibre web is formed which is fed to the cross feeder. This lays the web at right angles carding a uniform thin and opened parallel fibre web is formed which is fed to the cross feeder. This lays the web at right angles to the machine direction. In the aerodynamic method the fibres are blown onto an enclosed suction drum. In this case a random layed fibre web is formed which is fed to the needling machine. Here the consolidation occurs (fig. 2.32) with the help of thousands of needles (fig. 2.33a, 2.33b,

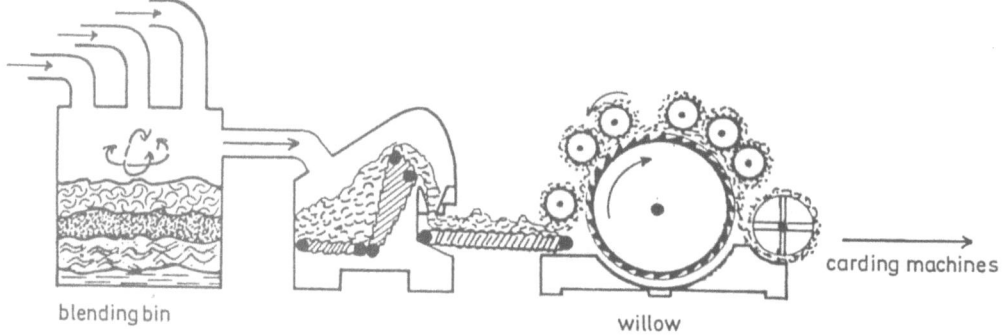

Fig. 2.30 Blending and carding [14]

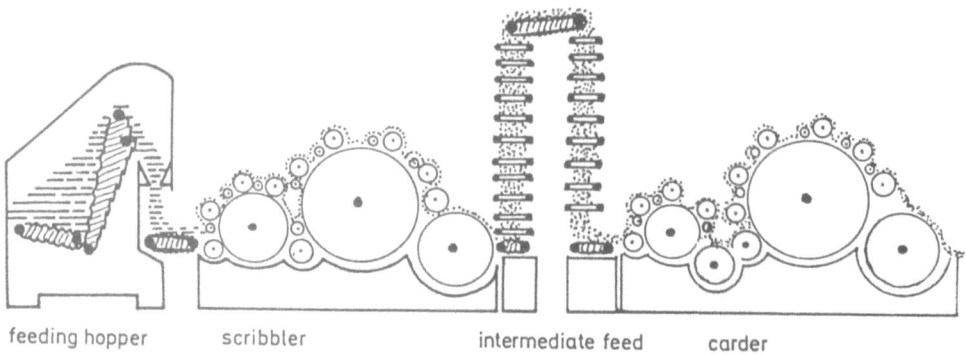

Fig. 2.31 Carding [14]

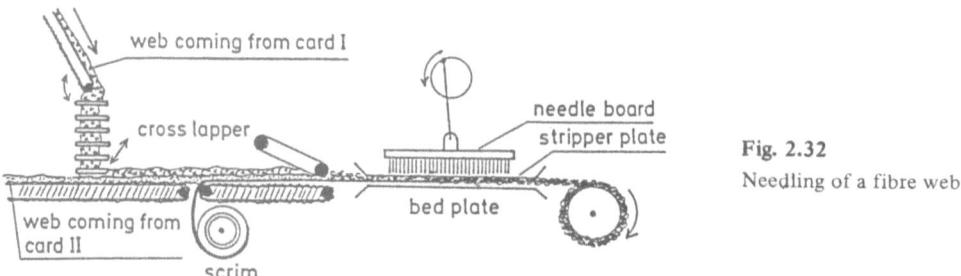

Fig. 2.32 Needling of a fibre web

2.34) which sit in a needle board that moves up and down. These needles have barbs. During needle insertion the barbs pull the fibres which lie more or less horizontal into a vertical position. When the needle is withdrawn it matts and knots the fibre web together (fig. 2.35). Needle gauge, shape and depth of the barb, number of insertions per sq. area and depth of insertion influence the desired density and characteristic of the filter medium. It is the skill of the needlefelt specialist to bring these variables under control

2.3 Textile Structures as Dust Filters

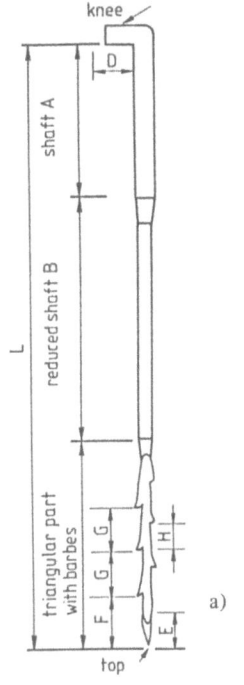

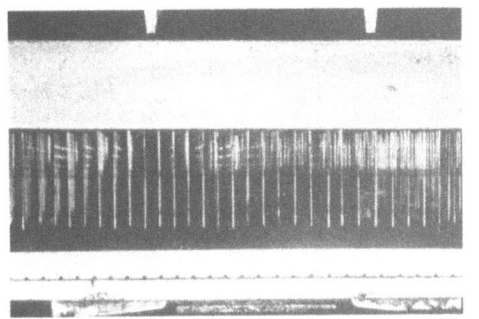

Fig. 2.33 Felting needle
a) Needle, b) Needle plate

Fig. 2.34
Needles with attached fibres penetrate the base plate of a needling loom

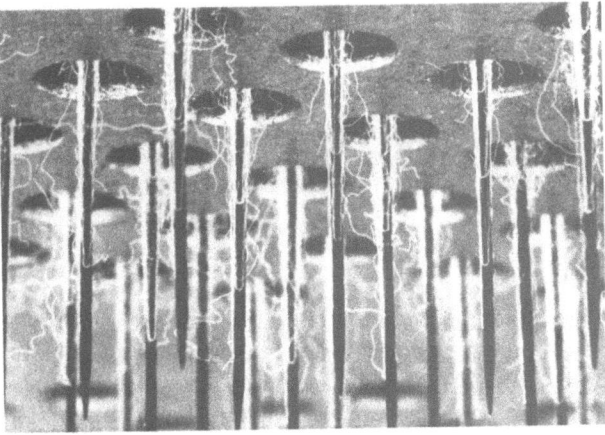

for a predetermined and reproducible product. High quality filter felts have several hundred stitch insertions per cm². The gauge and finish of the needle, its form and type of barbs are very important. With today's insertion speeds of 15–20 insertions/s the sharp-edged barbs of the kick-up needle (fig. 2.36) causes fibre damage in both the fibre web and the support scrim. For this reason needle manufacturers offer today a more expensive, finer and better finished needle in which the barbs are stamped (fig. 2.37). Such needles are more prone to break but make smaller holes and protect the fibre during needling.

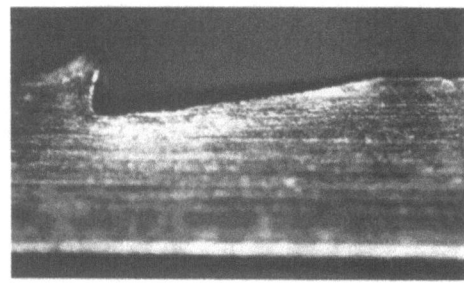

Fig. 2.36 Kick-up barb (enlarged 60x)

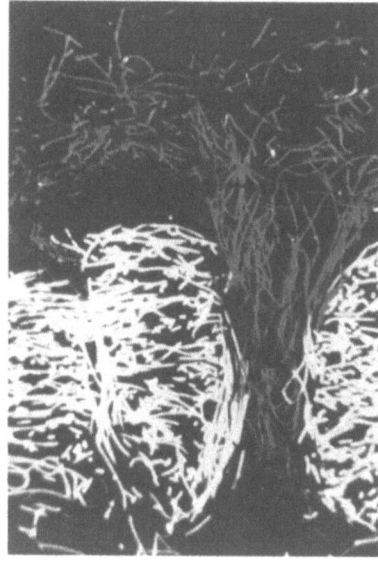

Fig. 2.35
"Insertion funnel" of a needle within the fibre web

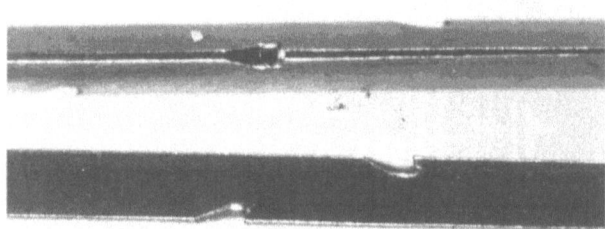

Fig. 2.37
Fibre protective non kick-up type needle (enlarged 20x)

After needling, a quality check takes place to test for technical specification. Broken needle points are detected electromagnetically and removed. Singeing removes extraneous fibres; this prevents dust accumulation which in turn adversely affects differential pressure. The filter medium is stabilised by thermofixation with conditions selected appropriate to fibre type. This makes them suitable for hot gas conditions. Many chemical finishing treatments can be selected to add fire protection, to improve repellency or cleanability, to improve flexibility or reduce friction. High pressure calendering under heat gives the finishing touches to a smooth and stabilized surface (fig. 2.38).

2.3.3 Wool Felt as The Forerunner

The latin word "filtrum" stands for "felt" as well as "filter". This could be evidence that in ancient times felts were used to filter liquids. Indeed felt from wool or animal hairs are among the oldest textiles which man has learned to produce. It is as old as the story of man. Many ancient myths and stories surround the origin of felts. One of the best

2.3 Textile Structures as Dust Filters

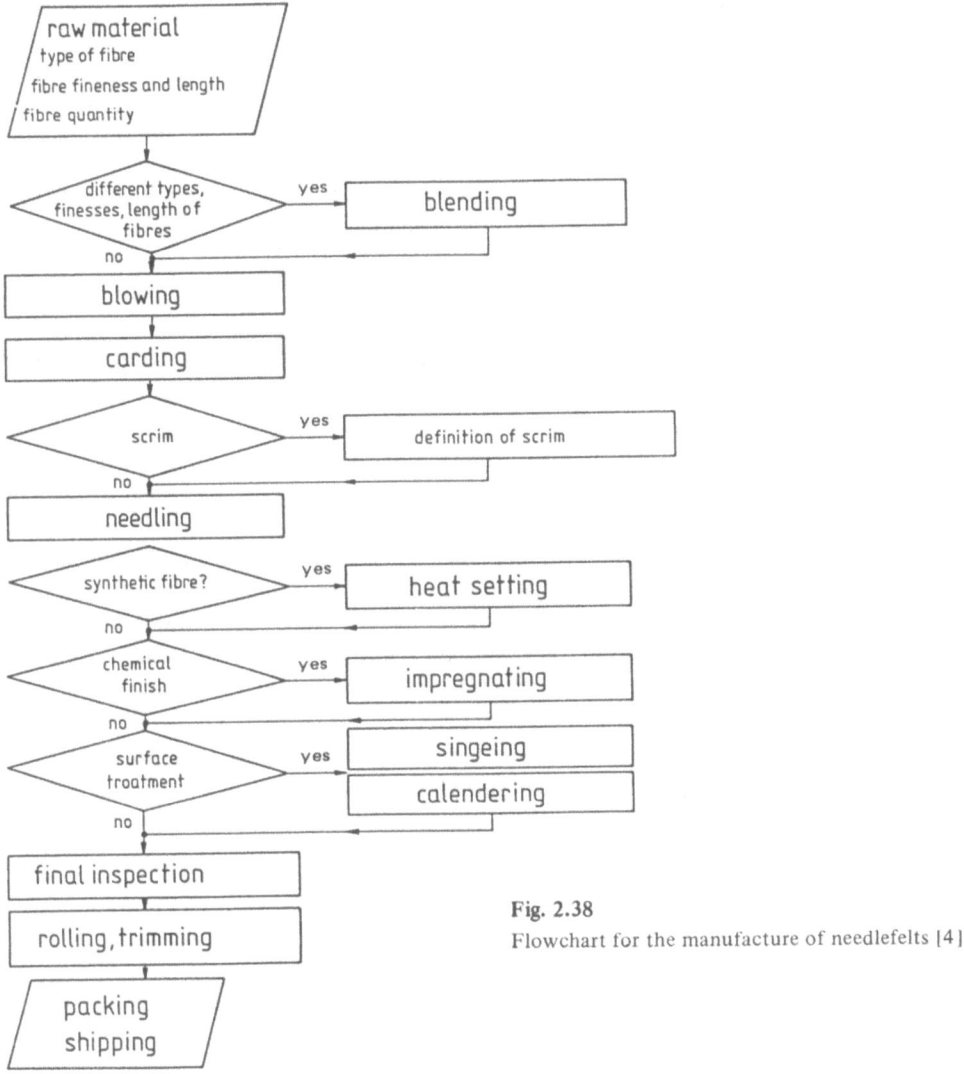

Fig. 2.38
Flowchart for the manufacture of needlefelts [4]

known tells of the biblical Noah laying out the floor of the Ark with sheep wool to give man and animal a soft bed as the waves tossed the vessel hither and thither. As the water masses engulfed his ship he saw with sad eyes the wool become wet and trampled on by all the animals. When at last the Ark landed and his whole animal kingdom breathed fresh air, he noticed with surprise that the floor covering had compacted. What was previously saturated through and through had become a flexible underlay by the wetness and stamping of the animals and the warm air in the Ark.

Heat, humidity, pressure and movement are the four components which work together to felt wool or animal hairs today as yesterday [15]. To appreciate the beginnings of felt making we can still see it among the nomadic Asians in the Steppes and in Mongolia. These tribes often live in home-made felt tents (Jurds). These tents are made of sheep wool, camel hair and goats hair or mixtures of these. They are worked with hot water with the addition of urine or fullers earth and by stamping with their feet. In Mongolia such felt tents are made by spreading the material with tensioned bow tendons, soaking in whey, forming into a lap and rolling on the ground with horses galloping over them. W. F. Reinig tells how the inhabitants of Kirghistan in the Asian steppes free wool from siunt by spreading it evenly on a felt mat, impregnating it with whey, forming it into a roll and rolling it for several hours with hands or under their arms [16, 17].

It is not too long ago that in Europe felts were made on a steaming table in which a narrow card web was felted by hand with a circular rolling movement. The desired density and compaction was achieved by rolling a felting roller or by striking it with strong blows, or working with the feet.

Even today wool and hair felts are compacted under conditions of warmth, humidity, pressure and movement [18, 19, 20, 21]. Animal fibres swell and the unfinite number of scales – visible only under a microscope – (ref. sect. 2.2.2) lock into each other. The fibres become matted and entangled. Wool or hair felt was the first nonwoven product not made in the conventional way of spinning, weaving, knitting or similar. In felt the basic element is the single fibre, not the thread or the yarn. Fibres lie ajacent, on top of, underneath, parallel, crossed, horizontal, vertical or at any angle. Each fibre is at the same time substance and binding agent. Between the fibres are the pores whose aggregate volume depends on the degree of compaction of the felt. These intersticies are three dimensional.

The process of felting – today produced on modern and efficient machinery (fig. 2.39) – is nevertheless always limited to wool or hair with all their inherent disadvantages. Wool

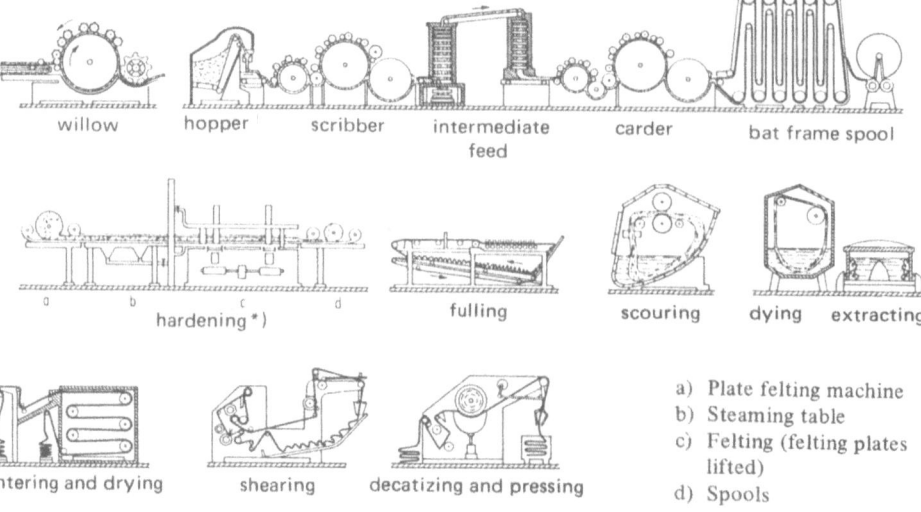

Fig. 2.39 The manufacture of felt (Courtesy of the German Museum in Munich)

2.3 Textile Structures as Dust Filters

felt thus fathered todays needlefelts. Felt can take a place of pride in being rediscovered in the needlefelt form with its three dimensional filtering characteristics. Felt, man's oldest textile, has "revolutionalized" filtration methods of the 20th century. This is only one example of modern dust collection technology where it is instructive to put our "technical progress" into a time perspective.

2.3.4 Nonwovens

In the narrow definition of Standards both felts and needlefelts fall into the category of "nonwovens", because they are not woven, but consolidated by mechanical means. A separate section will be devoted to them, because their filtration properties differ significantly depending on whether they are chemically or adhesively bonded, or spun-bonded in nature [22, 23, 24, 25].

There is a confusing array of nonwoven types on the market. They are grouped according to the method of manufacture based on textile- or paper industry technology, though there is considerable overlap between different types. It is not the aim of this exposition to deal with the full range of nonwovens on the market today. It is only relevant to point out those nonwovens which find application as filter media in industrial gas cleaning [26].

The basics are found in DIN 60000. The two areas of nonwovens and of structures produced by sewing or knitting are covered under DIN 61210 and DIN 61211. These Standards further subdivide the groups according to technical aspects. According to these Standards nonwovens are "textile structures of fibres in which cohesion is achieved by the fibres themselves". Such fibre webs are the starting point for the production of textile nonwoven fabrics. Depending on how they are compacted or bonded (with the exception of those needlefelts covered above) they are grouped as follows:

2.3.4.1 Nonwovens bonded with Adhesive Binders (fig. 2.40)

Nonwovens bonded with binders are textile structures made form fibre layers where internal cohesion is achieved with adhesive or by spotwelding of fibres.

2.3.4.2 Nonwovens that shrink

These textile structures gain internal cohesion by using fibre shrinkage of one or all of the components through heat or chemicals.

2.3.4.3 Nonwovens through swelling

These structures are held together through the action of swelling agents.

Combinations of the above manufacturing techniques are possible and practiced. Those nonwovens which use threads or other textiles for reinforcement, e.g. yarn sheets, fabrics, knitteds, nets of foils, are called reinforced nonwovens.

2.3.4.4 Spunbondeds

These textile structures are produced during the spinning of the fibre from either the melt solution or wet spinning methods, the fibre sheets are laid, blown or floated on a transport belt (fig. 2.41).

Fig. 2.40
Non-woven with binder

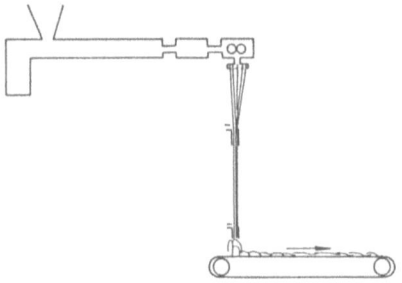

Fig. 2.41
Principle of spun-bonding

The above mentioned types of nonwovens play a role in dust collection. Their market contribution is small compared to the dominant needlefelts. Reasons for this lie mainly in their construction. Chemical components that are temperature sensitive or that can be hydrolysed limit their application. They are stiffer in handle and in some applications lack a reinforcement fabric. This affects service life and separation effect of the filter medium. All nonwovens irrespective of their origin and method of manufacture are three dimensional filter media and have high separation efficiencies on both their surface and within their depth. Tables 2.6 and 2.7 summarize technical parameters and relative price.

Table 2.6 Needlefelt characteristics and relative price (W. Schmid [14])

Needlefelt/support fabric "quality abbreviation"		Weight DIN 53854 g/m²	Thickness DIN 53855 draft mm	Density DIN 53855 draft g/cm³	Pore volume DIN 53855 draft %	Air permeability DIN 53887 l/dm²·min	Tensile strength DIN 53857 length (200 × 50 mm) N	Tensile strength DIN 53857 across (200 × 50 mm) N	Extensibility DIN 53857 length %	Extensibility DIN 53857 across %	Bursting strength DIN 53861 bar at 50 cm²	Dome height mm	Service temperature °C	Relative price per m² Polyester 400 g/m²
Wool/polyester	"LEw/PE"	400 / 500	1.6 / 1.9	0.25 / 0.26	81 / 80	190 / 150	500 / 550	400 / 450	26 / 28	24 / 26	5 / 6	22 / 20	70	1.8 / 1.9
Polypropylene/polypropylene	"PP/PP"	400 / 500	1.9 / 2.1	0.21 / 0.24	77 / 75	170 / 150	1020 / 1100	11 / 1160	10 / 11	12 / 13	11 / 12	20 / 19	95	1.55 / 1.62
Polyamide 66/polyamide 66 (Nylon)®	"PA/PA"	400 / 500	1.6 / 1.8	0.25 / 0.28	78 / 78	250 / 210	780	1100	13 / 13	18 / 20	10 / 12	20 / 24	110	1.31 / 1.47
Acrylic/acrylic (Dralon T)®	"DT/DT"	400 / 500	2 / 2.6	0.2 / 0.19	83 / 83	240 / 190	710 / 900	750 / 1250	13 / 14	9 / 11	10 / 12	16 / 16	125	1.37 / 1.51
Orlon/polyester	"O/PE"	400 / 500	1.8 / 2	0.22 / 0.25	80 / 79	220 / 180	1200 / 1440	1060 / 1260	15 / 14	19 / 18	14 / 14	20 / 17	125	1.37 / 1.51
Polyester/polyester	"PE/PE"	400 / 500	1.4 / 1.8	0.29 / 0.29	79 / 79	260 / 210	1400 / 1500	1200 / 1300	18 / 17	20 / 20	13 / 15	18 / 18	150	1 / 1.07
Aromatic polyamide (Conex®/Nomex®)	"NO/NO"	400 / 500	2 / 2.5	0.2 / 0.2	86 / 82	250 / 200	570 / 650	700 / 730	18 / 20	20 / 21	10 / 11	23 / 23	220	4.16 / 4.6
Polyetrafluoroethylene (Teflon®/Teflon®)	"TF/TF"	750 / 900	1.2 / 1.4	0.63 / 0.64	70 / 70	150 / 110	680 / 730	710 / 800	45 / 50	50 / 52	7 / 8	29 / 29	250	25.6 / 29.7
Fibreglass/fibreglass	"GL/GL"	950	2.5	0.38	80	100	2150	2850	1.7	2	12	7	220 (250)	6
Stainless steel (Bekinox)®	"B/B"	1500	2	0.75	91	125	1270	1090	5	5	6	6	450	60

Table 2.7 Woven fabric characteristics and relative costs (W. Schmid [14])

			Area Weight DIN 53854 g/m²	Thickness DIN 53855 draft mm	Density DIN 53855 draft g/cm³	Pore Volume DIN 53855 draft %	Thread Density DIN 53853 warp/weft No/cm		Thread Count DIN 60850 draft warp/weft dtex		Weave Type DIN 61101 draft
Polypropylene	PP		300	0.6	0.5	45	11.5	26	⌀0.3mm	0.3mm mono filament	2 * 2 twill
			550	1.2	0.46	49	21.5	9	311	1392 endless spinning	3 Layer-Fabric
Aliphatic Polyamide (Nylon)®	PA 6.6		260	0.8	0.33	71	16.5	19	250/2	870	2 * 2 twill
			395	0.7	0.56	51	12.5	12	940	940	plain weave
Acrylic	PAC	DT	185	0.7	0.26	77	46	19	220	220 multi filament	satin
			265	0.7	0.38	67	21	21	290/2	625	2 * 2 twill
			320	1.0	0.32	72	20	14.5	420/2	1000	2 * 2 twill
			370	1.5	0.25	78	28	30	290/2	625	double pick
Polyester	PE		130	0.32	0.4	71	25	26	⌀0.15	⌀0.15 mono filament	plain weave
			230	0.53	0.43	69	15	15	⌀0.25	⌀0.25	plain weave
			320	0.65	0.49	64	19.5	25	290/2	625	2 * 2 twill
Aromatic Polyamide	NO		170	0.38	0.44	68	35	23	245	308.5	3 * 1 twill
			290	0.75	0.39	72	23	20	487	710	2 * 2 twill
PTFE			485	0.48	1.01	52	19	15.5	440	440/2	plain weave
Fibreglass			290	0.34	0.85	67	21	20	680	680	twill
			310	0.8	0.39	85	21	12	680	140	twill
			440	0.92	0.48	82	19	13	1360	140	twill
			500	0.95	0.53	80	19	8.5	1360	140	cross twill

* Seize of specimen 100 × 30 mm
** The relative prices are based on the indications of 3 manufacturers [38, 42]

2.4 Market and Technology

Among all methods of dust collection, the fabric filter (fig. 2.42) has markedly increased its significance during the last 10 years. Figures available show a market share increase from 17% in 1964 to 40% today [121]. A 50% market share is now within reach. There are a number of very good reasons for the growing popularity of this type of dust collector:

1. Regulations are more stringent. The new German "Technical Manual for Clean Air" (TA Luft 1986) requires a reduction in particulate matter of earlier limits of 150 mg/m³ to a 3 level system of emission values of 5, 2 or 0.1 mg/m³, respectively, depending on the potential toxicity of the dust. Thus an overall reduction in dust levels is required [27].
2. Improved efficiency through higher admission velocities.
3. Greater separation efficiency compared to cyclone filters.
4. Better investment and lower maintenance costs.
5. A dry process; no waste water or costly waste water treatment problems.

2.4 Market and Technology

Air Permeability DIN 53887	Tensile Strength DIN 53857 warp/weft		Extensibility DIN 53857 warp/weft		Bursting Strength DIN 53861	Finishing Treatment	Service Temperature	Relative Price (**) to 1 m² polyester fabric
1/dm²·min	N	N	%	%	bar		°C	
WG (water gauge) 540	3580	1470	33	18	17.5	calandered	95	1.15
(at 10 mm WG) 10	3435	4500	31	33.5	26.5	calandered		1.88
250	1720*	790*	42.5*	40*	13		110	1.15
15	6865	3435	48	28	28.5	calandered		1.5
100	1275	880	24	18	6.5	grey		1.85
200	720*	800*	12*	23*	8.25	grey	125	1.46
175	820*	750*	25*	25*	9.25	grey		1.54
150	595*	1020*	15*	32*	10	raised on are side		2.1
800	758	830	41.5	38	7.5	grey		1.15
800	1420	1570	42	45	10	grey	150	1
180	1020*	1000*	42*	26*	13.5	grey		1.44
420	1470	830	40	24	10.5	grey	220	2.85
310	1860	1370	30	44	14.5	grey		2.88
5	820*	520*	50*	30*	9	grey	250	4.5
240	1750	1650			40	silicone or graphite finish		1.2
380	1750	1250			35	silicone or graphite finish		1.27
220	4500	2000			45	silicone or graphite finish	290	1.56
290	4500	2150				silicone or graphite finish		1.65

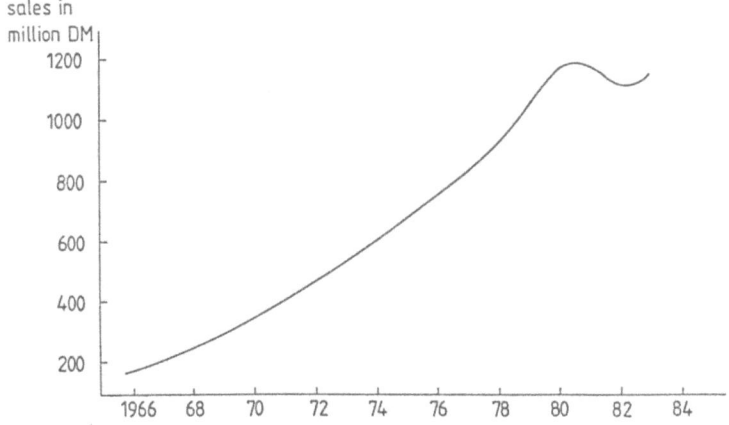

Fig. 2.42 Estimate of turnover for dust collectors manufactured in the GFR

6. New technology in effective filter media with chemical and temperature resistant treatments have opened new areas of application for dust collectors previously not accessible.

The success of the filter type dust collector in achieving results has drawn the supply industry to compete for greater market share in the plant hardware as well as filter media. In the Federal Republic of Germany at least 40 companies build bag and/or pocket filter plants. Among the "fashion items" offered is the jet filter with pulse jet cleaning. The "jet" age in filtration has arrived.

There is no doubt that the fabric filter leads the 80's ahead of the electric [28] or wet [29] or the mechanical collector [30]. With the increasing number of installations the demand for needlefelts has also grown.

Market estimates for the production of filter media (felt and fabric) in the German Federal Republic were 10 million m^2 per annum in the early 80's. By value the filter medium in a dust collector contributes around 15% of plant cost. Needlefelts are given a market share of around 85% of all filter media with the result that in the European market by now more than 40 companies are involved in manufacturing needlefelts for dust collection [31].

This has led to an overcapacity with consequent extreme competition. This may explain the paradox that with increasing raw material and finishing costs, prices are stagnant or even falling in the market. More than ever the qualities offered on this market must stand out through excellence and compliance to recognised standards within the European market. A guarantee of quality standard has become increasingly important in a time when –

- Emission values are to be lower than ever.
- Filter efficiency ratings are rising
- Long service life guarantees are demanded
- Both higher temperatures and more diverse gas compositions are making additional demands on the physical and chemical resistance of the filter medium

2.5 Quality Criteria

The dust separator wins or fails on its filter medium. For this reason the quality of the filter medium cannot be emphasized enough. Frequently we encounter that only price matters, quality is pushed aside by a slight of hand: "Fabric is fabric, and felt is felt". The consequence of this thinking is found latest when the dust finds its way into the filtered air, when the ominous cloud puff shows that the bags are no match for the pressure pulse, when the bags show premature wear, because they are unable to resist the physical or chemical attacks, when the filter elements contract or expand, because they have insufficient dimensional stability, or when "the rib cage outgrows the bags". This is how someone in jest described the shrinking of his filter bags. The result is premature change of bags, machine downtime and many hours of expensive labour to change the filter elements. As a rule these costs are manifold that of the original bags. It is curious that such costs are often completely overlooked by those who are most price conscious. Maybe it is because these costs are more difficult to assess and are

2.5 Quality Criteria

usually not quoted separately at the tendering stage. This apart, there are fewer and fewer operators willing to take on the dirty work of changing bags: and if they do, then at significant costs.

It thus appears proper and useful to set down unambiguously those quality criteria which should be taken into account when selecting a filter medium.

2.5.1 Area Weight

This is expressed in grams per m². It relates to the total fibre weight and is therefore a good value indicator. There is a relationship between the weight and durability. Thickness and density are proportional to weight. The following weight ranges apply to needlefelts:

Cleaning by shaking	300–350 g/m²
Cleaning by shaking or low pressure reverse clean	400–450 g/m²
Cleaning with low pressure reverse flow	400–450 g/m²
Cleaning with high pressure reverse flow	500–550 g/m²

DIN 53854 [125] specifies area weight with a tolerance of ± 5 %.

2.5.2 Density

The density of a nonwoven D_L is the ratio between area weight and thickness. It depends both on the fibre density D_F and the degree of compaction of the filter medium. It is an important indicator of the solidity of the filter material, e.g. its resistance to abrasive conditions. Density is also an important indicator of pore volume P_V. Some typical examples for filter fabrics are —

Polyester	fibre	D_F	1.36–1.38 g/cm³
	felt	D_V	0.28–0.30 g/cm³
Acrylic	fibre	D_F	1.15–1.18 g/cm³
	felt	D_V	0.22–0.24 g/cm³
Polypropylene	fibre	D_F	0.90–0.92 g/cm³
	felt	D_V	0.20–0.22 g/cm³
Polyamide	fibre	D_F	1.14–1.16 g/cm³
	felt	D_V	0.24–0.26 g/cm³
Aramide	fibre	D_F	1.38–1.41 g/cm³
	felt	D_V	0.22–0.24 g/cm³
PTFE	fibre	D_F	2.10–2.15 g/cm³
	felt	D_V	0.65–0.68 g/cm³

$$P_V (\%) = \frac{D_F - D_V}{D_F} \cdot 100$$

2.5.3 Air Permeability

The air permeability (fig. 2.43) is measured according to DIN 53887. It is expressed in liters per square decimeter and minute at 196 Pa (20 mm water gauge). Permeability testers are available commercially. The air permeability is the most important property of a filter medium, since it largely determines the differential pressure, face velocity rating, cleanability and separation efficiency.

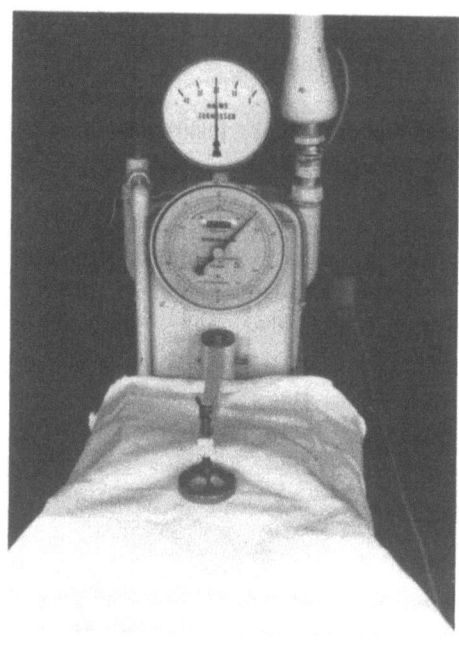

Fig. 2.43
Air permeability tester

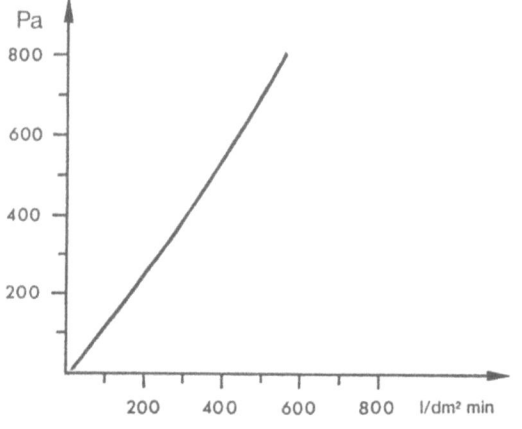

Fig. 2.44
Differential pressure of a virgin polyester felt 550 g/m^2
Air permeability 150 l/dm^2 min at 196 Pa (water gauge)
DIN 53 887   540 l/dm^2 min at 800 Pa

Depending on the type of cleaning the following air permeabilities are recommended for needlefelts:

Cleaning with shaking	400–600 l/dm^2 min
Cleaning with shaking + low pressure rev air	250–300 l/dm^2 min
Cleaning with low pressure rev air	250–300 l/dm^2 min
Cleaning with high pressure rev air	120–180 l/dm^2 min

2.5 Quality Criteria

For the same filter efficiency a needlefelt can have two to three times the air permeability of a woven fabric. This explains their higher face velocity rating.

Commercial tolerances for air permeability values are ±10 %.

2.5.4 Bursting Strength

DIN 53861 [126] describes the method of measuring bursting strength (fig. 2.45). It is expressed in bar with an associated dome height of a test area of 50 cm^2. A polyester felt with a polyester reinforcement scrim for jet cleaning burst at 18 bar and a dome height of 21 mm.

2.5.5 Breaking Extension

According to DIN 53857 [127] breaking extension is the extension of the specimen at the point of rupture (fig. 2.46). The specimen is 200 mm long and 50 mm wide. A load-extension recording diagram is an intimate part of the method and gives useful information about the work of rupture, and the extension characteristics in the lower region of the curve. The life of the filter medium depends significantly on positive or negative dimensional changes, it also influences its ability to be cleaned. If filter elements extend, they suffer premature failure through abrasion. If they shrink, their cleanability suffers. The filter medium blinds.

Typical values for a polyester jet quality with multifilament support fabric are 180 daN lengthwise and 140 daN widthwise with respective extension breaks of 19 % (machine direction) and 21 % across.

Tensile strength of a textile structure depends on its density, degree of thermosetting, staple length of fibres, but most importantly on the kind of reinforcement (with needlefelts). To withstand the high mechanical loading of filter media they should be reinforced with a multifilament fabric with a weight range of 140–150 g/m^2. Multifilaments are less damaged during needling, because the thread has relatively low twist (e.g. 60 turns/m). The filaments can move out of the way of the needles during needling. This differs for fabrics whose warp and weft threads are made from staple fibres which are cut into

Fig. 2.45 Burst strength tester

Fig. 2.46 Extension break tester

lengths of 40–80 mm. Here high twist increases the strength of the yarn and there is an increasing tendency for yarn damage during needling. This results in lower strength, lower bursting strength and reduced resistance to folding.

2.5.6 Abrasion Resistance

Abrasion resistance is measured according to DIN 53863 [128]. The test specimen has a size of 50 cm^2. Fabrics in the weight range 150–300 g/m^2 carry a 500 g load; over 300 g they carry 1000 g. The fibre loss after 500 revolutions is recorded. A typical polyester needlefelt with a density of 0.9 g/cm^3 should loose approx 0.5 g.

2.5.7 Additional Quality Parameters

Quality conscious needlefelt manufacturers give a whole range of additional specifications to ensure the quality of their filter medium:

Example – polyester needlefelt with support scrim 550 g/m^2

Area weight DIN 53854	550 g/m^2
Thickness DIN 53855	1.9 mm
Density	0.29 g/cm^3
Porevolume	79 %
Air permeability DIN 53887	150 l/dm^2 min 196 Pa (20 mm H$_2$O)
Bursting Strength DIN 53861	18 bar Dome height 21 mm (50 cm^2 specm)
Service Temperature	150 °C dry
Melting Point	250–260 °C
Self Ignition Temperature	508 °C
Regain DIN 54201 [129]	0.4 % at 65 % RH and 20 °C
Water Retention DIN 53814	about 4 % (Swelling value)
Acid Resistance	Good (e.g. resistant to 30 % hydrochloric acid and 50 % sulphuric acid to 50 °C
Alkali Resistance	sufficient (e.g. resistant to 25 % caustic soda)
Hydrolysis Reistance	Poor
Solvent Resistance	Fairly good, soluble in some phenols and hot nitrobenzene
Abrasion Resistance DIN 53863	0.115 g fibre loss (test size 50 cm^2, load 1000 g, 500 Rev.)
Creasing Resistance	Very good
Splitting Resistance	Excellent
Dimensional Stability	1 % (hot air at 150 °C)

2.6 Filter Medium and Cleaning Methods

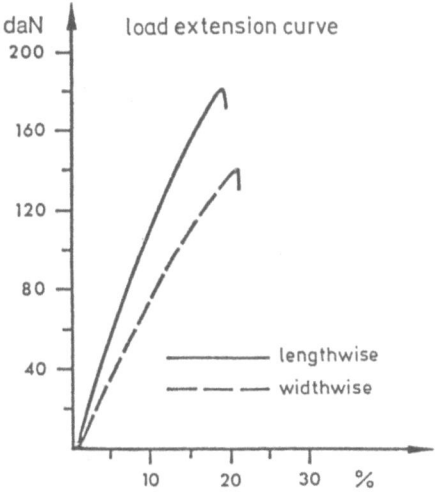

Fig. 2.47
Load-extension curve of a 550 g/m² Polyester needleleft
Strength peak: Machine direction: 180 daN
DIN 53 857 (Specimen size 200 × 500 mm)
Across: 140 daN
Extension at break: Lengthwise 19 %
Widthwise 21 %

Rotting Resistance	Good
Ageing	Very good
Electrostatic Behaviour DIN 54345 [130]	Specific Resistance $1.0 \times 10^{18}\ \Omega\,cm$ Durable antistatic finish possible, if desired
Width	200 cm Standard Width Can be delivered in wider widths
Breaking Extension	refer fig. 2.47

Treatments with specialized chemical finishes can critically affect quality parameters. It is worth devoting a separate section to the complexity of this subject.

Many manufacturers and users of filter media have their own quality control to continually monitor the adherence to specifications. It is usual to specify a tolerance for weight, thickness and air permeability. Tensile strength minimum values are given which are considered basic [135].

2.6 Filter Medium and Cleaning Methods

The story of the fabric filter begins in the year 1886 when the grinding mill engineer Wilhelm Friedrich Ludwig Beth was assigned Patent No. 38396 for a Suction Hose Filter with Automatic Cleaning [32].

Even the dust collector of the 80's possesses similar primary characteristics:

- The bag or pocket as filter body (fig. 2.48)
- Low or high pressure reverse flow cleaning as the cleaning principle: Usually in combination with pulse-initiated, but mechanically continued, cleaning shock [33].

Thus all jet filters are in reality both pneumatically functioning as well as mechanical shaker filters. Particles are shock accelerated and expelled with pneumatic assistance [34]. This type of cleaning is one of the most important construction characteristics of a

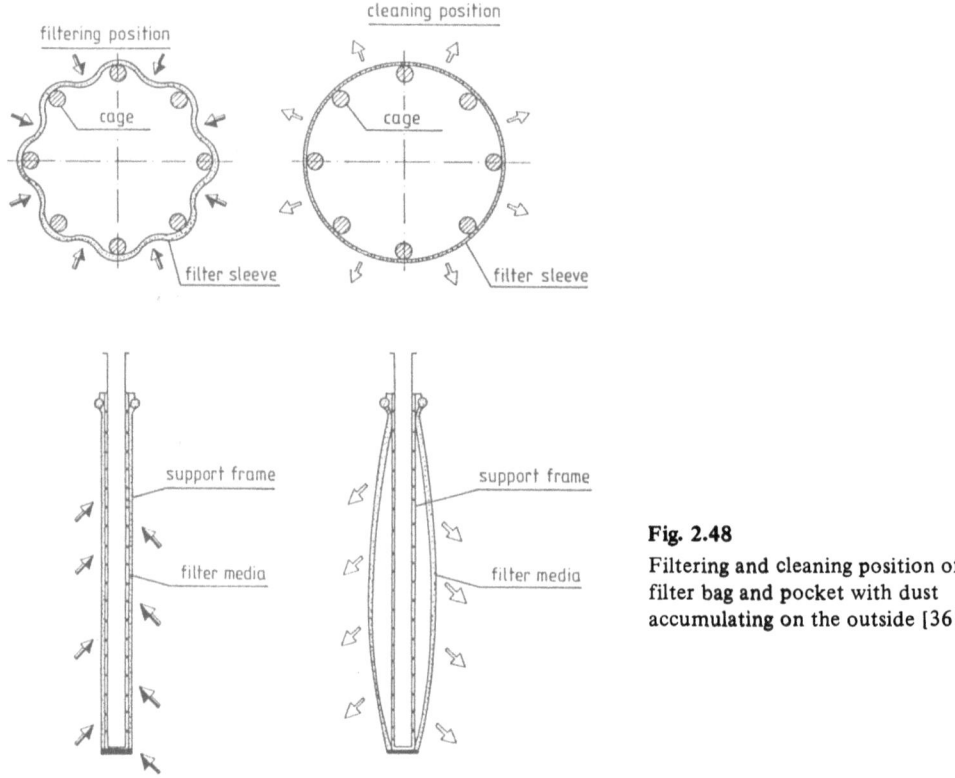

Fig. 2.48
Filtering and cleaning position of filter bag and pocket with dust accumulating on the outside [36]

dust collector. It determines the structure for the filter medium, the differential pressure, the air to cloth ratio. This in turn determines the size, space and energy requirements as well as the efficiency and competitiveness of a dust collector[1].

The advantages of filtration in the third dimension are beyond doubt [35]. The objection that needlefelts, because of their deeper layers, tend to blind is too generalized. It is more correct to say that in order to realize their potential, great care must be excercised in selecting filter qualities. The cleaning modes will largely determine the type to be selected as regards weight, density, flexibility and air permeability. With needlefelts the principle applies — what is deposited by air can also be removed by air. Needlefelts have proved that when properly constructed stable conditions can be reached, even in conventional reverse-cleaned dust collectors. As a general rule air permeability of needlefelts should be inversely proportional to the depth intensity of the cleaning mode. The build-up of a base coat on the surface, i.e. the third dimension filter layer, is less important with felts, because the felt itself is the third dimension. It is desireable on the other hand to achieve

---

[1] Constructional details are covered in a subsequent chapter by W. Flatt. In what follows those aspects are covered which are significant to the development and assembly of filter media.

2.6 Filter Medium and Cleaning Methods

a pore loading in the felt interior of up to 35 % of its own weight. This reduces the effective pore size and makes the filter media more homogeneous.

While fabrics may be used in shaking or reverse-flow filters, it is imperative to use needlefelts for jet filters. This applies also to combined shaking, low-pressure reverse flow cleaned filters as well as to filters with exclusive low-pressure reverse cleaning. The air pressure pulse forces the pores to open through which dust may leak between warp and weft. Dust collectors that are pulse-cleaned require the fine fibre structure of a felt which can cope with the buffeting movement when recoiling against the support cage.

When dealing with woven fabrics the particle separation obeys the law of the sieve effect at the beginning of each filter phase. All particles larger than the opening between warp and weft are retained. The gas stream is divided into a myriad of single streams. The active filter surface is simultaneously the free sieve area of the fabric, i.e. the sum of all openings between warp and weft. It is 40 % maximum. The sole purpose of the warp and weft threads is to give strength and dimensional stability to the filter medium. They are of limited significance for the separation process itself, because the gas stream takes the path of least resistance and streams past the yarn.

To enlarge the filter active surface of a fabric and to facilitate the build-up of a base coat, fabrics are sometimes brushed on the surface. While this decreases the openings between warp and weft, it also increases the tendency to agglomerate dust on the surface. This makes cleaning more difficult. The result is higher Δp and lower gas velocities. Thus the brushing of woven surfaces cannot be recommended.

Good separation effects can be achieved with densely woven fabrics, although with the price of lower air to cloth ratios and higher differential pressures. This results in oversized baghouses. An added difficulty is that in conventional construction of shaking or low-pressure cleaning the dust is deposited on the inside of the filter bag. Thus the cake is deposited within the smaller inside radius of the bag, before it is loosened through shaking, collapsing and reverse cleaning of the inside of the bag. This has advantages and disadvantages. On the one hand the dust is isolated, i.e. it is not affected by the neighbouring bag. On the other hand the dust release occurs in a constricted space. Baghouses with a bag diameter of 300 mm take account of this.

During back-flow, i.e. during cleaning from outside to in, a dense filter medium gives problems. The back-flow must overcome both the pressure loss of the filter medium plus the deposited dust. Not infrequently difficulties arise, because the reverse-flow pressure is insufficient. To achieve adequate filtering efficiencies fabrics with an air permeability in the region of 100 litres (DIN 53887) are chosen. If a needlefelt is selected, the same result can be achieved with an air permeability of 275 litres. Because of the lower drag and with the help of the smooth unstructured surface of the felt and added cake release properties it is easier to overcome the initial drag with more efficient cleaning. Experience has shown that in large projects where suppliers fit needlefelts to baghouses that they are more competitive, because they can increase air to cloth ratios by 30–50% with consequent smaller and cheaper installations.

With jet cleaned dust collectors the dust collects on the outside of the bag. The reverse cleaning is from inside to out. Menden [37] points out that this type of dust collector has displaced collectors with mechanical cleaning. The author lists among the significant advantages the following:

a) No moving parts of the mechanism.
b) The ability to use dense and separation active needlefelts which respond to intensive cleaning.
c) Achievement of high separation efficiency, even with fine dusts.
d) Increase in air to cloth ratios.

Flatt [38] goes even further to claim double to fourfold air to cloth ratios "compared with conventional dust collectors with the same bag life, same total flow resistance, and with same residual dust content". This refers to jet filters with single bag cleaning. Guethner [39] separates the cleaning process into 3 phases

a) Acceleration of the filter medium towards the outside.
b) Deceleration of the filter medium once it reaches a circular shape.
c) Acceleration of the filter medium inwards towards the cage after completion of the cleaning cycle and after the decay of the cleaning pulse in the bag interior.

Needlefelts are exclusively used in jet cleaned dust collectors. A weight of 550 g/m^2 and an air permeability of 150 l/dm^2 · min at 2 mb is preferred.

During the opening of the diaphragm valve — depending on construction between 20 and 150 ms — a pressure wave enters the interior of the bag or pocket (fig. 2.49). The filter elements resting of the cage are inflated from support. By the use of a centric jet or venturi tube the pulse enters the bag centrally. The pulse draws in secondary air as an injection jet of clean gas (fig. 2.50). In a fraction of a second, a pressure wave shoots

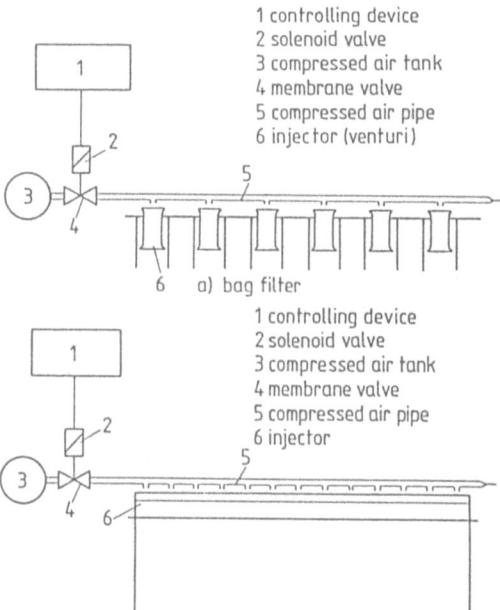

Fig. 2.49
Jet cleaning of bag and pocket filters [36]

vertically downwards to the bottom of the bag. A metal cap supporting the bottom of the bag intercepts this pulse resulting in a compression wave which is reflected back. Meantime more cleaning air enters from the top (fig. 2.51). Thus bags require special reinforcement near the bottom because of the periodic but brief pressure peak. With venturi jets (fig. 2.50) the jetstream pushes centrally down the bag, while with ring jets the main air movement is along the wall of the bag (fig. 2.52).

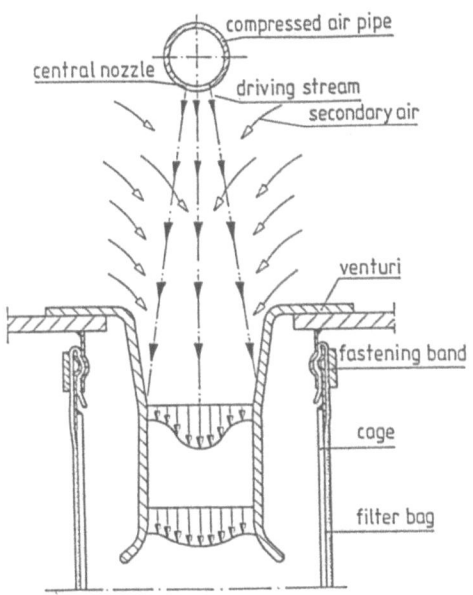

Fig. 2.50
Impulse jet to produce primary and secondary air induction [37]

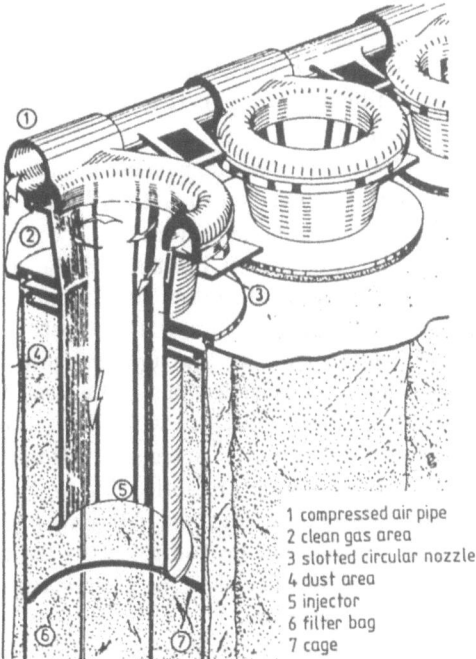

1 compressed air pipe
2 clean gas area
3 slotted circular nozzle
4 dust area
5 injector
6 filter bag
7 cage

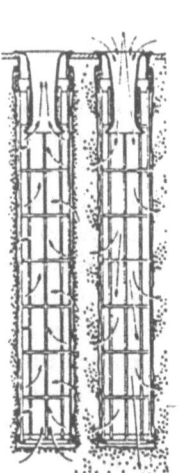

Fig. 2.51 Principles of filtering and cleaning

Fig. 2.52 Ring jet type [41]

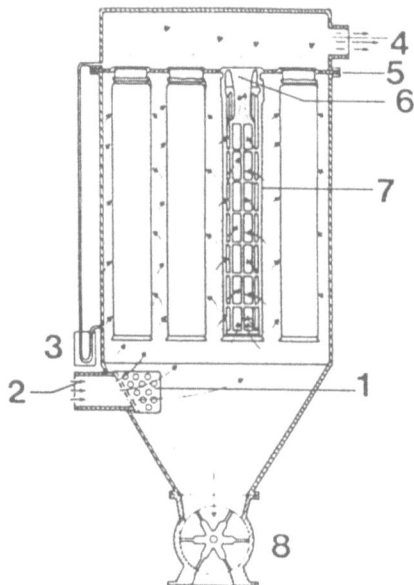

Fig. 2.53
Jet filter with exterior dust deposition on bags
1. Gas manifold
2. Flue gas entry
3. Differential pressure gauge
4. Filtered gas exit
5. Head plate
6. Venturi tube
7. Support cage
8. Dust lock

The dust cake on the surface of the bag is broken by the inflation of the filter element. The acceleration from a concave to a convex position (fig. 2.48) with sudden impact mechanically ejects the dust together with particles that have penetrated the felt. What is mechanically initiated is pneumatically continued by the cleaning air which flushes out the dust [41]. The reinforcement fabric takes on special significance in needlefelts for jet filters (fig. 2.53). It gives dimensional stability, strength, bending resistance and filtration efficiency.

It has been said that filter media in dust collectors that are jet cleaned are unstable [40]. One speaks of the "puffing effect." This tendency need not arise or can be reduced to a minimum, if suitable and properly constructed support fabrics in the felt are given sufficient attention. It has become the practice overseas to use very light scrims or no support fabrics at all. Either their significance is underestimated or for reasons of strong competition and to eliminate a cost factor, support fabrics are totally or partially ignored. It is overlooked that the support fabric has an important function in the separation efficiency of jet cleaned dust collectors. When a support structure is needled to fibres the fabric layer is enriched with fibres, resulting in an increased density zone within the reinforcement area. This density difference increases with greater thread density and weight of the support fabric. This acts as a barrier against penetrating dust particles. Particles penetrate into the felt only to the scrim, so that the dust build-up on and within the felt can be controlled to the depth of the density zone. Only particles that can get past this density zone are released into the clean gas exhaust once the filtration process recommences.

Experience with jet filters since the middle of the 60's has shown that dust penetration during the bag repositioning into the concave position — when it reforms around the support cage wires — are minimised, if the barrier in the inside of the felt holds. Separa-

2.6 Filter Medium and Cleaning Methods

tion problems with dusts that do not agglomerate can only be solved via a support fabric. The recognition of this fact can be used to advantage in the mechanical part of the cycle during jet cleaning. Dust particles that have penetrated to the denser middle zone are expelled more easily by the acceleration of the fabric from the "star form" into the circular shape, because they do not penetrate as deeply into the felt interior. Thus when specifying a needlefelt, weight and permeability alone in not sufficient, also weight and set of the associated support fabric should be included (fig. 2.54).

The Bubble Test (ASTM E 128/61) [44] gives a measure of the pore opening pressure. It is an indirect measure of the resistance of the felt to that pressure which acts upon the bag during the pressure pulse. A felt specimen is saturated in alcohol and is exposed to an adjustable air pressure. This pressure is recorded. When the first bubble appears on the felt, the opening pressure for the largest pore in the specimen is recorded. The average pore opening pressure is reached when many bubbles appear. Needle set, fineness, fibre fineness, shrinkage and thermosetting processes and most importantly the support weave are the significant influences on this pressure related pore opening.

Contrary to opinions occasionally heard in the market, a cleaning process has no or very little effect on the life of the filter element. Provided the needlefelts are properly selected to suit the particular installation. Thus it is not true that needlefelts have a shorter life in a jet cleaned dust collector than in a conventional collector. Practical experience contradicts this. In many installations jet cleaned needlefelt filters last between 16000 and 40000 operating hours even under extreme mechanical and abrasive conditions. It is also true that there are instances in which much shorter service lives – a few months in some cases – can occur. These problem cases will be dealt with in a separate chapter. Let it be sufficient at this point to say that this is caused by other factors than the cleaning method.

Chemical, hydrolytic, catalytic influences and not the least, poor quality and confusion over basic specifications or improper manufacture may account for such problems. With

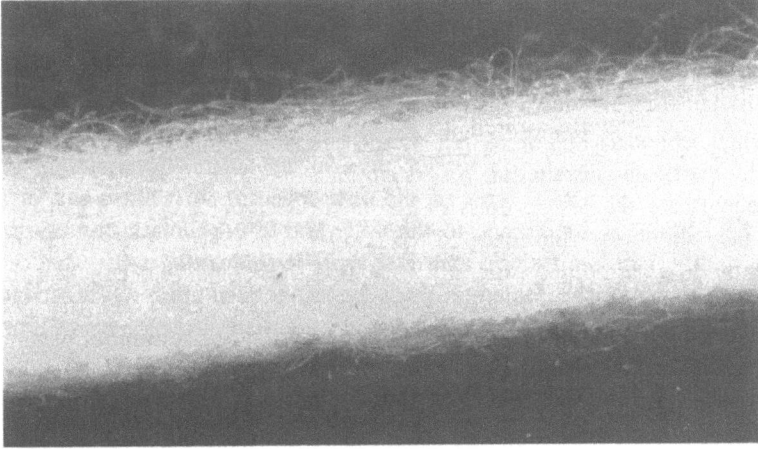

Fig. 2.54 Cross section of needlefelt with support scrim

shaking or vibrating filters it is possible to experience premature failure in the lower region of the bags, because the medium is insufficiently resistant to the periodic folding stresses. This is frequently the case in needlefelts where the support fabric is damaged during needling. As a rule substituting a multifilament fabric insert cures the problem. In many cases an additional reinforcement with a multifilament sock 100–200 mm long is a sure method to prevent premature creasing failure. With conventional cleaning systems one should remember that felts are stiffer than fabrics, so that the pulse induced wave motion from top to bottom is more difficult to transmit. Bag constructions with a small diameter have a stove-pipe like effect. In such cases the filter medium should be especially flexible to distribute the cleaning evenly over the length of the bag.

During discussion of cleaning modes and appropriate filter media there appear at times fundamental differences between American and European practices. It is surprising that in the USA generally a conservative approach is practiced with strong emphasis on fabrics and low gas velocities but very long service lives. There is also greater emphasis on the significance of the basic dust layer acting as a filter aid. Also their concept of the composition of needlefelts differs from the European view. Support fabrics tend to be lighter and more open. The surface of the felt (ref. chap. 2.8) is frequently left hairy to encourage a filter active cake. When looked at from the point of flue gas velocities, the European observer will see that the desire for compact construction as practiced in the space limited Europe does not find a parallel in the spacious USA. The requirements of the filter medium are thus lower. European qualities are often judged as "too good" in the USA, while American grades are viewed as inadequate by European plant manufacturers. As always the exception proves the rule.

2.7 Selection Criteria

The selecting of a suitable filter medium for a particular installation is among the most difficult tasks in arriving at a design concept. The following are decisive:

- Construction of the dust collector and its cleaning principle
- Composition and temperature of the carrier gas
- Physical and chemical properties of the dust to be separated

The assessment of these partitioning variables is set out into a series of detailed questions below, but first a fundamental statement:

A filter medium which is modern and up to date should be made to measure. It should be selected to suit the particular installation for its physical and chemical requirements. Filter bags and pockets are the textile dress of the dust collector. Just like a suit or a piece of clothing, they should be tailormade to the body and to the climatic conditions, i.e. temperature, humidity, gas composition and dust type. It frequently occurs that the technician dealing with the filtering problem is given precise details, often to the second decimal place, about the gas components, while the vital water and water vapour proportion in the gas is ignored: You always have water and, by the way, it never does any harm! This totally neglects the fact that water vapour is often the worst enemy for synthetic fibres.

Another prominent example is the particle size distribution of the dust. Difficult and costly analyses are made on particle size to the submicron level. Any mention, however,

2.7 Selection Criteria

whether the dust agglomerates or fluidizes is absent. A coarse non-agglomerating dust can be more difficult to separate than a fine agglomerating one. These are only two examples of often neglected and important variables.

A correct filter medium can only be selected, if a mutual cooperation exists between three experts:

- The construction engineer of the dust collector
- The process engineer of the plant
- The filter medium expert [43]

It goes without saying also that to achieve proper partitioning three factors work together and must harmonize:

- The dust
- The filter
- The filter medium

There are positive indications over the last years that mutual cooperation is becoming a reality. The filter media manufacturer as "third in the team" is consulted more and more often "to close the gate before the horse has bolted". This was not always so: Filter media specialists were previously only consulted when the filter elements, agreed to by both the plant manufacturer and the plant operator, did not measure up to their expectations, and a third party was sought to offload a complaint.

A specialist skilled in end-use requirements and concerned with impartial information will make a detailed examination of the separator which is to be equipped with a filter medium. He will determine origin, form, dimension, number of filter elements. He will investigate whether a preseparator has been installed and whether bags or pockets are in one or several chambers. His special attention will be given to the type of cleaning – the pressure in the pulse system, the cleaning air volume and the cleaning frequency. A significant question will be the expected air to cloth ratio (cubic meters per square meter minute).

His next series of enquiries will concern the operating conditions: is the plant housed in a building or in the open? Is the plant insulated or subject to outside temperatures? What is the average operating temperature, what peaks and frequency and at what intervals can they be expected? Does the plant operate in continuous shift or can plant shutdowns be expected? In this connection the moisture content of the gas is of interest, the water and acid dewpoint and the probability of their occurrence. Next is the question of gas composition: Are there potentially corrosive or absorptively bound gas components? If we are considering a combustion plant, it will be interesting to know the quantity of sulphur burnt per hour and the type of sulphur oxides formed. The next aim of questioning is the desired and maximum permissible pressure loss as well as the prescribed clean gas dust content in mg/m^3 and any guarantees pertaining to its achievement.

The third aspect is the dust level of the flue gas, its chemical composition, density fraction bands, hardness and abrasiveness. Important other variables that are worth investigating are tendency to agglomerate, i.e. its physical properties in relation to flow or possible adhesion or hygroscopy, when the dust is removed from the bags.

Now it is the filter specialists turn. He has to decide whether fabric, nonwoven or needlefelt is appropriate; the type of fibre especially in relation to anticipated temperature and chemical exposure. Depending on cleaning mode and the eventual residual dust content the area weight and air permeability should be set. Gas and dust may, in relation to fibre protection and cleanability, necessitate chemical treatments or special surface effects. Note must be taken whether bags are to be washed, whether there is a risk of electrostatic charging and what is the service life that is to be guaranteed.

The filter medium manufacturer has a rich palette of synthetic fibres to choose from spanning from viscose via PTFE to metal or even mineral fibres (ref. sect. 2.2) Unfortunately the choice of a suitable fibre is not always easy, because there are no fibres which do not have physical, chemical or commercial limitations. The result is a range of custom-made qualities to overcome the many potential limitations which can arise in the thermocatalytic chemical attack. Such a large quality range brings with it economic and organisational disadvantages which can only be overcome, if the chemical industry would finally be able to come up with a fibre that resists high temperature, stands up to corrosion, can withstand high physical stress and is cheap. Such a fibre is not yet on the horizon, so we must fall back on the inventiveness of the filter media specialist to fit a made-to-measure suit for the individual filtration problem. Since gas cleaning is moving into areas in which there is little experience to date, or where other separation principles have been used, we must expect at continuing challenge to solve ever new problems.

2.8 Filter Medium and Separation Process

In three dimensional filter media such as needlefelts and nonwovens several processes operate together to achieve particle separation from the gas stream. These processes are treated in detail by F. Löffler in chapter 1. He considers depth filtration in some detail as applies to the partitioning system in its initial state.

In contrast to surface filters such as fabrics, membranes and papers, the depth structure of three dimensional filter media is of special significance in the separation process [44].

Surface or Depth Filtration

At this point some categorical statements are appropriate concerning "surface filtration" or "depth filtration". The question of "either/or" is not the author's choice; it is also inappropriate in this form. A better formulation would be: "When surface- and when depth filtration".

When depth filtration with needlefelts and nonwovens in the 60's and early 70's began to dominate the market more and more it was inevitable that mistakes would be made in this euphoria. Depth effective filter media were used in unsuitable situations. Difficulties arose with sticky, agglomerating, hygoscopic dusts, for example in the food industry or in the separation of fine non-ferrous metal industry. Suitable needlefelt qualities were not coordinated with prevailing cleaning systems. Blinding occured with consequent short service life of the bags or pockets, high pressure losses and reduction in gas flow capacity.

New solutions had to be found. Among them were attempts to trap dust particles on the surface of the medium to avoid depth blocking. This was done by applying a coating to the surface of the filter media; the spreading and sintering of synthetic resins; the asymmetric insertion of a support fabric on the flue gas side and the flame-bonding of a semiporous membrane or foil. Such filter media did indeed succeed in preventing premature blinding, but failures still occurred.

2.8 Filter Medium and Separation Process

The technical arena is no different to life's stage, and one should not "tip out the baby with the bath". Both surface and depth filtration have their place, proveded the right techniques are used. A decision for depth filtration implies that huge filter surface areas are provided, made up by the sum of all fibre surfaces on and within the filter medium. It goes without saying that pore cleaning of the filter medium requires a depth effective cleaning system (as detailed in chap. 2.6). Depth filtration is the economic utilisation of the total available filter active surface of the medium.

In surface filtration the zone where the cake forms is transferred from the depth to the surface. Here the filter surface retains particles on impact. This can be advantageous where particles agglomerate, when they are tacky, fatty or moist or otherwise cohesive. One should remember, however, that a filter surface which is too dense leads to pressure drop problems. In felts with a heavy surface calender (eggshell finish) the small particles work themselves through the plasticised surface into the interior of the medium. During the reverse cleaning pulse they can no longer exit through the pore through which they entered. In this case the particles will build up beneath the surface layer on the intake side. The layer is build up progressively and leads to increased differential pressure and finally to complete blinding. The term "chicken effect" has been coined in analogy to a chicken that slips through a hole in the fence to the neighbour's garden, but cannot find its way back.

An interesting compromise was suggested to give the fibres themselves a repellent treatment instead of the felt surface. This allows access of the particles into the felt interior thereby enabling separation in the third dimension. The particles are not retained in the interior by high friction of the fibres. Both mechanical as well as jet cleaning pulses enable the particles to be easily detached from the fibre surface.

Silicones are suitable repellent treatments (fig. 2.55 and 2.56) as well as resin based on PTFE, silicones and fluorocarbon. These may be applied either to the surface of the medium or to the fibres themselves. Depending on what emulsifiers are used or the type of condensation product, impressive repellent effects can be achieved. When used correctly

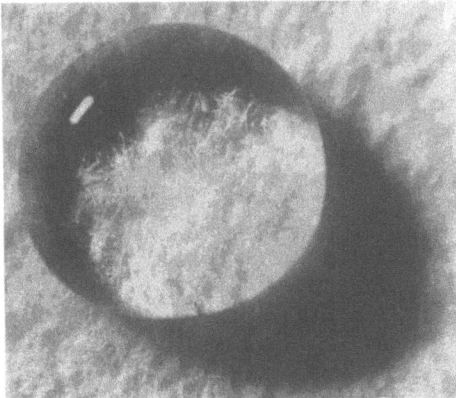

Fig. 2.55 Repellency effect of water

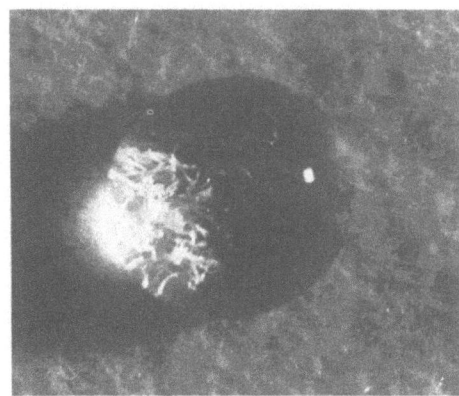

Fig. 2.56 Oil repellency

repellency can give additional fibre protection. Whilst encouraging cake and dust release, they are not essential and do not necessarily increase separation efficiency compared to untreated products. When such repellent products are applied correctly they resist wear and abrasion and can still be detected on filter media after many years of service. Wash-fastness should be checked with the manufacturer; treatments containing emulsifiers may have excellent repellency, but their wash fastness can be impaired by residual emulsifier.

Coatings or chemical treatments of filter media produce a fourth chemical variable in the interaction between dust, gas and fibre. This fourth element should be compatible with the fibre and chemically inert. It is inadvisable, for instance, to sinter a PVC or polyethylene granulate onto a polyester felt to make the surface easier to clean. The service temperature and chemical resistance of the filter medium will be limited to the durability of the coating – a chain is as strong as its weakest link. The same applies to all other fibre treatments (e.g. flame-resistance, antibacterial treatments, etc.) in which the thermal and chemical operating environment must be considered before choosing a treatment. This prevents difficulties which may arise due to decomposition effects associated with cohesion or contamination.

Luenenschloss and co-workers [45] show that the accumulation and separation of dust "occurs mainly on the face side of the carrier material (on the flue gas side)" (fig. 2.57 and 2.58). They underline the observation found in practice that the carrier material is

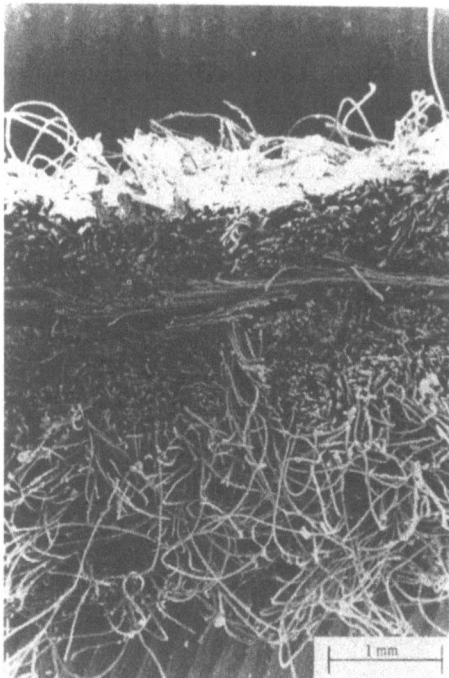

Fig. 2.57 Cross section of needlefelt with support scrimm and initial dust deposition

Fig. 2.58 Cross section of needlefelt after more dust exposure

2.8 Filter Medium and Separation Process

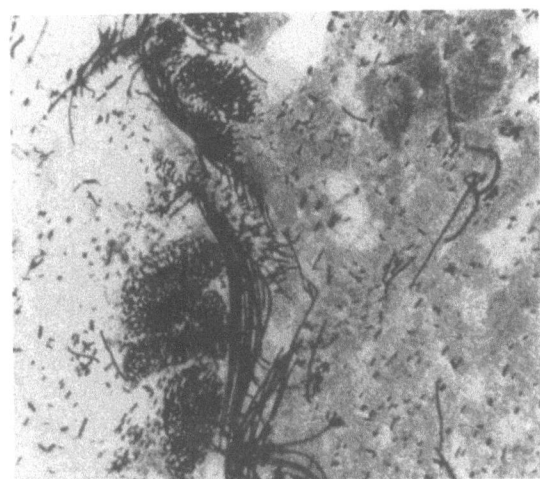

Fig. 2.59
Scanning electron micrograph showing dust penetration to interior support weave barrier [46]

a barrier to dust penetrating to the clean gas side (fig. 2.59). Photomicrographs and scanning electron microscope photos illustrate how fibres compact at the insertion points of the needles; the zone between these insertion points is less dense. They conclude that dust penetration does not occur at these insertion channels, but rather between them. This is further evidence for the filter efficiency of a support fabric, because the dust enters at places in the felt other than where needles are inserted into the support fabric.

During the course of their examination [45] scanning electron micrographs were produced which showed the deposition of a test dust "Quarz F 500" on polyester fibres (fig. 2.60). These showed that in all cases the particles attached themselves to the fibres. Evidently strong adhesive forces cause the dust to adhere to the fibre surface. This lends support to the idea of depositing a release finish onto the fibres. Their photographs show also that electrostatic forces play an important role in separation. Polyester has a specific resistance of $6.3 \cdot 10^9$ Ohm · cm (DIN 54345). Thus the purpose of an antistatic finish should not be limited to whether the filter medium can accumulate enough electrical energy to produce a spark, but should also consider the cleanability of the fibre medium when the conductivity of the fibre is increased. This is a wide field for research. It is interesting to consider in this context the behaviour of different dusts as it relates to their fineness and tendency to adhere. From many trials two are selected for illustration. These will show the strong influence of the dust properties in the design of a dust collector and in setting performance parameters [47]:

a) Test material: Hydrated Aluminium Oxide

    Particle size distribution   10 % less than 10 μm
                                         35 % less than 20 μm
                                         50 % less than 45 μm
                                         90 % less than 90 μm

Fig. 2.60
Dust deposition of polyester needlefelt interior. Additional magnification of inset above [46]

b) Test material: Acrylobutastyrene Powder (ABS Dust)

Particle size distribution 10 % less than 1 µm
20 % less than 6 µm
40 % less than 10 µm
95 % less than 20 µm

An acrylic needlefelt was used as a filter medium with a multifilament support fabric, final weight 550 g/m², air permeability 150 l/dm² min. These trials were conducted on a pulse jet filter type dust collector.

The particle size distribution shows that aluminium oxide is not particularly fine; it is however known for its high abrasiveness and its inability to form a cake. Aluminium oxide in its various forms is a classical example of a flowing dust. Contrary to current opinion that high dust deposit and high pressure drop goes hand in hand with good collection efficiency, in this case the percentage dust leakage into the filtered side rises with increasing differential pressure (fig. 2.61). ABS dust is much finer, however more prone to caking. Here the contrary is the case. With rising differential pressure, dust penetration decreases (fig. 2.62).

A similar unexpected result is indicated by the face velocity rating diagrams. With aluminium oxide the penetration increases with increasing air to cloth ratio (fig. 2.63); with ABS dust the penetration falls (fig. 2.64).

2.8 Filter Medium and Separation Process

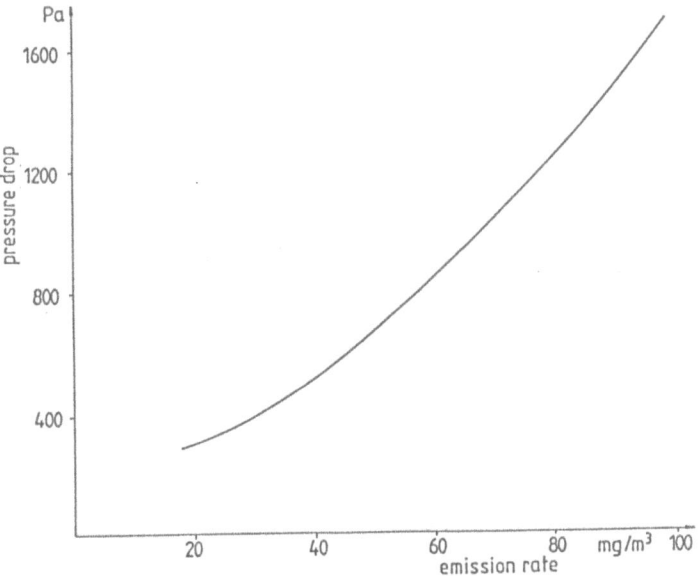

Fig. 2.61 Dust concentration of gas discharge as a function of the differential pressure – aluminium oxide dust

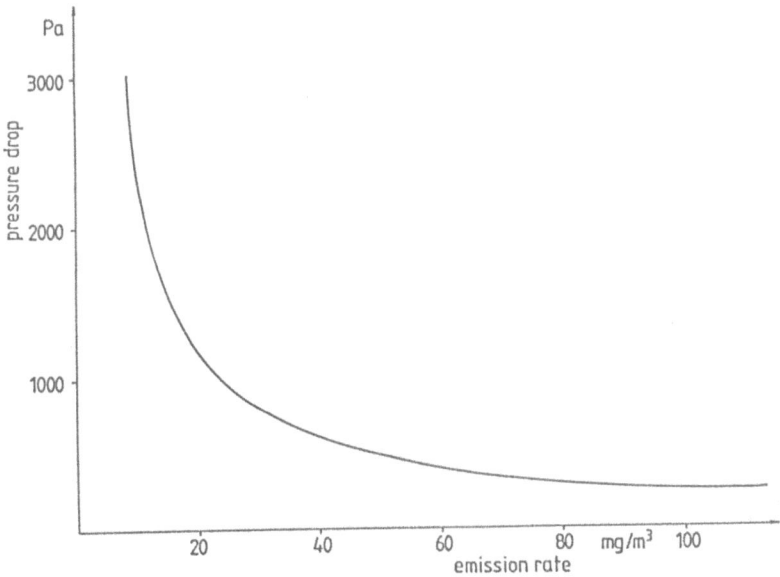

Fig. 2.62 Dust concentration of gas discharge as a function of the differential pressure – ABS dust

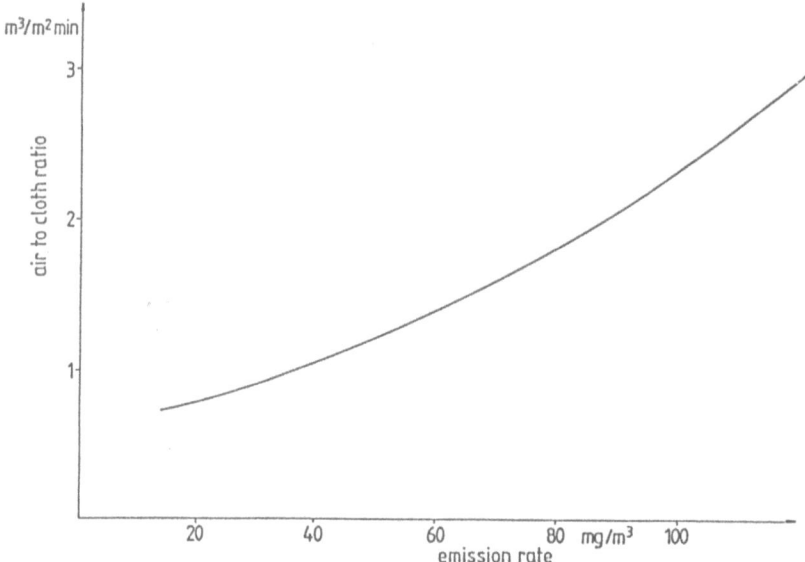

Fig. 2.63 Dust concentration of gas discharge as a function of the filter area rating – aluminium oxide dust

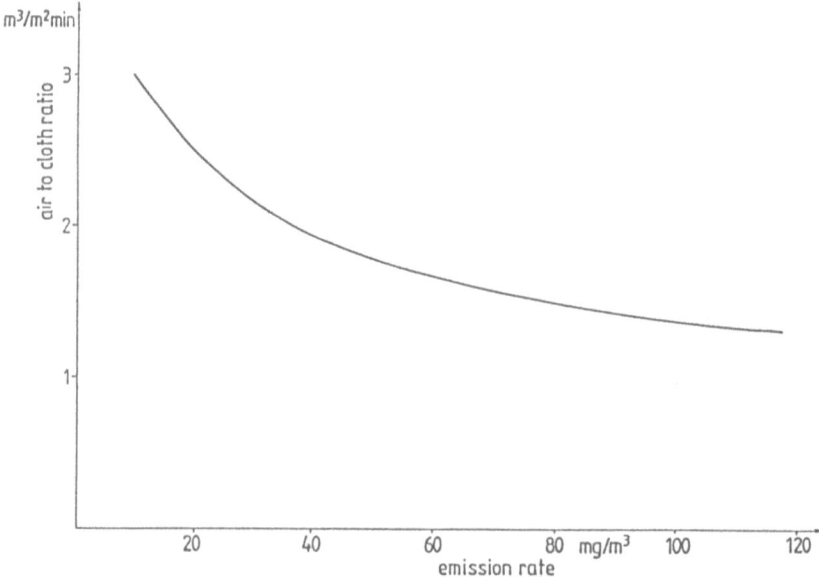

Fig. 2.64 Dust concentration of gas discharge as a function of the filter area rating – ABS dust

2.8 Filter Medium and Separation Process

If air to cloth ratio is kept constant at 2.2 m^3/m^2 min and the flue gas load is varied (fig. 2.65), aluminium oxide shows a delayed progressive rise in differential pressure. This is caused by a subsequent deposition of the non-caking dust in the felt interior. With ABS dust (fig. 2.66) the rise in differential pressure is linear to the flue gas dust

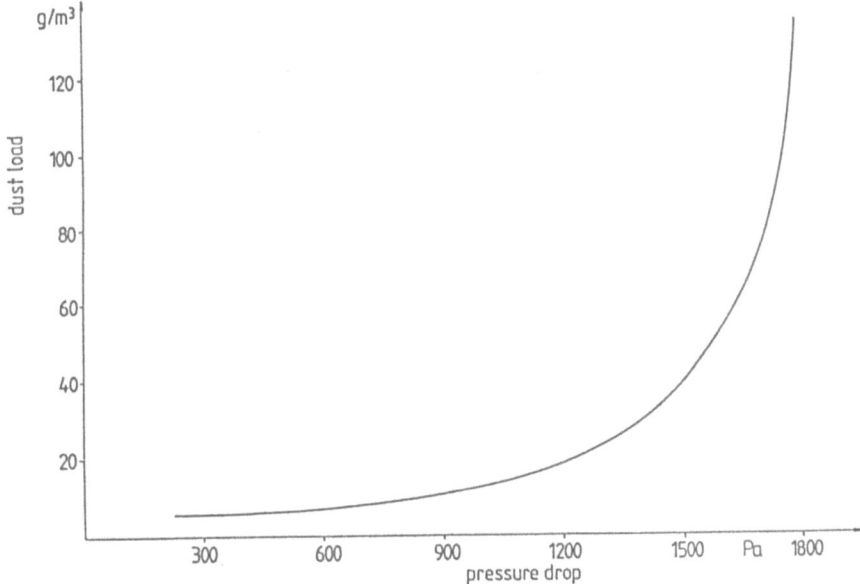

Fig. 2.65 Flue gas dust concentration as a function of the differential pressure – aluminium oxide dust

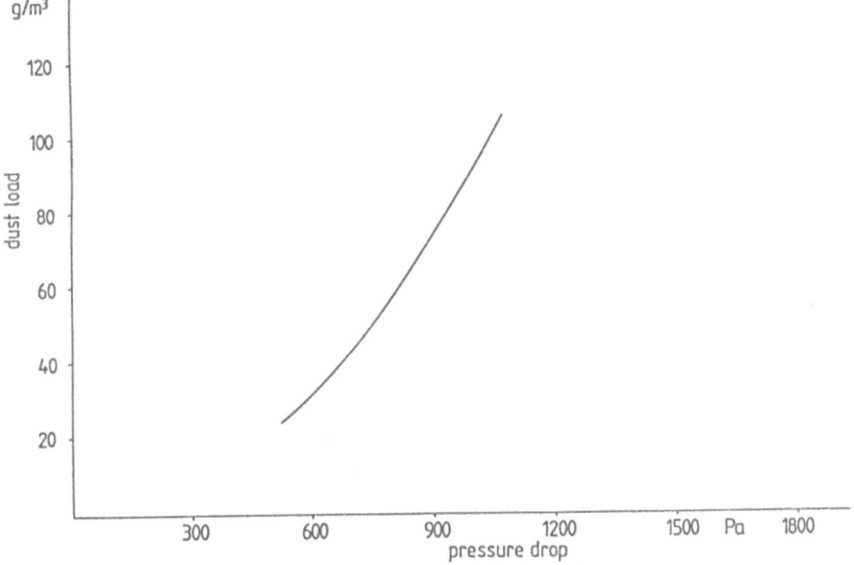

Fig. 2.66 Flue gas dust content as a function of the differential pressure – ABS dust

content. These extremes also become evident when considering filtration efficiency. While aluminium oxide shows excessive emission values (fig. 2.67) with increasing flue gas load, ABS dust emission decreases even slightly with increasing flue gas dust content (fig. 2.68).

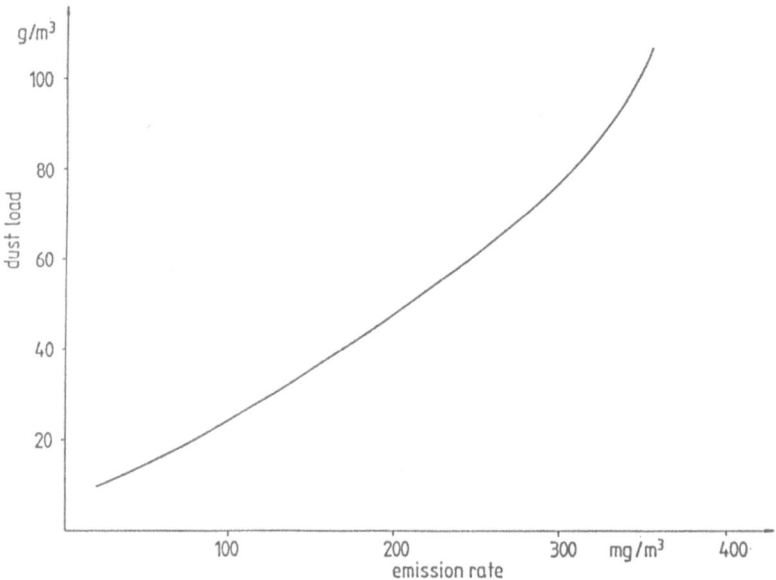

Fig. 2.67 Flue gas dust content as a function of the discharge gas dust content – aluminium oxide dust

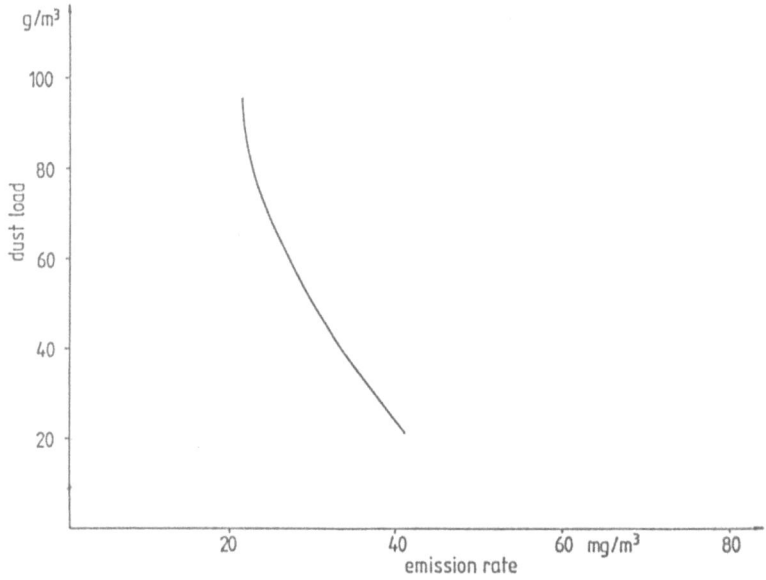

Fig. 2.68 Flue gas dust content as a function of the discharge gas dust content – ABS dust

2.8 Filter Medium and Separation Process

Menden [37] came to similar conclusions when he wrote: "When dealing with flowing non-caking dusts which do not produce an initial cake, increased emphasis must be given to the reverse cleaning of the filter medium. The drag on the filter bag is due solely to the dust particles which have penetrated into the three dimensional filter medium."

Based on suggestions of Bakke [132], Guethner [70] emphasizes different controlling factors for non-caking, i.e. flowing, and for caking dusts. The "injector gradient" (fig. 2.69 and 2.70) shows the relation between injector feed pressure and the cleaning air quantity. The injector must initially overcome the differential pressure of the burdened

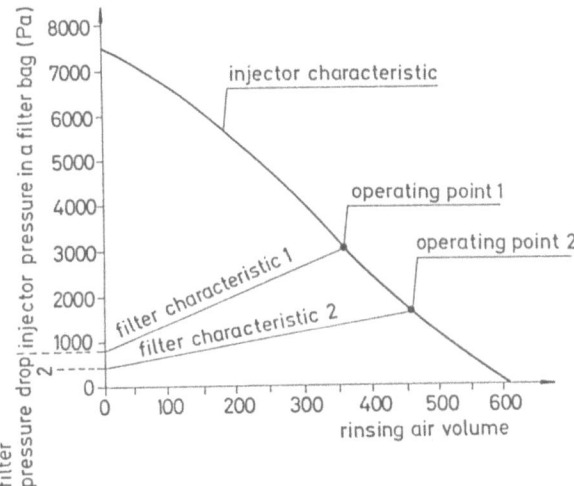

Fig. 2.69

Characteristic curve of venturi pump [37, 48]

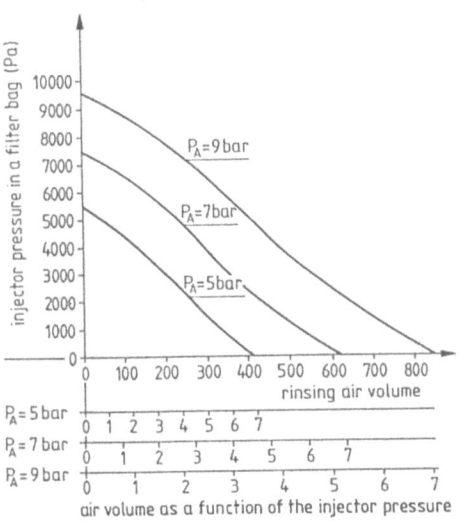

Fig. 2.70

Characteristic curve of a venturi pump as a function of cleaning pressure [37]

filter before air starts to move in reverse to the filtering direction. The point at which the reverse cleaning phase begins is the intersection between the injector characteristic and the filter characteristic curves. The resultant cleaning pressure in the filter bag is the difference between the cleaning pressure at this intersection and the differential pressure at which no cleaning air flows. Non-agglomerating dusts tend to form no filter cake: rather, the dust lodges in the inside of the filter medium resulting in reduced clean air requirements (operating point 1).

With caking dusts both Guethner [48] and Menden [37] conclude that the reverse air resistance is smaller than in case 1. This presumes that the filter cake will be removed at the first pressure pulse (refer "Filter medium and Cleaning Modes") and no longer represents a resistance to the cleaning air. This results in a lower operating point (operating point 2) an the line (lower total cleaning pressure, more cleaning air quantity).

The injector gradient line characterises the type of cleaning system and depends on the pressure in the pressure vessel. Fig. 2.70 illustrates the gradient line for one injector type at different storage pressures.

Meyer zu Riemsloh [41, 49] summarises different types of construction for air pressure cleaned filters and sets out a range of cleaning systems according to type of construction. With high mass velocity ratio (sum of jet stream and secondary air stream) he predicts an efficient cleaning. Eberling and Menden put the comparability of different systems into question [50].

It is a fact that a three dimensional filter medium needs a minimal cleaning air volume. The dust particles that have penetrated to the interior of the felt are loosened from the fibre during the mechanical part of the cycle and are then partially carried outwards. Supplementary pneumatic action must follow to blow these particles from the felt. With increasing cleaning volume (fig. 2.71) the air to cloth ratio can be increased within certain limits [37]. Flatt [38] puts great significance to the cleaning air and shows an exceptional increase in air to cloth ratio, if bags are cleaned individually.

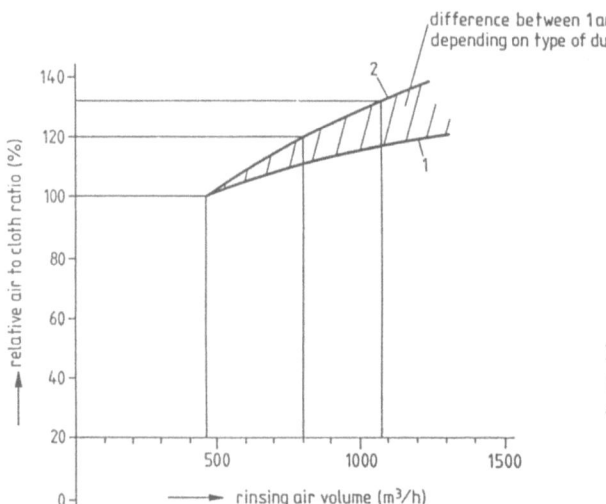

Fig. 2.71

Relative filter area rating as a function of cleaning air quantity [37]

Experience of the author and that of his colleagues is that the separation of dust particles is complex and that generalisations are not applicable. It is important to emphazise that the ever repeated assertion that the filter cake layer solves all problems is a gross simplification. The practical consideration of a dust in relation to its caking tendency, its electrical chargeability and polarity, catalytic properties and chemical reactivity are as important as the analytic data of its particle size distribution. It would be of great assistance, if test procedures and simulators were developed, that could determine simply and systematically the caking behaviour of a dust and that could be used to predict dust behaviour. This would be a step in the right direction to setting down the layout of a dust collector and would provide data not available to date for the proper analysis of a dust.

2.9 Chemical Attack on Filter Media

Chemical damage and hydrolysis are among the most frequent which affect the life and effectiveness of filter media. They are frequently not identified as a cause, because they hide behind latent mechanical damage which originates from previous chemical effects. Chemical and hydrolytic influences age the fibre; it becomes brittle and less flexible. It becomes more sensitive to creasing, it is more difficult to clean and finally it will blind. Hig drag, holes, abrasion marks or splits give a superficial appearance of mechanical damage or fatigue. They are frequently, however, initiated by an "invisible" attacking agent.

At elevated temperatures organic compounds undergo several reactions:

a) Oxidation: The reaction with oxygen from the air.
b) Hydrolysis: The reaction with water or water vapour.
c) Splitting off of smaller reactive molecules.

Some substances can initiate an acid or alkali catalytic reaction which can significantly accelerate the progress of a reaction.

Hydrolysis is the main enemy of textile filter media. By this is meant the action of moisture in breaking apart the molecular chains of a polymer or other organic and/or inorganic compound. Macromolecular compounds differ in their reactivity with water [51].

In the case of cellulose (Viscose, cotton) the oxygen bridges can be broken by the action of water. The reaction product is glucose. For this reason cotton is sensitive to hydrolysis (fig. 2.72).

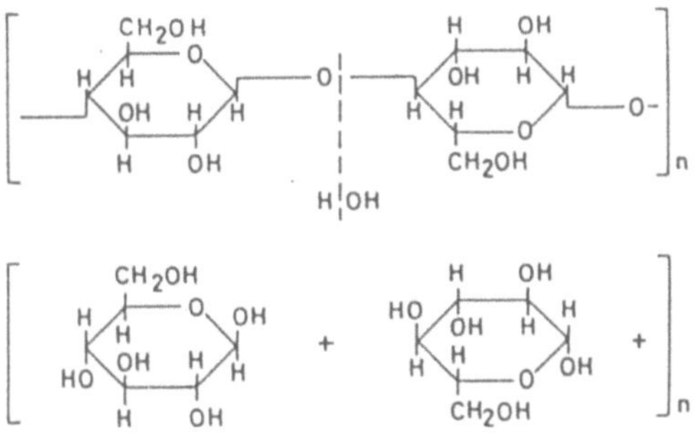

Fig. 2.72
Hydrolysis of cotton [51]

The acid amide bridges in wool can also be cleaved by water. As reaction products, amino acids result. Thus wool is similarly sensitive to hydrolysis (fig. 2.73).

Polyester (fig. 2.74) contains a chain ester formed by splitting off water during the reaction of an alcohol with an acid group. Under certain conditions this reaction can be reversed in the presence of water. For this reason polyester fibres can be hydrolysed.

The polyacrylonitrile chain is not as sensitive to hydrolysis, because its polymer chain is based on polyethylene However, it is possible for the carbonyl group to react with water which produces substantial changes in properties of the polymer (fig. 2.75). Rieber [51] warns that while the acrylic fibre resists hydrolysis, its propensity to swell is much increased.

With polyamide and polyurethane fibres, bridge members exist in the form of acid amide groups similarly to wool. They behave therefore in the same fashion to wool in the presence of water.

Polypropylene, PVC and PTFE do not hydrolyse because there are no points in the macromolecular complex which can be attacked by water.

Depending on whether water is present as water or gas, under pressure or not, or in liquid form without pressure the hydrolysis reaction proceeds in different directions. Foreign

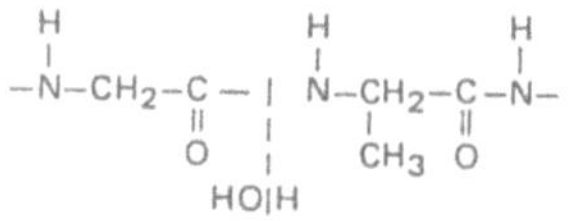

Fig. 2.73 Hydrolysis of wool [51]

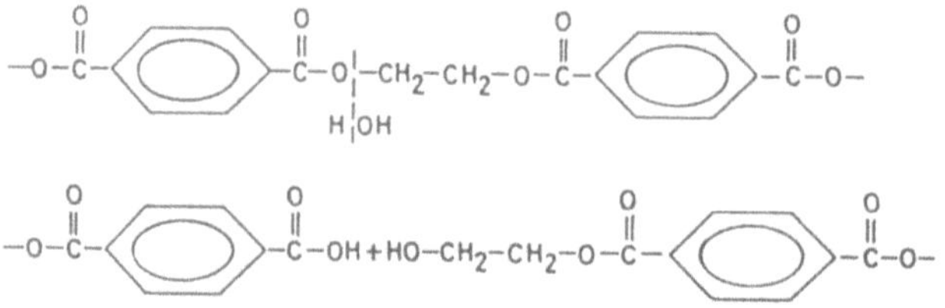

Fig. 2.74 Hydrolysis of polyester [51]

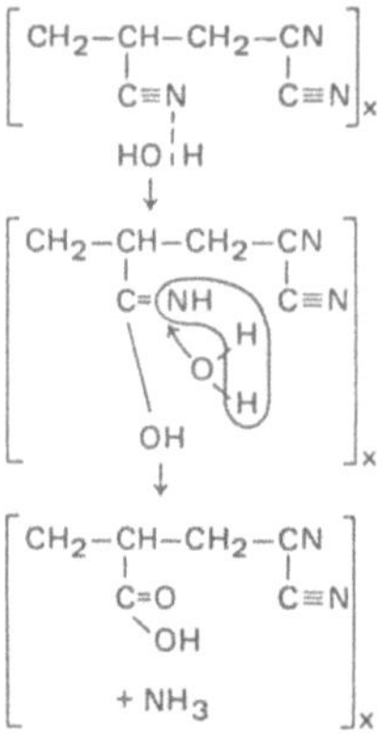

Fig. 2.75 Hydrolysis of polyacrylonitrile [51]

2.9 Chemical Attack on Filter Media

matter can act as catalyst to accelerate hydrolysis. Certain salts or organic solvents may also accelerate the reactions as well as acids and alkali. Swelling and solution effects occur which facilitate the reaction with water. Elevated temperatures increase the rate of hydrolysis. Rieber [51] theorizes "that a layer of water with a catalytic agent forms around the target group to make hydrolysis possible".

In the early stage, hydrolysis produces little change in fibre strength and extensibility. Progressively it causes aging which becomes apparent as a decreasing extension break and lower loop strength, as well as greater sensitivity to creasing. Mechanical stresses during hydrolysis accelerate the ageing without the associated chemical changes becoming necessarily evident. Decreasing loop strength and increased sensitivity to creasing is evidence that hydrolysis has affected the structure of the macromolecule.

At a humidity of 50 gram/m^3 air at 120 °C reacting for 7 months, polyester looses 50 % of its viscosity, i.e. it becomes unusable.

The extent of hydrolysis can be determined by chemical and physical methods. Chemical methods consist mainly in measuring molecular groups, i.e. COOH-, OH-, NH_2-groups [86, 87]. A further analytical tools is based on gel permeation chromatography: It compares the molecular weight distribution to undamaged material.

The specific viscosity measures the average degradation of the polymer. This method is suitable for polyester and polyamide, less so for acrylic, because hydrolysis affects the side chains in this case. Specific viscosity is an exact measure; it is therefore suitable for all fibres where the chain of the macromolecule is split by hydrolysis [51].

By determining the tensile strength and extensibility (fig. 2.76 and fig. 2.77) it is possible to detect whether hydrolysis has affected important properties required for the functioning of the product. During the first phase of hydrolysis the extensibility suffers. Only as reaction progresses is the strength significantly affected and the material ages increasingly. An excellent measure of strength loss caused by hydrolysis is the continued tear strength (carried out on a slit specimen). Also, knot or loop strength or bending delivers useful comparison data.

Hydrolysis is frequently confused with running a plant through the dewpoint. Some operators assume that hydrolysis is only initiated when the bags have become wet due to passing through the water or acid dewpoint. This assumption is quite wrong, because water vapour in gas form is much more corrosive than in liquid form. Passing through the dewpoint leads to baking and has other undesireable consequences. It can also happen

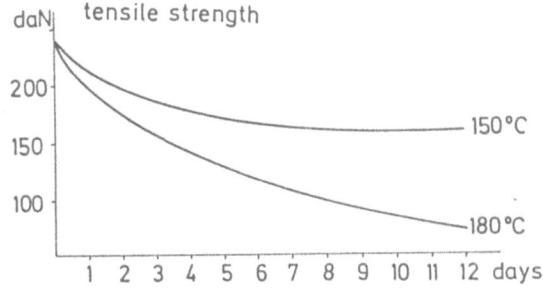

Fig. 2.76
Decrease of tensile strength of polyester needlefelt in saturated steam [52]

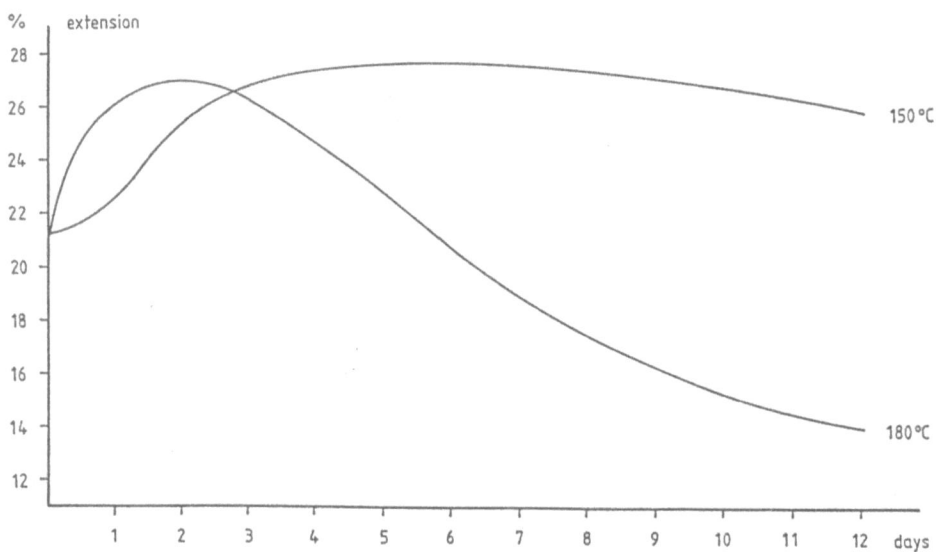

Fig. 2.77 Extension at break of polyester needlefelt in saturated steam

that acid or alkali contained in the dust goes into solution whereby hydrolysis is initiated. The possibility of hydrolysis can never be ruled out, if water, sulphur oxides, nitrogen oxides, hydrogen chloride, etc. are present in the flue gas, even if only in gaseous form.

Sulphur oxides are the most frequently encountered deleterious gas components in flue gases. When present as sulphur dioxide it forms sulphurous acid with water at the dewpoint. When sulphur trioxide and water react, sulphuric acid is formed which results in a dewpoint which varies with its concentration. Such problems occur in all those installations where light or heavy oil or coal is burnt.

Some cases are quoted from the non-ferrous metal industry (copper, brass, lead, zinc, tin) and from the roasting or sintering furnaces [55]. Without exception, the flue gases contain considerable quantities of sulphur dioxide and water vapour with additional smaller proportions of chlorine and fluorine. The dust contains oxides, sulphates, chlorides, and fluorides of a multitude of metals which in many instances behave as active catalysts eager to oxidize the SO_2 to SO_3. The dust may react with the gas; some potentially corrosive components may be absorbed on the dust; the fibre within the filter medium may swell so that gaseous attack can take place within the molecular structure of the fibre to initiate the depolymerisation process.

For the filter medium this can lead to a catastrophic chain reaction: From SO_2 some SO_3 is produced which together with water form sulphuric acid. This raises the acid dewpoint. At the same time the acid contains the above mentioned swelling agent to gain access to the fibre interior. Should the plant run through this acid dewpoint, the bags will become

2.9 Chemical Attack on Filter Media

wet and the dust will adhere to the walls of the bags. When conditions rise beyond the acid dewpoint the dust bakes onto the bags. The bags blind and are chemically damaged.

Should zinc be present (for instance in galvanized rings or support cages) atomic hydrogen will be produced from the action of sulphuric acid. Hydrogen is highly active in the nascent state. The bags become stiff through hydrolysis and can no longer be regenerated.

Even aromatic polyamides (aramides) are sensitive to hydrolysis at elevated temperatures, especially if alkali or acids are present in catalytic proportions.

Chemical modifications to polyester and more significantly to aramides have brought improvements during the last few years. The basis of such protection is to shield those members of the polymer chain which are sensitive to acid or alkali catalysed hydrolysis. This is a form of immunisation and strengthens the groups to withstand the splitting forces of the hydrolising agent. The treatment contains components which are integrated into the macromolecular structure during a secondary polymerisation. The fibre is not only coated, it is also modified [53, 54].

When such aramides are dissolved in suitable solvents they show a much higher specific viscosity when compared to untreated fibres. There is also an increase in insoluble residue, demonstrating the cross-linking effect of the treatment. This treatment is known as "Plus CS42" (fig. 2.78).

An even closer scrutiny must be made of the reactions of oxides of nitrogen on filter media made from polyamides, polyesters and acrylics. Huber/Joerg [56] list 3 probable reactions:

a) "A change at the chemical valence bond (depolymerisation by way of acid hydrolysis with polyamide).

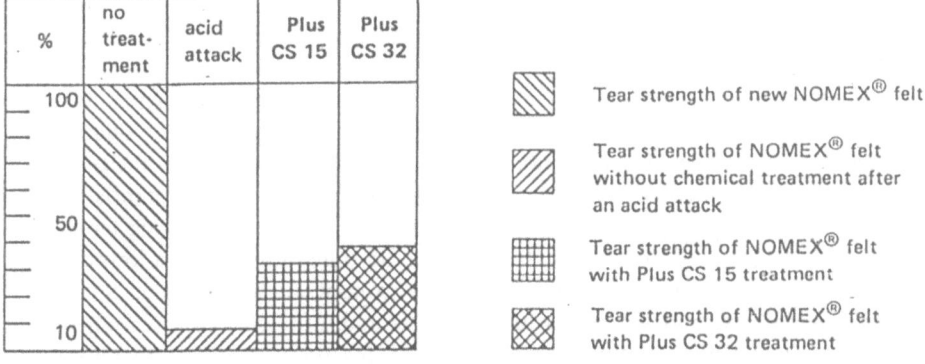

Fig. 2.78 Tear strength comparison of treated and untreated NOMEX® after hydrolysis attack

b) A diffusion of nitric oxides into the polymer which changes mechanical properties without changing morphology.

Since the polymer chains are not cross-linked, adjacent molecules are held together by electric dipole and higher order forces. These can be weakened by the infusion of nitric oxides.

c) A reaction can also take place via oligomers."

Nitric oxides react strongly exothermic. They are formed from the basis elements at very high temperatures (eg. the temperature of electric arcing above 3000 K). A characteristic of these oxides is their strong tendency to react with oxygen to form nitrogen dioxide. As soon as the colorless nitric oxide comes in contact with air it forms a brown vapour of nitrogen dioxide: This reaction is exothermic. Such exothermic reactions with this or other compounds within filter media are not a rarity. For this reason it is recommended, expecially with combustion or melting processes, to allow for a long delay line for the gas and dust (long pipelines) before they have access to the filter medium for cleaning. The gases must be allowed to react with each other and with any other catalysing dust component, before they pass through the sensitive textile matrix. If this delay is too short, combustion initiated attack or chemical degradation takes place on the filter elements. It should also be remembered that some chemical reactions may reverse on cooling to produce additional heat.

Not infrequently the dust collector is the scene of action of a variety of chemical reactions. The risk is ever present, if there is a potential for a chemical reaction between carrier gas and dust particles. Often this reactivity is latent and does not take place while the gas moves through the pipes at high speed, but rather when the moving gas slows to be forced through the stationary dust layer on the filter medium. This can produce oxidation, reduction, absorption or adsorption. It is prudent to choose filter media which stand up to these chemical and thermal conditions.

If provisions are made for an adequate delay between the source of the dust and the separation point, it will reduce the possibility of spark fly. It is also a precaution against pyrophoric dusts.

Pyrophoric iron oxide is a typical example. Iron oxides in the displaced lattice form (crystal lattice) react already at room temperature with oxygen in the air (purging air during cleaning) to produce large amounts of heat and ultimate glowing. Whether or not this special form of iron oxide is present depends on the degree of oxidation and the temperature conditions (tempering) during and subsequent to the smelting. It is therefore difficult to predict whether pyrophoric effects are likely. Potential energy reserves that may remain in the iron dust are of critical importance. During tempering, iron dust with a disturbed crystalline structure can be transformed into a lower energy state so that it can no longer generate heat when in contact with oxygen. Such cases have been recorded in cold air cupola furnaces. The addition of oxygen scavengers will deactivate the dust. Other methods are reverse cleaning with nitrogen to prevent any oxidative glowing of the dust on the bags, alternatively running the filtering plant at higher temperature (at least 180 °C) which tempers the dust and lowers its potential energy. A filter surface which can be cleaned easily to release the dust quickly will reduce the deposit time of the dust on the bags or pockets to a minimum.

The reactivity of dust cakes on filter media is used industrially on a large scale today. Dust collectors can serve both as dry absorption reactors for the gas and also for dust separation. According to Kaeppeler [57] dry absorption "can fundamentally be used for all gaseous contaminants for which an absorber is available". The required reaction time should not be more than 2–3 seconds.

Calcium carbonate and aluminium oxide may be used to remove:

Sulphur dioxide/trioxide (SO_2/SO_3)
Hydrogen fluoride (HF)
Hydrogen chloride (HCl)
Hydrogen sulphide (H_2S)
organic acids

The above contaminants arise during the following processes:

Electrolytic refining of aluminium
Ceramic ovens
Enamelling kilns
Coal or oil fired power stations
Combustive waste disposal and/or pyrolysis plants
Sedimentation sludge combustion
Brick kilns
Non-ferrous metal roasting and smelting furnaces
battery incinerators

In the last mentioned process the lead oxide (PbO) is chemically very active and acts as a dewpoint depressant absorber to SO_2, SO_3 and HCl.

We cite tests conducted on the fly ash in a filter type separator behind a coal fired boiler. It was found that the fly ash was much more absorbent that had previously been assumed. It was shown that absorption of SO_2 took place on the surface of the filter bag by carrier gas and dust particles reacting together. The relatively large surface area of the amorphous fly ash particles greatly assists this. It was observed that the fly ash tested alkaline in the raw gas stream, while after cleaning from the filter bags it gave an acid reaction: A practical example of chemical absorption that can take place on filter elements [6].

2.10 Hot Gas Filtration

Filter type separators are moving increasingly into areas where gases need to be cleaned at high temperatures. Low emission values and cost effective investment pressures play their part. It is interesting that heat resistant filter media have been available for more than 10 years. Thus media manufacturers have intuitively pointed the way in which future development of filter type separators was to move. Since suppliers of plant generally wish to include a cooling tower with their plant, this development did not necessarily find their approval. The present energy situation has vindicated those that have invested in the development of technology for hot gas applications [6].

Hot gas filtration in the region of 180–600 °C can be justified for the following reasons:
a) Energy can be saved and put to other useful purposes [58].

b) A cooling tower can be dispensed with to save capital cost and maintenance [59].
c) Processing conditions can make it necessary to separate dust at high temperatures, for instance dust separation above the condensation point of a product, or to encourage or prevent thermocatalytic reactions [60].

When recycling energy from flue gas the heat exchanger is a form of interference. Dust contaminates the heat exchanger leading to abrasion and corrosion damage and leaving deposits which are detrimental to the efficiency of the heat exchanger. The heat transfer medium must be separated from the heat carrying gas. Flue gas permits only indirect heat exchange.

According to Dornieden [61] efficient heat exchange requires turbulent airflow. This necessitates small ducts to achieve high velocities. Assuming a flue gas dust content of 5 gram/m^3 as the output of a combustion or roasting furnace, 500 kg dust per hour or 12 tons per day pass the heat exchanger at an estimated gas volume of 100 000 m^3. It is inevitable that this leads to complications.

H. C. Guertler and R. de Bruyne [59] suggest a direct heat exchange method using filtered hot gas. The hot gas heats a suitable packing material which is subsequently flooded with cold air. Dust collection plants are often seen unjustifiably as a necessary evil. The possibility of recovering valuable products and of energy recycling is often neglected. On the other hand possibilities must exist for utilising this energy once recovered. Such possibilities could be remote heating, steam generation or pre-driers. The cost of a filtration plant would be quickly amortized by the energy saved.

There are further incentives. C. N. Davis [62] shows that the diffusion dependent deposition of aerosols on porous bodies at elevated temperatures increases, because the Brownian motion increases, leading to more collisions with the fibres. Pich and Binek [63] came to similar conclusions. The higher viscosity of the air at lower density compensates to a large extent for the higher gas volume at elevated temperature.

Experience shows, however, that the separation is more difficult at higher temperature with conventional media. While the air to cloth ratio decreases, the leakage increases because of the thermal plasticity of synthetic fibres. This is especially true of dusts that do not agglomerate.

To ellucidate this the extensibility of filter media was measured at elevated temperatures (table 2.8). As shown, tensile strength decreases with increasing temperature for all synthetic fibres. Fibres of the acrylic family show excessive extension, so that detrimental changes in geometry and separation efficiency must be expected at higher stresses and temperatures.

While glass fabrics have always enjoyed a strong position in baghouses, particularly in the carbon black- and steel industry, and in the dust collection of boiler rooms – in the USA particularly in the cement rotary furnaces –, the jet filter has had to rely on more expensive alternatives for a long time. For temperatures around 300 °C mineral fibre felts appeared on the market in the early 70's under the name of PYTROTEX® and reached a measure of acceptance [66, 67]. Apart from metal fibre felt this appears to be the first three dimensional filter medium suitable for jet filters made from inorganic fibres. It was manufactured in several types and contained as main component fibreglass and/or white asbestos and/or metal fibres. To increase strength, density and service life temperature resistant binders were added [131].

2.10 Hot Gas Filtration

Table 2.8 Extensibility with increasing temperature

		Trevira® felt with PE fabric 600 g/m²	Nomex® felt with Nomex fabric 550 g/m²	Acrylic felt with Acrylic fabric 600 g/m²	Woolfelt with PE fabric 500 g/m²
22 °C					
Tensile Strength N	lengthwise	1750	1702	1088	553
	across	1235	719	655	371
Extensibility %	lengthwise	16.1	32.8	15.6	49.3
	across	16.2	23.3	26.5	35.0
80 °C					
Tensile Strength N	lengthwise	1442	1544	863	440
	across	1118	679	495	336
Extensibility %	lengthwise	15.9	30.7	20.3	47.9
	across	16.9	21.8	38.5	51.0
140 °C					
Tensile Strength N	lengthwise			364	
	across			298	
Extensibility %	lengthwise			85.1	
	across			90.0	
150 °C					
Tensile Strength N	lengthwise	1203			
	across	813			
Extensibility %	lengthwise	16.1			
	across	16.5			
220 °C					
Tensile Strength N	lengthwise		1074		
	across		465		
Extensibility %	lengthwise		29.4		
	across		20.3		

Table 2.9 Heat stability of textile filter media [64]

Polypropylene	90 °C	(100 °C)
Acrylic	120 °C	(125 °C)
Polyester	150 °C	(160 °C)
aliphatic Polyamide	110 °C	(120 °C)
aromatic Polyamide	180 °C	(220 °C)
Fibreglass		
a) felt	220 °C	(250 °C)
b) fabric	250 °C	(300 °C)
Polytetrafluorethylene	260 °C	(280 °C)
Mineral fibres	300 °C	(350 °C)
Metal fibres X CrNiMo 18/10	400 °C	(450 °C)
Inconel	550 °C	(600 °C)
Quartz	800 °C	(1000 °C)

Tighter regulations in the use of asbestos and the high cost of metal fibres have divided the product range so that felts are at one end of the scale and pure metal fibre felts on the other extreme. The fibreglass felt is made in a range 700–900 g/m².

It consists of a fibreglass woven base onto which a fibreglass web is needled to one side. A high proportion of binder which is temperature resistant is required to ensure cohesion. Typical specifications for a fibreglass felt are:

Weight	850 g/m²
Thickness	2.6 mm
Density	0.33 g/cm³
Air permeability (DIN 53 887) [85]	120 l/dm² min 2mb
Pore volume	87 %
Burst strength	14 bar, 10 mm dome
Tensile strength warp	230 daN
weft	250 daN
Extension at break warp	4 %
weft	2 %
Dimensional stability at 200 °C	99 %

Binders which as a rule are PTFE or silicone resins inevitably stiffen the filter medium which is therefore only suitable for air pressure cleaned dust collectors. The support cages should have double the number of rods (distance between rings max. 20 mm) to reduce the concave/convex movement. Fibreglass requires a more intensive reverse air cleaning, because the primary cleaning through acceleration of the particles is largely absent. Due to the lower permeability it is important to have sufficient air pressure, so that the clean air volume injected into the interior of the bag can penetrate sufficiently. Fibreglass felts aim to reach the large market of boiler and power generation dust collection where temperatures between 180–200 °C are common, and where assurances are required that the plant is capable of running safely above the acid dewpoint temperature with high sulphur content fuels. Fibreglass felts have achieved significant popularity in a short time. Extensive experience has been collected on their special requirements so that failures have become rare.

For the sake of completeness it should be mentioned that in the USA a fibreglass fabric (500–700 g/m²) with a TEFLON B coating competes with fibreglass felt. This has given good service in pulse jet dust collectors. The manufacturers claim higher physical durability for their product. Despite undisputed successes the weight of experience, however, points in the direction of felts for the future, because of their higher filter capacity rating.

Fibreglass filter media have additional advantages besides their high filter efficiency:
a) Non flammability.
b) Low sensitivity to hydrolysis.
c) Good acid resistance (with exception to hydrofluoric acid).
d) Good resistance to alkali and organic solvents.

For a temperature range above 300 and up to 600 °C two alternatives are available:

2.10 Hot Gas Filtration

a) A metal fibre felt from X 5 CrNi 188 and X 2 CrNiMo 1810 (Material ref 4301 and 4404). For low pressure or high pressure cleaned dust collectors. These media are durable up to 400 °C with peaks up to 450 °C [59, 63, 6, 69].

The fibres used in these needle felts have a fineness of 8–12 µm. For this reason they have a very large filter active surface. To achieve high strength, low extensibility and high dimensional stability they are supported by a continuous filament woven fabric made from the same material. We are dealing therefore with a textile material of a three dimensional nature made from 100 % stainless steel [65].

b) By developing a fibre with the same order of fineness from Inconel it has become possible to produce filter media with properties similar to above with added temperature stability (service temp. 550 °C, peak 600 °C) and additional chemical resistance.

Filtration trials with metal fibre felts led to the following conclusions:

a) Needlefelts from stainless steel fibres behave like normal textile filter media.
b) They are easier to clean than organic filter media, they require less effort for cleaning (pressure, frequency, air volume).
c) They build up less differential pressure.
d) They can tolerate higher dust loading and gas velocities. Compared to conventional polyester felt the required filter surface area can be halved at room temperature.
e) They do not become statically charged with consequent easier cleaning.
f) Because of the rough fibre surface, metal fibre felts achieve very high separation efficiencies.
g) They can be electrostatically charged and may be used as electrodes in a combined electrofilter and fabric filter.
h) If carefully tuned to the cleaning mode, metal fibre felts are good dust accumulators so that gases that are prone to chemical absorption can be bound within the dust.
i) Metal fibre felts have good resistance to radio activity.

The temperature resistance of metal fibre felts does not compare to that of the solid metal. The fibres are manufactured by the wire drawing method and possess a very large surface area which is advantageous for filtration, but presents a large attack surface for chemicals. Verplancke [68] predicts a number of different applications, e.g. in dust collection of clincer coolers, for the energy recycling of ferrosilicate plants, behind electric arc furnaces, and in the nonferrous industry. An air to cloth ratio of 180 m^3 is in trend with increasing face velocities. A differential pressure of 16 mbar underlines the easy cleanability of metallic filter media.

Filter media of quartz fibres or modified fibreglass are in the pipeline to reach temperatures up to 1000 °C and higher. Potential applications are in the coal liquification and gas cleaning for turbines. Developments in this area are proceeding at full speed. The application of a fabric filter in a red or white hot glowing pulse jet filter is not far off. Ceramic man made mineral fibres are also being tested. Should the next edition of this book eventuate, results of current trials and research findings will be known. An exciting progress in the present state of the art is foreshadowed [131].

2.11 Filter Medium and Static Electricity

Static electricity is one of the most elusive causes of explosions. During the period from 1900 to 1972 281 explosions were recorded in the USA of which 9 % were attributed to electrostatic causes [69]. A similar statistic was assembled for the German Federal Republic by the former Dust Research Institute in Bonn (now National Institute for Safety of Work – BIA). Of 291 dust explosions 10 % were attributed to static electricity. Official documents record that in 20 % of cases the cause was unknown. It is impossible to say what percentage of this 20 % may have been initiated by static electricity. The VDI 2263 of August 1969 page 3 [70] contains this significant comment: "Should a report conclude that the cause of a fire or explosion was a suspected self-ignition of accumulated dust or 'in all probability as a result of static charging', it is neither an acceptable nor an adequate explanation."

In the Federal Republic of Germany approx. 300 dust explosions occur every year with damages in the millions. A large proportion originate in the food and animal feed manufacturing industry.

For an explosion to occur 3 conditions must be fulfilled: Apart from a finely distributed dust and oxygen, a source of ignition with sufficient energy is required. To initiate a dust explosion the ignition source must release a quantity of energy above a critical value. According to Vensteenkiste [69]. The following minimum ignition energies for dust/air mixtures have been documented:

Dust type	Minimum ignition energy dust/air mixture mJ
Potato starch	25
Powdered sugar	30
Rubber	10
Polyethylene	10
Polystyrene	15
Aluminium	20
Brown coal	30
Hard coal	30
Sulphur	15

Franke [71] gives the following values for coal dust/methane combinations:

Dust type	Min ignition energy mJ Methane content by volume in percent volume			
	0	1	2	3
Coking coal	135	30	10	3
Gas coal, type A	4	–	0.4	0.32
Gas coal, type R	59	12	2.3	0.41

2.11 Filter Medium and Static Electricity

Most literature references [71–79] give minimum ignition energies for dust/air mixtures between 5 and 800 mJ. There are also research findings for dusts such as epoxy and resins, esters and dyestuffs where the minimum ignition energy lies below 3 mJ. Ignition energies below 1 mJ have been recorded [69] with hybrid mixtures of dust and air in combination with very small amounts of flammable gases.

Franke [71] has investigated the relationship between minimum ignition energy and flammable gas content in hybrid mixtures of coal dust with methane. From the above table it is evident that for only 3 % methane in air the minimum ignition energy for coking coal dust falls from 135 to 3 mJ. It is not impossible for hybrid mixtures to form in plants designed solely for pure dust/air mixtures as a result of a machine malfunction. An overheated bearing can produce flammable oil vapours in sufficient quantities to form hybrid mixtures [69]. There are differing opinions among experts whether electrostatic charges can build up sufficient energy on filter media to ignite a dust/air mixture. This question has become acute in filter type dust collectors behind pulverised coal drying plants in the cement industry. The problem is similarly serious in the areas of food preparation, timber, plastics and chemical industry which have used filter medias for years and where precautions for the rapid discharge of static are taken.

The author does not wish to participate in a controversy which is of more academic interest. He will leave this to experts. From the point of view of the filter medium a more practical approach is called for which focusses on the following points:

a) Filter media made from synthetic fibres have a high specific resistance and inadequate electrical conductivity.
b) Unlike wool and cotton they do not take up moisture so that their conductivity changes little with use.
c) Dust separated in high speed, high production grinders can carry high electrostatic charges. This can be exacerbated by friction in the duct work and low conductivity of the dust.
d) Some dusts originating from plastics or resins also have low conductivity.
e) Temperature and moisture content of the gas affects the conductivity of gas and filter medium.
f) During the separation process friction effects occur between fibre and/or the deposited dust. This can contribute to charge generation.
g) It has been demonstrated that cleaning efficiency is influenced by electrostatic charging.
h) Actual experience has shown that spark discharges in dust collectors leave traces within the filter medium.
i) As explained above, a significant number of explosions remain unexplained or traced back to probable static electricity.

These points have convinced the author to emphasize the importance of filter media equipped with antistatic treatments and to forecast their increased importance in future years. Many cases in the useful application of antistatic filter media since the end of the 60's lend strong support to this view.

Several techniques are common to conduct electrostatic charges to earth; these will be discussed in detail. It should go without saying that filter type dust collectors equipped with antistatic devices should be earthed. Nevertheless examples are know where dust collectors, exemplerary equipped, have been mounted on vibration pads of rubber or plastic which are nonconducting!

The old method of sewing a copper braid to the seam of the filter bag is still alive. This is quite useless and even dangerous with synthetic fibre media. The braid conducts electricity from within its close proximity only, while the rest of the circumference of the bag, because of the high specific resistance of the material, remains unprotected. If the braid breaks, there is danger of forming a capacitance with eventual spark discharge. Even the metal braid which connects the filter element to the housing has only a localised effect, even if the bags are pulled over the support cage.

Similarly ineffectual are chemical finishes which are deposited on the surface of the fibres. These rely on the principle that the substance either intrinsically or by its ability to absorb water increases the conductivity of the fibre. A disadvantage of such treatments is their limited durability to abrasive dusts, their potential reactivity with gases and dusts and a limited temperature stability which lies below that of the fibre.

Of greater interest are suggestions to modify the synthetic fibre by coating or adding graphite or other carbonaceous substances to the spinning solution. With this technique specific resistance can be reduced to the order of 10^7 to 10^8 Ωcm depending on the percentage of added "carbon fibres". Such fibres can be added to the support fabric as well to the felt layer. Other developments tend into the direction of metal coating on synthetic fibres where the coating is deposited on single fibres as well as on the finished filter media. Experiments have been recorded in the USA where a carbon containing substance is deposited by a printing method on the surface of the filter medium. It is questionable, however, whether conductivity can extend into the three dimensional structure.

The addition of fine stainless steel fibres has achieved great success in filter media (needle-felts and fabrics) to prevent electrostatic charging. The fibre can be produced in a fineness range $8-12$ μm which allows a homogenous fibre distribution on the surface and within the depth of the textile structure. By adding metal fibres to the warp or weft threads at regular intervals, the manufacturer is able to control the conductivity in both the length and width direction. With needlefelts metal fibre additions between $3-7$ % depending on the specific resistance of the carrier fibres are usual. Woven fabrics may require even less. Although problem situations are not known, it is important to point out that a uniform fibre distribution is of utmost importance. Single, isolated fibres can form a condenser and may cause a spark discharge.

The stainless steel alloy ASI 316L which is used by a Belgian fibre manufacturer [78] has a very high corrosin resistance and the following composition: $8-12$ % Nickel, $16-18$ % Chrome, $2-3$ % Molybdenum.

The measurement of surface resistance and core resistance is made according to DIN 54 345 [130]. A circular electrode is used at standard conditions of temperature and humidity.

With three dimensional filter media the core resistance, which expresses the conductivity across the thickness of the material, is regarded as important, since with jet filters especially there is contact with the wire support cage which provides continuity between entry side of the three dimensional medium and the support cage on the clean gas side. Both,

2.11 Filter Medium and Static Electricity

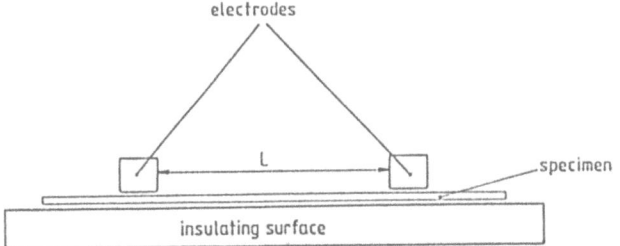

Fig. 2.79
Measurement of surface resistance according to DIN 54 345 [133]

surface- and core resistance [133] are of importance in the antistatic behaviour of a filter medium. Surface resistance gives an indication of its ability to distribute the charges, while core resistance gives a measure of the electrical contact with the support elements and the discharge of the potential.

Surface resistance values (fig. 2.79) (DIN 54 345 [130]) of needlefelts with stainless steel fibre additions [69]:

Fibre type	Area weight g/m^2	stainless steel %	Surface resistance Ω cm
Polyester	500		> 10^{12}
Wool	480	0	> 10^{12}
Polyester	600	1.79	7.85 · 10^2
Polyester	500	2.03	3.56 · 10^3
Aromatic polyamide	480	2.23	9.24 · 10^3
Wool	400	2.32	7.63 · 10^2
Fabrics			
Polyester		0	> 10^{12}
Polyester		2.00	2.06 · 10^2
Polyester		2.28	1.34 · 10^2

Core Resistance (DIN 54 482) [133] of Needlefelts with Metal Fibre Additions (fig. 2.80)

	Antistatic Ω cm	Untreated Ω cm
Polyester	1.1 · 10^3	6.3 · 10^9
Polyester siliconized	1.0 · 10^3	3.8 · 10^9
Acrylic	1.8 · 10^2	5.0 · 10^9
aromatic polyamide	4.6 · 10^2	9.2 · 10^7
Polypropylene	1.3 · 10^6	1.0 · 10^{18}
PTFE	1.0 · 10^5	1.0 · 10^{18}
Wool with polyester support scrim	1.1 · 10^6	1.2 · 10^8

The Labour and Industry Department in Germany have set a safety limit of 10^9 Ωcm for surface resistance. The Mining Offices issue certificates for filter media in the range 10^5 to 10^7. In the free market it is common for plant operators and builders to request core resistances of less than 10^5 Ωcm and even down to 10^3 Ωcm.

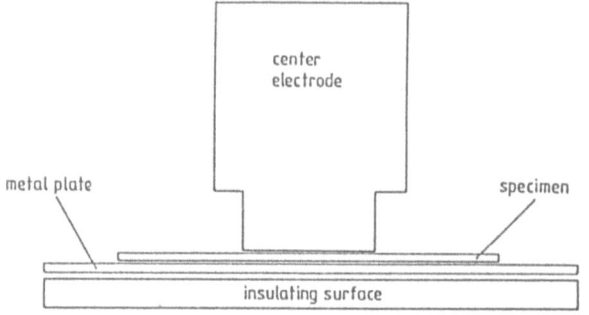

Fig. 2.80
Measurement of core resistance according to DIN 53 482 [132]

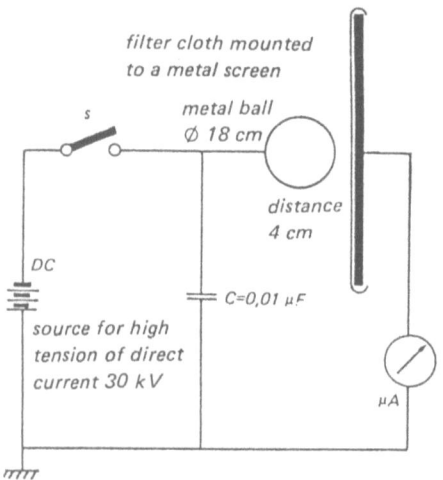

Fig. 2.81
Corona effect of stainless steel fibres experimental set-up [69]

According to Vansteenkiste [69] it is probable that stainless steel fibres discharge through a corona effect [135]. The stainless steel fibre in the filter media discharges not only the dust cake on the filter surface, but also neutralises the electrostatic charges that accumulate on the filter surface. This can be demonstrated in a simulated situation (fig. 2.81). A metal sphere is brought close to a filter specimen. The filter medium is mounted on a welded wire mesh simulating the conditions of a filter in practice. The distance between sphere and specimen is 4 cm. When switch S is closed the 0.01 µF capacitor is charged through a DC high voltage source. With switch S closed and an energy source of 30 kV, 4500 mJ in energy can be stored. This is 100 fold greater than the ignition energy for an explosive flour/air mixture. After charging, switch S is opened. At the same time the metal sphere is moved towards the specimen at 1 cm/sec. With an untreated filter medium a powerful spark discharges when the sphere is 1.7 cm from the filter. The total energy of 4500 mJ is released at once at this distance. With a filter media containing 1.79–4.48 % stainless steel fibres the sphere can be brought into contact with the specimen without a spark. During the course of this movement this stainless steel fibres continuously discharge the voltage through the corona effect. The total energy quantity of 4500 mJ is gradually

2.11 Filter Medium and Static Electricity

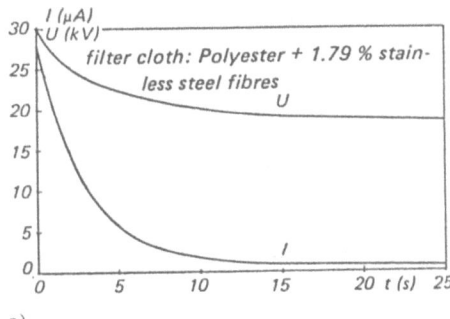

Fig. 2.82
Electrostatic effects during dust separation
a) Corona effect of stainless steel fibres. Discharge current of voltage field [69]
b) Explosion proof pressure filter

dissipated. At no point is the critical energy potential of 3 MW/m reached. The explanation for this phenomenon is the brush discharge effect: if a stainless steel fibre is exposed to an electrostatic field, a high voltage gradient arises on the surface. As soon as this electrostatic field has reached a critical level on the fibre surface the air in the vicinity is ionized; in turn the ionized molecules conduct the electrostatic chage. The corona current was recorded for a sample containing 1.78 % stainless steel. The initial charge of the capacitor corresponds to 30 kV at the moment of opening the switch. Within 25 seconds the corona current decreased from 28 to 0.3 mA (fig. 2.82a).

From the voltage curve it is possible to calculate the energy reduction within the circuit and to derive the corresponding voltage at the sphere at any one moment. Curve U shows the change in potential of the sphere as a function of time. For a given distance of 4 cm the voltage is immediately reduced from 30–19 kV: The electrostatic field of 750 kV/m is reduced to 470 kV/m, the ignition energy therefore dropped from 4500 mJ to 1900 mJ. If under these conditions the sphere is brought within 3 cm of the filter specimen, the electrostatic field rises to 630 kV/m. The corona current flows again and reduces the field to 470 kV/m.

The final voltage at 3 cm distance is 14.1 kV. Table 2.10 shows the residual potential in relation to distance as the sphere approaches the filter specimen. The voltage is reduced to nearly zero by the brush discharge effect of the stainless steel fibre.

Fig. 2.82b shows an explosion protected pressure filter.

Table 2.10 Corona effect of stainless steel fibres. Voltage in relation to sphere — distance filter medium with stainless steel fibre addition [69]

Distance between sphere and test specimen. 1.9 % stainless steel fibre mm	Residual sphere voltage kV
40	19
30	14
20	9.5
10	4.7
5	2.3
2	0.9
1	0.4

2.12 Flammability of Filter Media

There is an increasing demand to equip needlefelts, nonwovens and fabrics for use in dust collection with flame retarding finishes. In order to reduce the flammability of textiles, substances are applied which inhibit the formation of flammable distillation products. As the following table shows, textile fibres differ widely in their ignition and self-ignition temperatures:

Fibre	Ignition temperature °C	Self ignition temperature °C
Wool	325	590
Cotton	350	590
Polyester	390	508
Acrylic	250	515
aliphatic polyamide	390	510
Polypropylene	375	495
aromatic polyamide	490	675
PTFE[1])	–	–

[1]) PTFE sublimes and decomposes from 290 °C onwards with an evaporation rate of 0.0002 %/h. This increases dramatically above 400 °C. PTFE does not burn in air; combustion will take place, if oxygen content in greater than 40 %.

Inorganic fibres such as asbestos, glass, ceramic fibres or the man made mineral fibres, also carbon — which is produced by the pyrolysis of cellulose or acrylic fibres — cannot be ignited and are thus non-flammable. Fibres which do not decompose under heat or which produce few decomposition products during pyrolysis are also non-flammable. Fibres which decompose but are not self-igniting such as aromatic polyamides do not yield high decomposition products during heating, so that the heat energy released during combustion does not sustain burning (negative energy balance) [80].

Fibres with very low ignition temperatures are easier to ignite and are therefore more flammable than those with higher ignition temperatures. Besides ignition temperature the LOI value (lowest oxygen ignition) as reported by Einsele [80] is of significance.

2.12 Flammability of Filter Media

This measures the lowest quantity of oxygen in an oxygen/nitrogen mixture that just maintains combustion.

Fibre	LOI value
Wool	25.2
Cotton	19.0
Polyester	22.0
Acrylic	18.0
Polypropylene	18.6
aliphatic polyamide	20.0
aromatic polyamide	28.2

Fibres with low LOI values such as acrylics and polypropylene burn easily.

Decomposition energy and the quantity of heat released during burning in air are two additional important parameters which characterize the burning behaviour of fibres.

Under the present state of technology mixing flammable and nonflammable fibres does not achieve significant flame-retardant effects. Fibreglass simply acts as a "diluent" for flammable fibres without decreasing their flammability. The fibreglass, acts as a lattice and prevents dripping so that a web containing fibreglass and synthetic fibres in fact burns more readily than the pure synthetic web. Pure textiles structures of fibreglass, metal or other mineral fibres do not burn. If a binder is used for fibreglass, it must include a flame-retardant with comparable temperature resistance.

The fire risk is related to the heat released during combustion which in turn determines the flame temperature. Flame temperatures reach 690 °C for polyester and 941 °C for wool.

The choice of flame-retardant additives is exceptionally large. The patent literature on this subject is enormous and difficult to summarize. According to Einsele [81] the oldest flame-retardant originated from inorganic chemistry. An English patent from the year 1735 uses alum, iron-sulphate and borax as a flame-retardant. In 1820 ammonium sulphate and borax was used for the first time as a flame-retardant. Today's inorganic products are based on mono- and diammonium phosphate as well as ammonium sulphate, ammonium chloride and ammonium bromide. These inorganic compounds have the advantage that they are cheap. They give good flame suppression and reduce afterglow. They have low washfastness and tend to crystallize at the surface. Products of the basis of ammonium phosphate show limited temperature durability. The reagent evaporates at 80 °C and when ignited smothers the flame. When applied to non-absorbent fabrics 20–30 % weight increase is necessary and the handle, cleanability and flexibility are adversely affected.

Polyester fibres are mostly treated, with phosphor or halogen compounds, eg. bromine compounds. These are available as commercial flame-retardants and are fixed with melamine or polyvinyl acetate resins. The same substances may be used to treat polyamides.

Polypropylene is very flammable and produces low molecular saturated and unsaturated methyl groups during thermal decomposition. Halogen compounds, phosphor compounds and aliphatic brominated alcohols and aromatic bromine compounds are effective. These must usually be attached with the help of a binder and require supplements of 20 % and higher. Phosphor combinations and halogen/phosphor compounds are most frequently used to flame-retard synthetics that absorb little moisture. Washfastness, high weight and toxicity make these treatments not entirely satisfactory for filter media. It is recommended to carefully choose among the large assortment of flame-retardant compounds to ensure suitability in later use [80, 82].

The burning behavior of textiles is assessed by the vertical strip method (DIN 53 906) [83]. This Standard prescribes the determination of the burning time and afterglow as well as the degree of damage in a vertically suspended specimen lit by a flame. This vertical arrangement is a severe test and this procedure is used to differentiate between those textiles which are more difficult to burn.

The principle characteristics of the burning behaviour of textiles are defined under DIN 54 300 [84]. The following data are derived from this test: Burning time is the time from the moment of withdrawal of the ignition source to extinction of the burning specimen. A further measure is the time of afterglow: It expresses the time between flame extinction and when afterglow ceases. To determine damage to the specimen it is torn along its length and the length of tear recorded.

2.13 Filter Media in Use

Once the installation comes on stream it is the moment of truth for the filter medium in the dust collector [122]. Here it will show under hot, humid, abrasive, corrosive and dusty conditions with all the highs and lows whether the right choice has been made and whether the filter medium harmonizes with the plant design and whether it will withstand the onslought. It will become apparent whether the polymer chains – the building blocks of the fibres – hold together within the filter medium or whether they will be chemically, hydrolytically, oxidatively or catalytically split apart or broken. Here it will be decided whether the dust will lodge on the surface or whether it will penetrate into the depth of the material, whether the cleaning mode can generate sufficient energy to expel the dust so that it can be collected and removed.

Here all theoretical arguments and simulated experiments about dust load, face velocity, settling velocity, bulk density, fineness, agglomeration tendency, differential pressure must stand trial. Here it will be verified what on paper appeared an elegant solution, i.e. whether bag diameter, length, pressure distribution within the jet filter, clean aire pressure, cleaning air volume, cleaning frequency and jet distribution within the filter medium, its gas permeability, its density and its external dimensions harmonize. Here it will show whether the distance between bags, the flue gas entry, the distance of the dust collector from the dust source allows the medium to "breathe" and deposit evenly and remain free of undesireable physical and chemical reactions. Here proof must be delivered whether the medium can be taken through the dewpoint temperature without start-up preheating and without warm air conditioning and, if the plant is left to run in a wet state, whether the filter bags become "concrete pipes" or the filter pockets become armourplated building sheets.

The filter medium separates not only dusts; it also separates philosophies. The dust collector divides wheat from chaff. The so often repeated sales promise "no problem for us" can turn out to be a real one.

2.13 Filter Media in Use

Indeed, the success of the fabric filter and its growing popularity can surely be taken from the fact that it delivers what it promises. If problems do arise, it is usually that either not all important processing parameters were known, or their significance and contribution were underestimated at the time. Sections 2.5 and 2.7 of this book are designed to avoid overlooking important factors.

These sections and others clarify where the limits are und where problems arise, if the limits are exceeded. These limits may be set by the fibres which are the building blocks of the filter media. We must accept these limits as given by the fibre industry with all their strengths and weaknesses. Failures can come from misjudging the basic filtration problem or in material faults, i.e. the area of human failures. Quality control "ad nauseum" within the framework of the quality parameters outlined in section 2.5 is the solution.

Filter media manufacturers that involve themselves in their products place great emphasis on a continual monitoring of the filter medium during operation, or to study those parameters that deviate after a period of operation. Such a laboratory based examination keeps track of the condition of the fabric or needlefelt after a period of use; it also permits inferences about the physical and mechanical conditions within the dust collector.

For this purpose it is useful to remove a filter element in the dust-laden condition so that the dust load and permeability in the caked condition can be studied. This gives information on the amount of dust which the media has absorbed. This in turn allows conclusions about the proportion of dust deposited in relation to the vigin filter material.

The air permeability in the caked condition indicates the degree of blinding. It is often overlooked that even a much lower air permeability, when measured at 2 mbar differential pressure (DIN 38 857) [85] may still permit a high air to cloth ratio at 10 mbar. Thus, because of the approximately linear pressure loss curve, the air permeability of a dirty filter of 15 l/dm^2 min at 2 mbar corresponds to a face velocity of 7.5 m^3/m^2 min at 10 mbar differential pressure. Such a face velocity would be very exceptional. It is not true to say that such a filter is blocked.

When the filter element has been removed it is blown down with a hand held nozzle until no more dust can be removed. This procedure indicates the pneumatic cleanability. It shows whether the dust has baked on as a result of humidity, or whether it is difficult to remove because of greasiness, stickiness or water absorption. It further gives an indication of how efficient the cleaning system is and whether the elements could be regenerated through a simple dry cleaning.

The next phase of dust removal is washing in a drum-type washing machine. This indicates how much dust remains in the filter medium that could only be dislodged by washing. A final weighing will show any residues when compared to the virgin bag. One may also compare the air permeability in the washed condition with the original value. This may detect tenaceous residues caused by baking. At any rate, it is important to know the difference between the dust removed dry as against that removed by the wet treatment.

The next stage moves into the area of textile characteristics. This allows inferences about fatigue of the material or chemical damage. Reduced extensibility and dome height in burst strength testing indicates embrittlement which in turn points to hydrolytic damage. The pH value of an aqueous extract of the filter medium is important, since it indicates possible chemical attack. This value should be compared with an aqueous extract of the dust collected from both the flue gas side and the clean gas side. This comparison will show whether chemical reactions (adsorption) has taken place on the filter surface.

The redox potential of an extract is similarly important. It will indicate whether a solution is chemically oxidative or reductive. The redox value is a measure of whether a solution takes up electrons (reduction) or releases them (oxidation). An rH value below 20 indicates reducing properties, 20–30 is neutral, while values above 30 point to oxidising characteristics. Caution is advised in interpreting these measurements, because redox potentials are not only affected by substances going into solution from the dust or carrier gas, but also from remnants of destroyed fibres.

If further examination is warranted, we enter the much more difficult area of polymer analysis. Temperature induced damage in acrylic fibres are indicated by yellowing. The solubility in dimethylformamide is reduced. Heat damage of acrylics is indicated by reduced staining with Metythelene Blue [86]. Polyamides and polyesters are soluble in cresol [51]. The specific viscosity measures probable hydrolysis damage. If heat damage on polyamide or polyester is suspected, this can be demonstrated by an increase in carboxyl endgroups. Furthermore, heat damaged polyamide gives a red reaction with when dissolved in Guajakol. We refer to section 2.9 [87].

The above mentioned tests can give a basic indication which greatly facilitate the much more complex polymer analysis. Polymer chemistry has reached a degree of sophistication that can detect hidden causes of fibre damage. It has to be decided from case to case whether this is econimically justifiable. If however a hitherto unknown attack agent on a synthetic fibre can be elucidated, the effort is well worthwhile in avoiding future serious consequences.

After these introductory remarks some typical examples will be selected from different industries to show dust collectors and their textile media — needlefelts, nonwovens or fabrics — under actual use conditions. It may even be possible to overcome the bias that fabric filters are a necessary evil, and to give the dust collector an image it truly deserves. Where the health of man, animal and plants are being protected and where scarce resources are preserved, it is not justice to consider it simply as a cost factor. There are many examples in the field of product recovery where filter separators pay back their cost in a short time. Additionally, dust collectors and their filter media make an essential contribution to the quality of life in our industrialized society.

A few examples will be presented of the lesser known applications in product areas and environmental protection such as — dust separation in ceramic sintering, magnecite works, aluminium refining, lime works, sugar mills, fertiliser manufacturing, waste incineration, gold mining, dyestuff manufacturing, steam generation, steel works, cement works, uranium recovery plant, food processing, steel casting, PVC spray driers, furniture factories, aluminium manufacturing, corn mill, titanium dioxide plants, dried milk production, chemical works, carbon black plants, flour mills, plastics manufacturing, copper smelters and lead furnaces, the manufacture of chalk, coal mines, asphalt mixing plant, textile mills, production plants that use asbestos, coal pulverisation and many other production facilities in the modern diversified industrial world.

2.13 Filter Media in Use

New frontiers were made in the dust removal in electric arc furnaces with gas burner support (UHP-process) where dust collectors, in baghouses with low pressure reverse cleaning and also in compact filters with jet cleaning, have achieved new standards of efficiency in removing the dust from gas volumes in excess of 1 million m^3. In a specific installation (bag dimensions: length 10 000 mm, diameter 300 mm) the air to cloth ratio could be increased to over 80 m^3/m^2 hr with the help of a specially treated polyester needle felt. In this case dimensions and investment costs for the collector were considerably less than a system using woven fabrics (fig. 2.83).

The jet filter has proved in practice that it is a viable alternative for such large gas volumes. This stands in the face of common opinion that jet filters are only economic for smaller gas volumes. Below are specifications of a filter installation commisioned in 1978 for a West German steel mill:

Type	Pulse jet filter
Total gas volume	1 mill m^3/h
Total filter surface area	6720 m^2
Air to cloth ratio	2.48 m^3/m^2 min
Number of filter chambers	10
Number of filter bags	3360
Operating temperature	Max. 120 °C
Number of fans	2
Fan capacity	2 × 538 000 m^3/h
Power rating	1 × 1250 kW
Voltage	10 kV
Fan speed	1000 Rev/min
Cleaning	Pulse Jet, 3 Screw Compressors

A specially treated polyester needlefelt of 500 g/m^2 was used. The differential pressure operated at 14 mbar and clean gas dust level was below 10 mg/m^3. The bags have been in use for 4 years at the time of writing this report [88].

A further successful installation of a fabric filter is in the non-ferrous metal industry (fig. 2.84). A few years ago a plant was commissioned to remove dust from a rotary furnace of a scrap metal smether which was in close proximity to a residential area in a West German city. The filter is run at 180 °C and is kept hot during idling with a steam heated circulation at 130 °C to prevent it dropping below the acid dewpoint of 112 °C. The jet cleaned installation is equipped with 246 m^2 of TEFLON® needlefelt with an area weight of 840 g/m^2. The air to cloth ratio is 1.85 m^3/m^2 min. PTFE was the filter media of choice, because it enabled the plant to run safely at a temperature above the acid dewpoint, thus requiring less cooling. Also, the corrosive agents in the carrier gas (SO_2, SO_3, Cl, H_2O etc.) require a chemically resistant medium [89].

A year earlier not far from the last mentioned plant, a filter type dust collector with conventional cleaning was also equipped with TEFLON® fabric. 700 bags (1400 m^2) of TEFLON® woven fabric cleaned the anode furnaces of a copper smelter. Extensive pilot studies showed that only TEFLON® could stand up to the chemical attack associated with the smelting of recycled copper. The gas contains significant

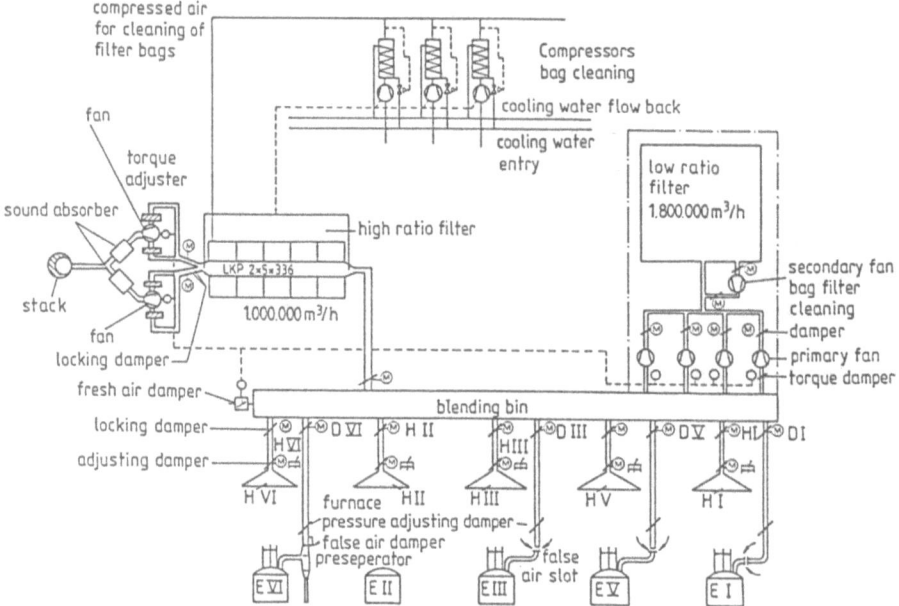

Fig. 2.83 Flowchart of dust collection in a steel mill [88]

Fig. 2.84 Dust collector in a brass smelter (Works photo Beth, Luebeck)

2.13 Filter Media in Use

quantities of sulphur oxides and water vapour. In addition hydrochloric acid from the PVC insulation and fluorine is present. The dust contains oxides, sulphates, chlorides, and fluorides of many metals (copper, brass, lead, zinc, tin, heavy metal) which become active catalysts oxidising SO_2 to SO_3 which raises the acid dewpoint. An added complication is that zinc chloride acts as a swelling agent for acrylics so that hydrolysis of the fibre interior is likely. PTFE filter material ensured an operating temperature of 180–200 °C and gave a wide safety margin above the acid dewpoint peaks. Although TEFLON® is considerably more expensive in the initial investment, it has proved to be economic with low maintenance and to be an environmentally acceptable solution with a service life of at least 5 years and nearly unlimited chemical resistance [55, 90].

Just as dust may catalyse, it can also absorb (refer section 2.9). For instance lead oxide is a by-product of non-ferrous metal smelting. It can bind significant quantities of sulphur dioxide and trioxide so that the acid dewpoint can be reduced or entirely suppressed. This came to light in lead smelting and in dust collectors behind lead acid battery incineration plants. Over the past years several dust collectors went into sevice behind lead rotary furnaces that utilise this property of lead oxide. It should be noted that both lead oxide and zinc oxide are regarded as compacting, fine and not easily stripped dusts. As long as great care is taken in choosing the correct media, very good emission levels can abe achieved. Careful attention must be given to select surface finish and added chemical and repellent treatments for the filter medium to keep the differential pressure loss to a minimum. Clean gas contents below 5 mg/m^3 are achievable with polyester felts of weigths between 500–600 g/m^2 – this applies to jet cleaned dust collectors and pocket filters at 120 °C and face velocities of 1.3–1.8 m/min.

Absorption plays an important part also in the dust removal from primary smelting furnaces in aluminium refining (fig. 2.85). Significant quantities of fluorine gas are generated during the process and are absorbed on to expanded alumina particles that are deposited on the bags. This allows dust separation and gas cleaning to be combined in a single process of dry scrubbing. The deposit of alumina on the bags has to be carefully controlled to maintain a differential pressure between 18–20 mbar at admission velocities of 1.5–1.8 m/min. Temperature ranges between 120–140 °C. A polyester needlefelt 500–600 g/m^2 in combination with a jet filter has proved effective. In low pressure cleaned dust collectors the weight can be reduced to 400 g/m^2. Despite the abrasiveness of aluminium oxide filter lifetimes of 3–5 1/2 years have been recorded.

As a rule dry scrubbing installations are used exclusively where environmental protection is essential. Additional costs are involved with a sizeable increase in plant size to remove the unwanted material. In some cases the collected waste can be returned to the production line to at least partially offset the protection cost.

During aluminium electrolysis fluorides can be recovered by absorbing it on aluminium oxide and returning this to the furnace. This reduces the need for fresh flux. A reactor preceeds the filter to achieve maximum absorber saturation [57].

Fig. 2.86 illustrates how the fresh adsorber from silo 1 is metered at point 2 and fed into the flue gas 3. To maintain the temperature within the maximum permissible range the gas temperature can be reduced by indirect cooling in a heat exchanger 4. This prevents possible baking on to the conducting surfaces of the heat exchanger. Only part of the adsorber is added to the hot gas stream at the entrance to the gas cooler. The remaining

Fig. 2.85
Primary dust collection plant in aluminium smelter

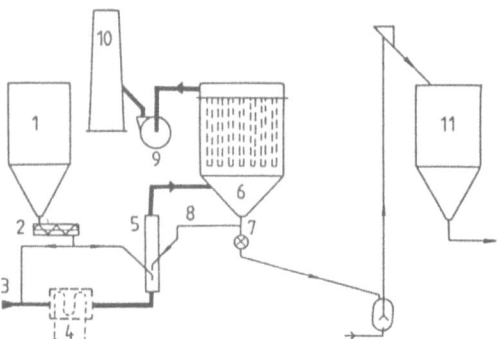

Fig. 2.86
Basic design of a dry absorption plant [57]
1 Fresh absorber silo
2 Adsorber metering
3 Flue gas
4 Heat exchanger
5 Reactor
6 Fabric filter
7 Absorber discharge
8 Recirculation
9 Extraction fan
10 Chimney
11 Silo for saturated absorber

2.13 Filter Media in Use

adsorber is added at reactor 5. Chemical absorption takes place in the reactor and also in the ductwork leading to the dust collector 6. The reaction continues on the outside surface of the filter bag. The dust particles holding the adsorbed waste are separated from the gas stream. The hopper collects the dust that has fallen off. The cleaning cycle must be adjusted so that only a small adsorber layer rests on the surface. This layer acts as a buffer for the filter, should the adsorber feed be interrupted. It also acts as a final gas scrubber. The saturated adsorber is removed from the system through an extractor and fed to silo 11. The adsorber can be recirculated to ensure full saturation 8. The filtered gas is removed via fan 9 and released into the atmosphere by the stack 10 [57].

Ferrosilicone is a magic word for dust removal experts, not only when discussing the most appropriate removal system, it also raises differing opinions. Ferrosilicone is an alloy of iron and silicone. It is produced by smelting in an arc furnace. The fine submicron amorphous dust is produced by the oxidation in contact with oxygen of the air. This fine dust occurs in low concentration and is prone to form dense cakes with consequent blinding. Matured installations work with low velocity baghouses equipped with fibreglass fabrics, also with NOMEX® both on pilot and full scale with low pressure cleaning [91, 92].

Recently a series of large installations with jet cleaning have come on line. NOMEX® needlefelts, usually with chemical protection treatments, were the media of choice. Sulphur dioxide and trioxide in the presence of water as well as other catalysing components in the flue gas have led to chemical damage of the filter media. Additional chemical protection has proved desireable in many cases. Besides stressing the physical requirements of a filter medium it is equally important to give due attention to the chemical behaviour. Despite difficulties, jet cleaned filter type dust collectors can reach clean gas emission level below a few mg/m^2 of submicron ferrosilicone dust at face velocity 2 m/min and average pressure loss of 12–20 mbar.

Filter type dust collectors have recently been installed to remove flue dust from lead glass furnaces. Installations of this type that have gone into the German Federal Republic, Great Britain and the USA have performed satisfactorily in reducing emissions. In all cases the recorded clean gas emission levels lie substantially below the legal requirements. This has also brought economic benefits. Expensive lead oxide can be efficiently recovered and recycled for the manufacture of lead glass with associated substantial savings.

Fabric filters in the glass industry in the German Federal Republic use several techniques. Pocket, horizontal and vertical bags are common. Air to cloth ratios, originally very low, have now been raised to more than 1.5 m^2/m^2 min for jet cleaned systems. Needlefelts of the acrylic type, treated aramides, PTFE (Teflon) and combinations of Nomex and Teflon are used for filter media. After initial difficulties the filter element life has been increased to at least 2 years. 3–5 years can be expected from Teflon. In some cases alumina or chalk is injected to absorb fluorine. This brings a major part of the noxious gas within the minimum emission requirement before entrusting it to the atmospheric environment [93–95].

Dust removal plays a significant part in the textile industry. A high socioeconomic value is placed on protecting mankind from effects of any kind that threaten health or life.

This is especially relevant for those industries that cannot avoid producing large amount of waste during the course of their manufacture and where dangerous dusts accumulate. This applies not only to the now stringent limits in the asbestos industry, but also in cotton spinning mills. "Byssinosis" has received much attention in this industry. The fine dust penetrates the breathing organs and irritates with asthmatic symptoms.

In News Sheet No. 10 of the Geran Research Society "the maximum workplace concentrations 1974" [96] cotton dust has been classified as Group H (danger of skin absorption). Cotton dust is given a maximum permissible level of 1.5 mg/m^2 in the German Federal Republic.

Her Majesty's Factory Inspectorate in England has already recognized these symptoms for a number of years [97]. The Shirley Institute in Manchester is engaged in developing techniques for dust removal in cotton mills.

The approach is to tackle the problem at the source (fig. 2.87) [99]. The H.M.F.I. has carried out its own studies with an extraction system where small volumes of air are drawn off at high velocity [98]. The is known as "Low volume – High velocity" (LV-HV) [101].

The air that is withdrawn is cleaned by a fabric filter [100]. The dust builds up an active fibrous filter aid on the filter medium so that as a rule good filter efficiencies can be expected. With suitable dust separators and appropriate media exhaust emissions lie below 1 mg/m^2. However, the dust cake is composed of very short fibres and does not drop easily from the filter medium (fabric or needlefelt) during the cleaning cycle. To keep the differential pressures within the range of 10–15 mbar (fig. 2.88) the filter requires intensive rapping or shaking, if mechanical means are used, or depth intensive reverse cleaning, if jet pulsing is used [102].

In the textile industry, shaking filters are frequently encountered where the dust is deposited on the inside of the bag. Jet cleaned filters with external deposition on the elements and higher face velocities are in the offing. The dust is deposited on the periphery of the bags and the combined mechanical and pneumatic forces during

Fig. 2.87
Automatic dust removal on a card condensor (Works photo Jungbauer, Augsburg)

2.13 Filter Media in Use

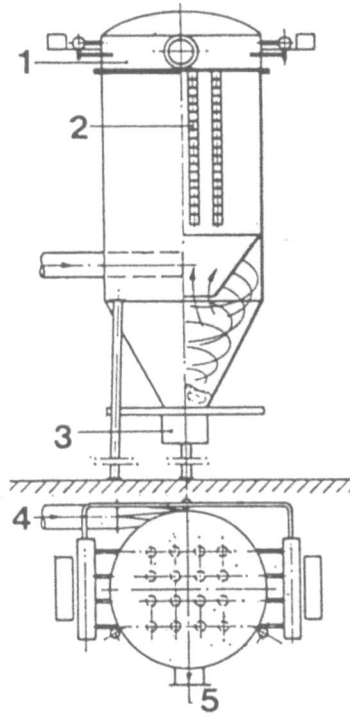

Fig. 2.88
Collection chamber with built-in pressure cleaned filters (System Jungbauer)

cleaning make it easier to dislodge the fibrous dust which tends to cling. Jet cleaned dust collectors allows the use of denser three-dimensional filter media with cake release treatments. A good performer has been a polyester needlefelt of 400–500 g/m² and air permeability of 250–150 l/dm² min at 2 mbar. With an air volume of 100 m³/h the air to cloth ratio in a closed evacuated system is 4.3 m³/m² min. Differential pressure lies in the range 20–40 mbar. At first sight this value appears high. However, such a system is entirely cost effective and technically representative given the peculiarities of an enclosed system and the high air to cloth ratio; also, if it is considered that the kinetic vacuum during valve opening can generate an air velocity in the ductwork of up to 30 m/sec at 400–500 mbar below atmospheric pressure. The filter medium has been shown to give a service life in excess of 5 years [101].

Dust collectors mounted behind asphalt mixing plants for road paving involve quite different service conditions. The dust is generated at the mixing drum: It is an abrasive rock dust mixed with water vapour removed from the mixture together with sulphur dioxide and sulphur trioxide from the fuel burning. Flue gas temperatures lie between 110–180 °C. Any unskilled start-up or close-down passes the water and acid dewpoints.

A filter medium for this end use has to survive mechanical, chemical and thermal shocks. In Central and Southern Europe bag or pocket filters of acrylic (Dralon®) have given excellent service, while in Scandinavia and the USA filter elements from aramides (NOMEX®) are preferred for temperature reasons. The different rock type and its

Fig. 2.89 Dust collector behind asphalt mixer

absorption capabilities permit the use of NOMEX without premature hydrolysis. Filter media reach service lives of 1–3 years at air to cloth ratios of 1.8–2.0 m^3/m^2 min (fig. 2.89).

Dust collectors behind coal grinding and drying plants have increased considerably in importance since the forced transition from expensive crude oil to coal (fig. 2.90). In the cement industry particularly, rotary furnaces are no longer heated with oil but with pulverised coal. A dust collector behind a pulverizer has to cope with huge dust levels of several hundred grams per cubic metre at temperatures between 90–120 °C. Water dewpoint varies between 50–55 °C. The possibility of spark generation from rock in the grinder and electrostatic charging is ever present. Three sources of heat are used to dry the coal dust:

a) The hot inert gas of the rotary furnace.
b) Heated gas from a separate coal or oil fired burner.
c) Indirectly heated air from a kiln cooler.

The gas composition is an important factor in the choice of filter medium, because water, sulphur oxides and nitrogen oxides (from the rotary furnace) significantly affect the life of the filter elements and the emission levels of the fine agglomerating coal dust [52].

Jet cleaned dust separators are mainly used. The face velocity lies between 1.2–1.5 m/min at differential pressures of 10–12 mbar. Although fine coal dust is difficult to separate from the gas stream, emission levels between 10–50 mg/m^3 have been reached with filter type dust separators. The chosen filter material play a critical part in this demanding separation. Good results have been achieved with acrylic needlefelts of 500–600 g/m^2 with an air permeability of 80–150 l/dm^2 min at 2 mbar. Here the importance of a

2.13 Filter Media in Use

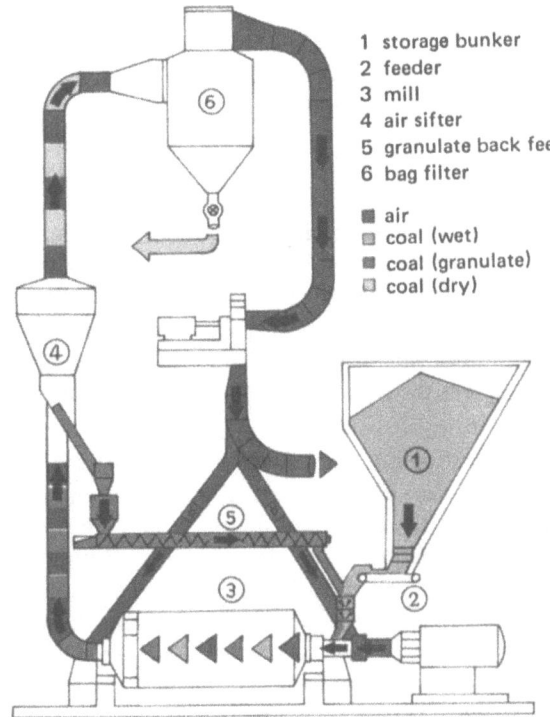

1 storage bunker
2 feeder
3 mill
4 air sifter
5 granulate back feed
6 bag filter

■ air
▫ coal (wet)
■ coal (granulate)
▫ coal (dry)

Fig. 2.90
Flowchart of a pulverized coal drier

densely constructred support weave has been proven to give high separation efficiencies, stable dimensions and to attain an econimic service life of the element. Although it is controversial whether sufficient ignition energy can build up on the filter to initiate a spark, the use of an antistatic treatment is an insurance (ref. section 2.11). It should be remebered that synthetic filter media have high specific surface and core resistance. Metal fibre additions to the felt and/or support weave reduce the discharge time and reduce the danger of sparking; they improve cleaning and contribute to the density and cleaning ability of the filter medium.

Funke [103] points to the great importance on the fabric filters in the cement industry. Bag filters find application here with conveyor systems, cement grinders, drying furnaces, grate coolers, crushers, granulators as well as silos and loading stations. In a survey [103] of the service life of filter media in this industry 71 % of filters were engaged in filtering clinker and cement dusts. Approximately half of the surveyed dust collectors had exhaust gas temperature above 50 °C. The above author found air to cloth ratios in the range 30–150 m^3/m^2 h with an average of 85 m^3/m^2.

94 collectors were found in 11 cement works where "more than 90 % used needlefelts". The woven fabric proportion was below 10 %. In plants where shaking is used to clean filters a woven fabric was used at the bottom end. The authour determined fibre proportions to be 95 % acrylic, 9 % wool, 7 % polyamide and 5 % NOMEX. Area weights

ranged between 350–600 g/m² with 520 as the mean. Needlefelts held the major share at 500–600 g/m².

The table of service live of filter media below is informative:

40 % 5000–10 000 h
38 % 10 000–20 000 h
22 % more than 20 000 h

Obviously, the construction of the dust collector and the cleaning mode influences the service life of the filter elements in the cement industry. In this survey, bag filters with mechanical cleaning tend to have a life in the mid to lower range. Jet cleaned bag filters are somewhat recent in this industry to draw any conclusions. The survey intimates "significant longer service life". Thus a longer service life is foreshadowed for filter elements in pulse jet cleaned separators – "on the average 15 000 h were achieved" – (fig. 2.91).

Exciting developments have taken place in treating hot gases from clinker coolers in the cement industry. NOMEX® has been used to date for gas filtering with jet cleaning at peak temperatures of 180–190 °C after cooling. In some cases multicyclones are found whose separation efficiencies are no longer adequate. Verplancke [65] advances a cleaning of the exit gas from clinker coolers at 250–350 °C with short peaks up to 450 °C. Under such conditions the gases would not be cooled before passing through the dust collector. For the bags a metal fibre felt was used with an air to cloth ratio of 180 m³/m² h and a differential pressure of 16 mbar. Bags were cleaned with a jet pulse every 6 minutes at 6 bar cleaning pressure. The maximum flue gas content was 100 g/m³ and final emission 20–50 mg/m³ with metal fibre filter bags.

This is only one example where filter type dust collectors can survive high gas temperatures. Such temperatures are crucial to many processes. There are examples where a plant must run above the sublimation or condensation point of a product. A case in point is given from the authors experience in North America where high sulphur content

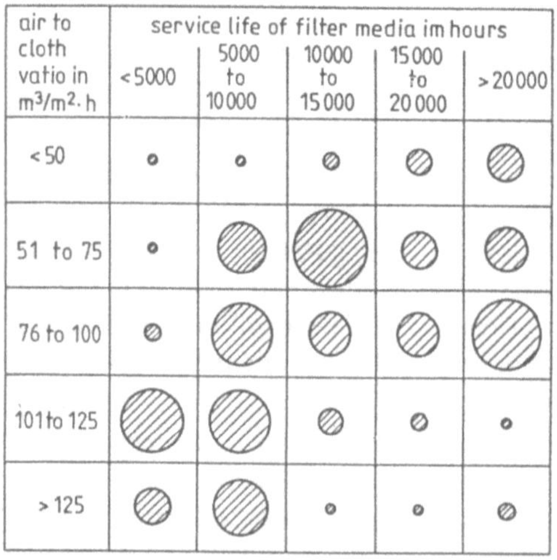

Fig. 2.91

Service life in relation to air/cloth ratio in the cement industry [103]

2.13 Filter Media in Use

flue gases from a precious metal roasting process were to be filtered. The flue gas temperature had to lie above 317 °C, because it contained arsenic trioxide whose recovery at high purity was a great cost incentive. A jet filter equipped with metal fibre felt withstood this temperature. Impurities were filtered while the arsenic trioxide passed through the separator in gas form. After cooling the arsenic trioxide was recovered in a further filter equipped with TEFLON® needlefelt with 97 % purity; this was sufficient for pharmaceutical use. This is a classical example where a properly conceived application of filter type dust collectors can generate both economic and ecological benefits.

In America and Australia interesting developments are in progress with the dust removal from oil or coal fired steam generators and power plants (fig. 2.92). The fabric filter is becoming more and more competetive to the electrostatic precipitator, and not just for low sulphur coal. More stringent emission controls, variable separation performance of the electrostatic precipitator when burning variable sulphur fuels, the necessity to condition sulphur oxides to keep emission level low and more attractive cost benefits give the filter type dust collector the upper hand. Today several alternatives for gas cleaning for power plants are available:

a) Baghouses with low pressure cleaning clothed in fibreglass fabric treated with PTFE, silicone or graphite. Temperatures of 200–260 °C are possible, though face velocities are limited to 0.6 m/min. The pressure differential depends on processing conditions and may present difficulties. This type of installation predominates in the USA. It is founded on the proven performance of fibreglass fabric bags in the carbon black industry operating on the furnace principle [12, 104–106, 117].

b) Low pressure reverse air cleaning baghouses equipped with lightweight PTFE needlefelt. It tolerates 200–260 °C and reaches face velocities of 1.2 m/min. Initial costs for these filters are significantly higher, but are compensated by higher air to cloth ratios. This method also has its origin in the carbon black industry.

c) Low pressure reverse cleaned baghouses for low sulphur coal (0.4–0.5 %) equipped with acrylic bags (Dralon T®) for maximum temperature 120 °C. For a woven fabric the face velocity lies between 0.5–1.0 m/min. This can be raised to 1.3 m/min for a needlefelt. If the acid dewpoint reaches 110–115 °C, it will exceed the operating range

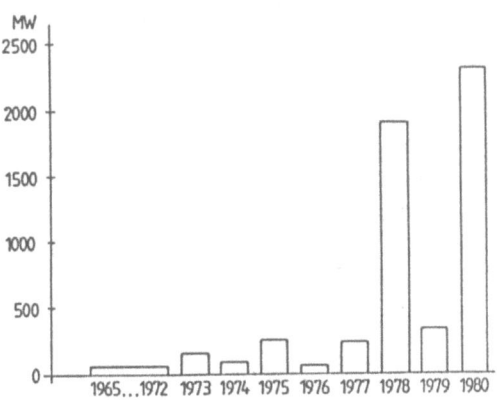

Fig. 2.92
Dust collector installed in power plants since 1965 [109]

USA	approx 85 %
Canada	
Denmark	each 2–4 %
Australia	
Fed. Rep. of Germany	
Total	approx 5500 MW

for acrylics, so that media with higher service temperatures should be considered. Baghouses of enormous dimensions equipped with acrylic fabrics are installed in Australia [107].

d) Jet cleaned dust collectors equipped with acrylic needlefelts for coal burning with low sulphur content. Such installations are found in Denmark, Canada and Australia with bag lengths 3–6 m. Face velocities range 1.2–2.0 m/min. These designs are similar to e) and function very well, if the acid dewpoint can be kept below the fibre temperature of 120 °C [108, 109].

e) Jet cleaned dust collectors with TEFLON coated fibreglass fabrics at 600–900 g/m^2. The applicable temperature range is 220–260 °C. Face velocities 1.1–1.3 m/min. Off-line cleaning is recommended for this design (filter turned off during the cleaning cycle) to protect the fabric and to allow the flue ash to sink and to keep the ΔP to a minimum with correspondingly lowered face velocities [110, 111, 114].

f) Jet cleaned dust collectors equipped with PTFE (TEFLON®) needlefelts of 850 g/m^2. Temperature ranges 220–260 °C. Face velocities, depending on construction and type of cleaning, between 1.5–1.8 m/min. A number of such collection plants are in use behind boilers [112].

g) Jet cleaned filter type dust collectors equipped with recently developed RYTON to operate at 170 °C, peaking at 190 °C (ref. section 2.10). This filter media is resistant to steam, sulphurous and sulphuric acid and permits face velocities of 1.2–1.8 m/min.

A further development which combines dust collection with simultaneous desulphuring of flue gases has become popular in USA and Europe (fig. 2.93) [113]. The flue gas is passed through a spray drier into which calcium hydroxide is injected. This forms calcium sulphate or gypsum for which there is a demand in the building industriy. This process has the advantage that the plant operator need no longer be troubled by sulphuric acid dewpoint, if he uses sulphur containing fuels. Thus, the filter media choice for this type of dust collector is determined solely by the temperature and the water content. Baghouses with jet cleaned filters are used. The filter media is an acrylic felt with a temperature range 110–120 °C. These temperatures lie considerably above any possible water dewpoint [115].

A further option may be added to the above where lime is added to the coal to obtain an alkaline and absorptive fly ash (fig. 2.94) [116].

In general it appears that those systems will survive which offer the following advantages:
- High air to cloth ratios for compactness of construction.
- High temperature performance to be able to run above any sulphuric acid dewpoint.
- Combining dust separation in high efficiency filters with simultaneous desulphuring.

Many fabic filters can be found in the chemical industry: e.g. for grinding organic or inorganic compounds, for the recovery of powdered products from drying ovens, behind pneumatic conveyors, mixers, conveyor belts, settling tanks, for the ventilation of silos or assistance in a multitude of complex and advanced processes [118]. As an example a dust collector is chosen behind a drier in the chemical industry. Following an initial

2.13 Filter Media in Use

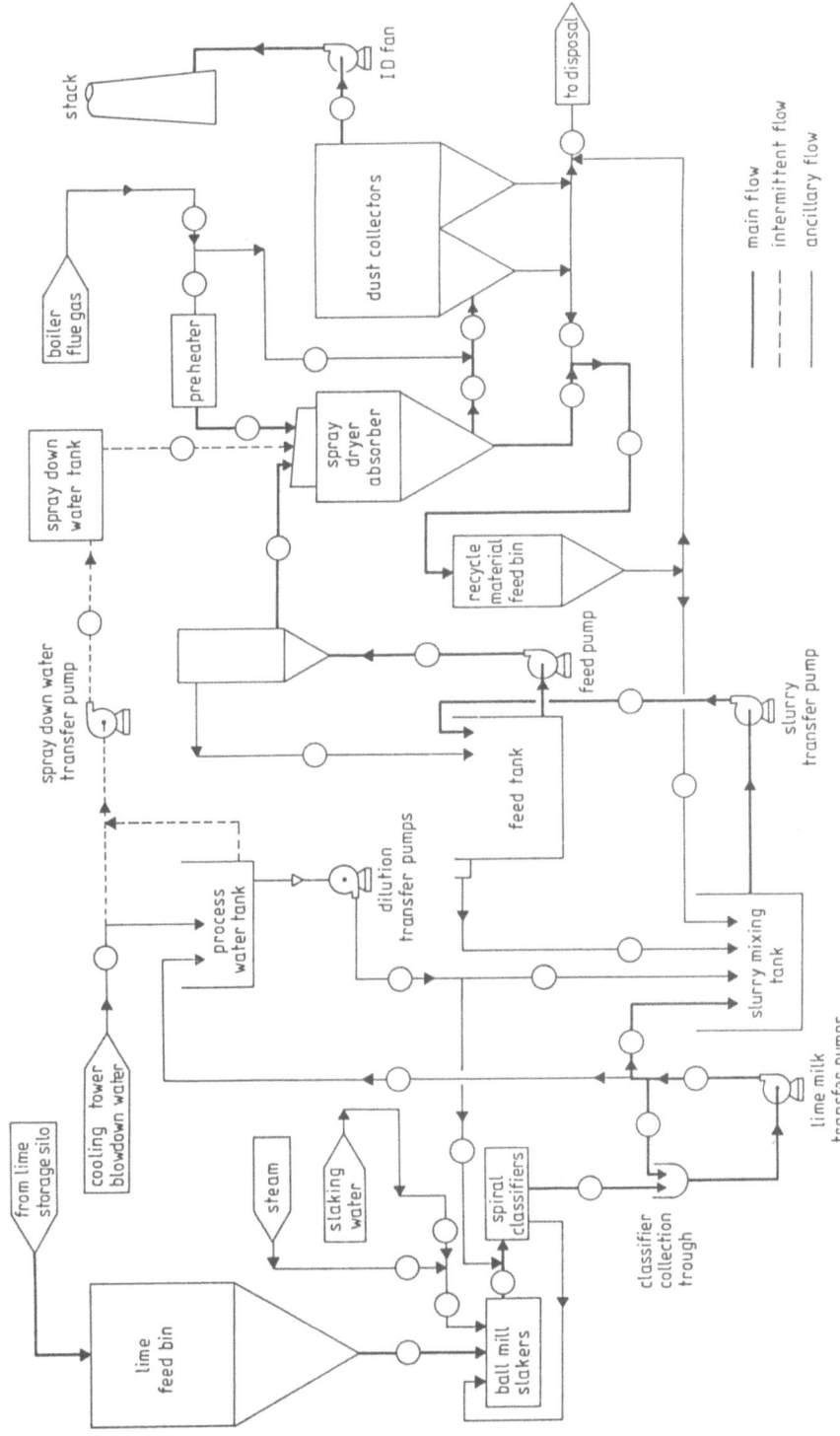

Fig. 2.93 Flowchart of an American sulphur dioxide dry scrubbing installation [115]

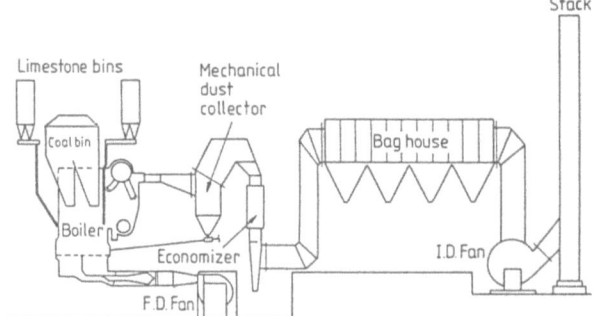

Fig. 2.94
American suggestion for binding sulphur oxides to alkaline fly ash by adding lime to coal [116]

pretreatment the fine dust portion reaches a pressure filter of which the following specifications are part of:

Capacity	17 000 Nm^3/h
Filter inlet temperature	above 120 °C
Flue gas dust content	18 g/Nm^3
Particle distribution	100 % < 10 μm
	50 % < 3 μm
Bulk density	318 g/m^3
Dust characteristics	agglomerates easily
	corrosive
Water dewpoint	65 °C
Cleaning action	Jet filter
Cleaning pressure	6 bar
Cleaning pulse timing	150 ms
Cleaning air demand	90 Nm^3/h
Cleaning frequency	2.7 min
Cleaning mode	in Rows
Air to cloth ratio	70 m^3/m^2 h
Differential pressure	8–15 mbar
Filter surface area	330 m^2
Filter medium	Acrylic 500 g/m^2
	Air permeability:
	150 l/dm^2 min 2 mbar

2.14 Warranties for Filter Media

To be precise, the textile material of a dust collector should really be considered a replacement part. Infortunately it has become accepted practice to make exceedingly high warranty demands on it. While a bearing manufacturer would not warrant his part for more than 3000 hours, a filter media producer is often obliged to garantee periods of 16 000 or even 24 000 operating hours. Why is this so?

Competition, prestige, and marketing pressures have softened suppliers in the filter media industry, and they have yielded to unrealistic exspectations of warranty periods.

2.14 Warranties for Filter Media

Others have followed suit. By now this has become common practice in Europe and overseas when large projects are put to tender; it applies equally to the filter media. Plant manufacturers pass up their own warranty obligations to subcontractors. These requirements are threatening to become excessive.

The wish of every plant builder and operator is understandable and legetimate to obtain assurances in regard to characteristics of the components he is about to purchase. However, this should be kept within limits that are realistic and sensible. Warranties have their place. They are educational, they raise discipline, they encourage more frequent and tighter control during manufacture; they separate the quality conscious tenderer from the haphazard one.

The following undertakings are usually required from filter media manufacturers:

- Adherence of the filter media to agreed and specified tolerances.
- Separation efficiency.
- Service life of the filter elements.

A basis for such warranties should be a detailed specification of the exspected performance parameters which should include the dust collector with the anticipated operating conditions. The compliance to such parameters, e.g. temperature, differential pressure, face velocity etc. should be capable of being monitored with pen recorders.

Frequently it is laid down that the differential pressure be maintained within certain limits. It appears to the author at times that the demands made on a filter medium are unrealistic, because of the many variables that can affect cleaning within the plant (ref. section 2.6) and because many of the processing variables cannot be laid down in advance. A "Statement of Warranty" is common in the industry where plant builder or operator set out all data including filter specifications. If these data are all-embracing, no one will have any objections. If, however, the medeium, in order to save cost, is specified at its lowest weight limit and construction and it is then directed to the media manufacturer, he has no option but to either lose the business, if he does not wish to be part of the warrantly, or to undertake the thorny road of demanding higher specifications with associated higher price at the tendering stage. Much diplomatic skill must be excercised to bridge this conflict between sales opportunity and technical realities.

VDI Standard 2260 [119] makes technical warranties an integral part of delivery contracts. These should extent to.

a) Fitness for use and specified performance characteristics of the dust collection plant (performance warranty).
b) Materials (parts warranty).
c) Safety provisions and legal requirements.
d) Miscellaneous requirements by special arrangements.

Material warranty refers to the suitability of the basic subcomponents and subsequent manufacture into the final product. All warranted values should be capable of measurements [120].

According to VDI 2260 [119] the performance garantee as referred to in point a) lays down the quality of construction as well as proper functioning of all subcomponents needed for the plant. First in line is the property of the collectors to achieve a given filtration efficiency (residual pollutants, degree of filtration) both of which must be

Staubforschungsinstitut
des Hauptverbandes
der gewerblichen Berufsgenossenschaften e. V.

Berufsgenossenschaftliche und
vom BMA bezeichnete Prüfstelle

53 Bonn, den 29. März 1976
Langwartweg 103
Telefon: (0 22 21) 54 92 90

Prüfzeugnis

Nr. 675211/00276

über die Prüfung von filterndem Material zur Verwendung
in Einrichtungen zum Abscheiden gesundheitsgefährlicher
Stäube mit Rückführung der Reinluft in die Arbeitsräume

1. Prüfling

Bezeichnung/Typ	Needlona PE/PE 401
Name und Anschrift des Herstellers	Bayerische Wollfilzfabriken
Monat und Jahr der Herstellung	Januar 1976

2. Name und Anschrift des Antragstellers:

Bayerische Wollfilzfabriken K.G.
Postfach 1120
8875 Offingen

3. Datum der Prüfung: 26.3.76

4. Art der Prüfung:

Materialprüfung — ~~Materialnachprüfung~~ *) gem. ZH 1/487, Abs. 2

5. Angaben des Antragstellers zum Prüfling

Art des filternden Materials:	PE/PE	
Flächengewicht:	426	g/m²
Temperaturbeständigkeit:	---	°C
Luftdurchlässigkeit nach DIN 53 887:	750	m³/m² · h

6. Prüfverfahren

Nach DIN 24 184 „Typprüfung von Schwebstoffiltern" Ausgabe 1974, Abs. 5.13
Bestimmung des Durchlaßgrades D_3 mit Prüfaerosol 3 (Quarzstaub).

Konzentration Prüfaerosol 3:	350	mg/m³

*) Nichtzutreffendes streichen.

Dieses Prüfzeugnis besteht aus 2 Seiten; es darf nur vollständig und unverändert vervielfältigt oder veröffentlicht werden.

2.14 Warranties for Filter Media

Staubforschungsinstitut des Hauptverbandes der gewerblichen Berufsgenossenschaften e. V., Bonn

— Seite 2 zum Prüfzeugnis Nr. 675211-00276 vom 29. März 1976 —

7. Prüfergebnisse

Flächengewicht	427	g/m²
Luftdurchlässigkeit nach DIN 53 887	742	m³/m²h
Mittlerer Durchflußwiderstand bei 0,033 m/s Anströmgeschwindigkeit	30,7	N/m²
Mittlerer Durchlaßgrad $\overline{D}$ gegen Testaerosol 3 nach DIN 24 184	0,60	%
Relative Standardabweichung der Einzeldurchlaßgrade	19,8	%

8. Bemerkungen zur Prüfung

---/---

9. Beurteilung

Nach dem Ergebnis der Prüfungen sind die Anforderungen an die Abscheideleistung des filternden Materials nach ZH 1/487, Abs. 2   erfüllt — nicht erfüllt *)

Aus diesem Ergebnis können keine Rückschlüsse auf die gesamte Einrichtung zum Abscheiden gesundheitsgefährlicher Stäube mit Rückführung der Reinluft in die Arbeitsräume gezogen werden.

10. Besondere Hinweise

Die Prüfung und Beurteilung erfolgt nach den „Grundsätzen für Prüfungen und Untersuchungen im Staubforschungsinstitut".

Die Prüfergebnisse beziehen sich auf das der Prüfstelle vorgelegte Prüfobjekt und alle identischen Produkte. Die Identität wird von der Prüfstelle nicht überwacht.

Dieses Prüfzeugnis ist bis zum 31. Juli 1983 gültig.

Bonn, den 4.8.1980
Berufsgenossenschaftliches Institut
für Arbeitssicherheit
Berufsgenossenschaftliche und
vom BMA bezeichnete Prüfstelle

(Dr. G. Riediger)

Staubforschungsinstitut
des Hauptverbandes
der gewerblichen Berufsgenossenschaften e. V.
Berufsgenossenschaftliche und
vom BMA bezeichnete Prüfstelle

i. A. (Dr. Riediger)

Fig. 2.95 Specimen of a test certificate

measureable. The type of material to be separated, the envisaged filtration techniques, the operational and regulatory requirements will raise different aspects for setting the warranty parameters.

Service life is the time during which a filter medium maintains its required functioning with due allowance for proper maintenance to retain the expected efficiency of the separation plant. Service life is measured in operating hours, but may also be split into operating time and idle time. VDI 226 [119] defines operating time as the time during which gas passes through the collection plant. As "life" that time is understood when complete replacement of the filter medium becomes necessary.

The quoted VDI guidelines also define the concept of availability. This expresses the probability that the system is in a functioning state at any point in time (DIN 44 042) [123]. It is expressed by the formula

$$V [\%] = \frac{t_B + t_S}{t} \cdot 100$$

- t_B is the sum of actual operating hours of the collector.
- t_S is the sum of idle hours of the collector which are beyond the control of the contractor. Included here are such things as unsufficient throughput or downtime of an upline plant.
- t is the sum of operating hours plus idle time as agreed by the contractor and the plant operator.

According to Engels [120] operation begins with the start-up of the plant and ends with shut-down. This also applies to the filter medium. Proper operating instructions for this are essential. The plant operator has to ensure that the plant is run according to these operating instructions. Engels postulates that specified parameters should be adhered to. This includes the selection of appropriate and proven materials, and the satisfactory functioning of all subcomponents that are part of the installation. Testing and acceptance certificates are desireable.

Such certificates may contain the mean filtration efficiency and its dependence on relevant parameters. One of these issuing authorities is the B.I.A. (Berufgenossenschaftliches Institut für Arbeitssicherheit, St. Augustin near Bonn, West Germany). They can also encompass testing of electrostatic behaviour (Testing Authority – Westfälische Berggewerkschaftsklasse – Bergbauversuchsstrecke – Institut für Explosionsschutz und Sprengtechnik, Dortmund). The effects of decomposition products from plastics on "Filterselbstretter" are of relevance in underground mining, as well as other safety or hygiene related properties for underground use, testing of the burning behaviour or the effect of combustion products (fig. 2.95).

In legal terms the concept of "Garantee" has the meaning of "unequivocally assured attribute". One should therefore always speak of a warranty, never of a garantee in an industrial situation. One should also take note of paragraphs 459 and 633 of the German Law.

A possible formulation for a warranty statement for filter media is given in the following example:

2.15 Assembly, Accessories Support Elements Maintenance

a) It is warranted that best possible material has been selected in good faith and according to best experience and proven performance.

b) In the event of the filter medium becoming inoperable within X operating hours, or within X tons of material throughput, but maximum within a time period of X years, it will be replaced at no cost within the usual commercial conditions, provided the plant has been operated, demonstrably so, under proper supervision and within the agreed operating and maintenance requirements.

c) An extension of the warranty period beyond the replacement of parts is not applicable. With filter media one may expect service lives of more than 2 years at minimum. Warranty period beyond 2 years should be avoided for any normal replacement part [124].

It is recommended that a maintenance plan for filter and plant be set down [120].

2.15 Assembly, Accessories, Support Elements Maintenance and Care, Washing and Cleaning of Filter Fabrics

Once the filter media, made from either needlefelt, nonwoven or woven fabric, is expertly made and has passed its performance tests a further important step begins — that of the assembly into the actual filter element. This can take the form of bags, pockets, cassettes, frames, inserts or whatever various forms and dimensions the inventive filter designer may draw up. This step is just as important as the carefully controlled manufacture of the filter medium.

The market for filter media is split into plant builder, bag maker and end-user. The bag maker is one of the most important members of this market, because he is the intermediary between the media manufacturer and the plant operator. Some plant manufacturers operate their own textile and bag making facilities; similarly, large media users may operate a bag making section. A number of filter media manufacturers maintain their own sewing rooms so as to be able to direct their marketing efforts into areas where they can better compete for the end-user, i.e. the operator of dust collectors, with other bag makers. Some manufacturers of filter media choose not to get involved in the manufacture of finished products. Their main customers are builders of installations and bag manufacturers. As a rule the filter bag maker buys the raw material in roll form and sells the filter elements within its own framework of pricing determined by market forces. Commission bag makers also exist in this market.

Making filter elements requires textile expertise, in-use experience and great individual care. Most of the filter bodies in use today are still assembled by sewing. Special sewing machines with long arms which are fast and robust produce double or triple seams. Bag bottoms are inserted, sealing strips attached, rings encased with tapes, "collars" with turn-ups and steel rings inserted; quick-fastening rings, draw cords or string pockets attached, "socks" are attached to increase bending resistance, reinforcement patches are placed where parts may wear or loops or straps are attached for easier handling (fig. 2.96).

It is possible to weld needlefelts and nonwovens made of synthetic fibres because of their thermoplastic nature. This saves labour, economises on thread and avoids possible errors in selecting matching sewing threads. Welding is possible for polyester, acrylics, poly-

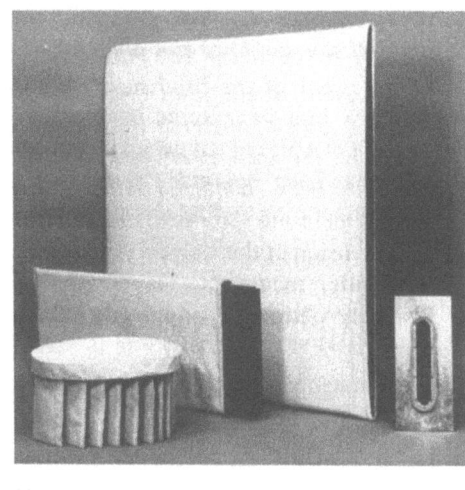

a)   b)

c)

Fig. 2.96
Examples of different bag and pocket type filter elements

propylene and aliphatic polyamides. These fabrics are welded by heat contact in which the work pieces are plasticized by an electrically heated wedge or a hot air jet.

For several years in the USA circular woven bags have been on the market which form a seamless bag. It has met with strong competition from the local European industry. It is made in a limited range of standard diameters and is no match for the large variety of bag diameters. The rise of the needlefelt with its higher filter efficiency rating has provided added competition.

It is technically feasible to produce seamless needlefelt bags. The procedure is still complicated and has not matured sufficiently to be adapted for filter use and is not yet of any commercial significance for dust collectors. However, seamless needlefelt tubes have been produced in large quantities for cartidge and candle filters in dry and wet filtration.

Starting with a technically appropriate filter material for the element the following quality factors are important during manufacture:

2.15 Assembly, Accessories, Support Elements Maintenance

a) Dimensional Accuracy

Bags or pockets that are too small sit too tight or exert too much tension in the guide. When inverted over support cages in jet filters, they become as tight as a drum. The felt cannot make sufficient convex/concave excursions during the cleaning cycle so that the dust cannot be shaken off sufficiently for collection and removal. Bag or pockets blind quickly. If filter elements are made too large, they sit too loose. They hang slack in their mounting, form creases on the floor and suffer abrasion. The cleaning process is disturbed, because the wave motion during shaking cannot propagate over the whole length of the bag. There is a danger that the bags oscillate to such a degree during cleaning that they abrade against each other.

If pocket filters are oversized, they have too much movement in the frame. Filter surfaces touch and valuable filtering surface is lost. If this occurs in jet cleaned filters, the bags have too much room to move on their support cage. They abrade on the support cage wires, they are pushed too far into the spaces between the ribs of the cage during the filtering cycle. While this greater slack may improve cleaning, it increases stress. As a rule of thumb, for jet cleaned dust collectors the outside diameter of the bags can be taken from the external diameter of the support cage plus 10 mm tolerance plus 14 mm width for the seam turnover. Both length and diameter must be accurate to ensure proper functioning.

b) Quality of Seams

The sewing thread must be the same fibre type as the filter medium. Any mixing of threads should not be tolerated either deliberately or inadvertently. If a polyester thread is used for an acrylic material, such a thread would hydrolise in moist, wet conditions: The seam bursts! Each synthetic fibre (ref. section 2.2) has different chemical and thermal properties, and the thread should be compatible with the properties of the medium.

For extra safety filter elements are sewn with double seams, jet filters have even triple seams. This applies to all seams that receive extra stress such as the length or floor seams. Specialised sewing machines are used where the sewing head has multiple needles to ensure parallel distance between seams. Even then the operator must be careful to allow sufficient clearance from the edge of the material. The choice of needles is important; thick or blunt needles cause holes through which dust can leak.

Occasionally one meets filter bags where the side seam takes the form of a spiral or where the filter bag is shaped like a banana. The problem of such "banana bags" may lie either with the medium where both edges are of differing lengths or with differential feeding during sewing. In any case such bags are not to be recommended, because their wear and tear is non-uniform and the uneven seam interferes with cleaning. It goes without saying that quality control during assembly is a most important function. One needs only to consider the catastrophic results of a "missing seam". The more sensitive the filter material the more care should be excercised in tailoring. Textiles produced from fibres that are prone to physical stress such as fibreglass are more difficult to assemble than the more robust and hard-wearing polyamide or polyesters.

c) Meticulous Care in choosing Accessories

Accessories for filter bags are stiffening or spacing rings, flaps for hanging up, hooks, springs, strapping tapes, screws, spring steel circlips, cords, tubular seams with draw tapes, and similar. All are important — to reinforce the bag or pocket at exposed points, to strengthen, to seal, to fix the turn-up, to suspend it from the top floor plate, or to attach it to the bottom flange and to keep under tension. For these accessories the same applies that has been said about seams: Textile accessories must match the filter media in chemical and temperature performance, i.e. they must have equal or better durability. Hollow seams and seals must not become brittle to endanger the dust seal. Metal parts also must withstand the anticipated chemical conditions to prevent abrasion of the filter media at contact points, and to prevent oxidation or other undesirable chemical reactions.

Based on the anticipated conditions, exposed metal parts can be steel or its various coated forms with plastics, lacquer, zinc or cadmium or stainless steel. It is also important to ensure that there is no possibility of a galvanising reaction with gases; e.g. formation of zinc chloride which damages acrylic, or worse still a catalytic zinc release of nascent hydrogen with catastrophic results (ref. section 2.9). It must be established that plastic coatings can withstand the heat so that they do not peel and expose the medium to rusting steel.

For obvious reasons it is not possible to enumerate all eventualities which may influence the assembly of filter elements. Quality conscious filter element makers draw upon their extensive experience and will select their material accordingly. Nevertheless, it is recommended that the specification for the filter elements in given equal emphasis to that of the filter medium, because the complete filter assembly must function inside the collector when exposed to the gas and dust and must not be in conflict with the material from which the filter material is made.

Screwdrivers and pliers are the filter materials, worst enemies. Media and plant manufacturers go to great length to produce bags or pockets which do justice to the various requirements, and they are confident nothing more can happen. The bags are mounted, but dust seeps. The project engineer thinks: "Well, this will disappear once the material has settled. The pores have to take up some dust to become homogeneous." If after 3 days dust is still leaking, he rings the filter media supplier: "The bags leak." (This is the usual way in which such a complaint is voiced.) The supplier hastens to the installation to find that heavy hands have been at work. The fitters slipped a few times with their screwdrivers. The bags did not take kindly to this.

These faults are well known, still they occur regularly. In wise anticipation some contractors have excluded such damage from their warrenties. Through this careless handling filter elements are put out of action before they have even come in contact with the dust.

If they have survived what should have been their last hurdle, further dangers lie in wait. The welded join of the support ring can break so that the frayed end burrows into the fabric like scalpels. During welding of the support cage for a jet filter a sharp burr can form, this scratches the inside of the bag or pocket, like a nail penetrating the sole of a shoe. The flange of the floor plate of the support cage is sometimes more like a knife

2.15 Assembly, Accessories, Support Elements Maintenance

than a protection cap, and the length wires of the support cage can swing like the strings of a double bass and abrade the felt when they have become detached from the transverse rings by improper welding.

Thus, filter bags or pockets have no easy task. To perform flawlessly in modern jet filters they require first quality support cages that are free from faults, accurate in dimension, without sharp corners or edges, not sensitive to corrosion, and durably welded. This type of wire construction is now readily available. Nevertheless there are still some suppliers in the market that have not yet discovered that they are saving at the wrong end when using unsuitable support cages (fig. 2.97a and 2.97b). Support cages for jet cleaned dust collectors are commonly made in diameters of 115 mm (4 1/2") to 150 (6"). 10 vertical support wires attach to the circumference with a wire thickness of 3–4 mm. Support cages for larger diameters have correspondingly thicker wires. For the more sensitive fibreglass and PTFE filter media it is recommended in the interest of longer service life to increase the number of vertical wires sometimes to twice the number. The horizontal support rings are 200 mm apart. If these support rings are indented between the vertical wires to form a star shape, they exert less pressure on the filter material. This protects the filter medium. For special cases support cages are made from wire mesh with a pitch of 15-25 mm. This support cage construction relies entirely on the pneumatic effect of cleaning. Covering of the support cage with a fine woven stocking either underneath or on top of the actual filter bag has not proved successful. Dust accumulates between both textile layers which leads to plinding. The floor plate of the support cage should always have a steel plate with turned edges. This dampens the descending pressure pulse during cleaning so that it does not hit the floor of the bag with full force. Even if the boot has been sewn with a triple seam, the top seam is the most stressed. The pressure wave causes

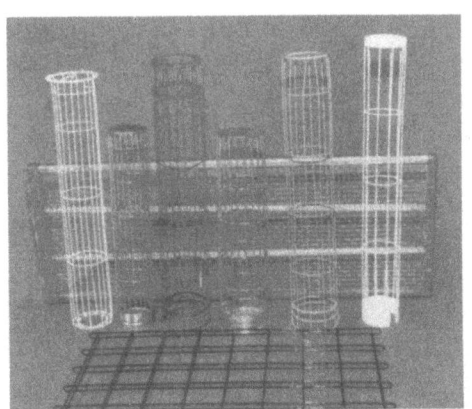

a) Support cages for conventionel bag or pocket filters

b) Typical support cage for pocket filter

Fig. 2.97 Support cages for jet filters

a bulging of the floor and leads to distortions in the seam area with the risk of dust seepage and bursting of the floor.

A disposable or single use cage which is changed with the bag has been suggested in the USA. In the more roomy USA such features may be acceptable. In Europe it would exacerbate the already huge disposal problem. Because of the popularity of the jet filter several alternatives are being tried to integrate filter bag and support cage into a single unit. Porous sintered metal candle filters are on trial.

Bag changes are costly and objectionable. They are therefore receiving increased attention in modern constructions of dust collectors. Several new systems have come to light in which the bags can be changed with a minimum of human interference. One method is to attach both support cage and bag to a flange on the top plate, another uses the snap ring principle to clip the bag into the top floor plate while the support cage is suspended over shouldered wire ends. It is important to achieve a tight seal between bag and housing to stop leaks and to prevent vertical rubbing contact by proper stacking of the bags in the filter housing (fig. 2.98).

The chemical industry has special requirements such as in the production of wash powder when quick and thorough cleaning is necessary without contaminating the following production run. In cooperation between plant manufacturer and operator the technique of C.I.P. (clean in place) has been developed. The plant is converted into a "scrubbing chamber" whereby large quantities of water are sprayed into the filter with simultaneous pulsing. Other constructions divide the dust collector into exchangeable bag batteries which can be lifted out and exchanged (fig. 2.99). The filter is cleaned outside the chamber by a series of floodings. It is best to dry the bags on their support

Fig. 2.98 Bag filters with removable cages. Bags are clip fastened to the floor plate (Flaekt)

cages. This should be done at 40–50 °C with uniform reverse air flow. This is a fast and fibre protective method. Precondition is that the media manufacturer has delivered a dimensionally stable product, which by todays standards should be garanteed to ± 0.5 %.

Should such washing processes not be possible, because of construction or processing reasons, the bags must be removed individually and washed in drum type washing machines. Some high quality media manufacturers have marketed fully washable needlefelts. It is recommended to initially remove excessive dust in situ with mechanical and pneumatic cleaning. This is done best through extensive use of the cleaning cycle with dust feed shut off immediately prior to a plant shut-down.

It should be remembered that textiles are sensitive to abrasion. Rough handling leads to surface damage and roughing. Only drum type washing machines can be recommended, whereby the filter media should be protected from friction against the drum wall. This can be done by placing them in oversized bags of very open synthetic fabrics. The rule applies: "A lot of rinsing, very little movement". Dust that has penetrated into the interior of a needlefelt must be removed by pore depth intensive rinsings.

Lukewarm water with the addition of commercially available detergents suitable for synthetic fibre is most appropriate for water soluble and easily loosened dirt. If this alone is not successful, the temperature may be raised to 50–60 °C. Soaking the bags overnight makes washing easier. A nonionic wetting agent ($1-2$ g/l) as well as a detergent suitable for synthetics (3 g/l) is added at the wash cycle of the drum laundering machine.

Alkaline contaminants such as cement or lime can be removed with acids. This is carried out with a pretreatment of $1-2$ h soaking in a bath of max 50 °C with $1-2$ ml/l acetic acid. Acid contaminants from anionic minerals are removed with alkali. Soaking in

Fig. 2.99
Removal of filter cartridge from a pressure filter filtering epoxy (Micro Pul)

1–2 ml/l ammonia is recommended. Temperature again should not exceed 50 °C. This is followed by the normal wash cycle as described above.

For hardened contamination such as bitumen, organic solvents are required. Here again a presoak helps. In this case a warming of the soaking bath should be avoided to prevent evaporation of the solvent. Health and safety precautions should be observed and the work area well ventilated. Washed filter elements can be dried in the open air or in industrial driers at max. 95 °C. As already mentioned it is recommended to dry in the separator itself to avoid the formation of folds and creases in the filter media which may be fixed in. Also, regular pulsing keeps the pores clean during drying and removes any residual dirt or washing agent. If the filter media has a chemical finish, e.g. a silicone treatment, it can be reapplied to the bags during the last rinsing bath according to instructions of the manufacturer.

Some companies maintain a service division for filter bags. In this case, bags are removed after a predetermined time, cleaned, examined for damage and repaired. However, companies that undertake this are becoming rare, because Department of Labour and Industry regulations regarding work safety and transport of dusts and the disposal of rejected filter materials require special handling which becomes costly.

Like other things, filter bags require just as much maintenance and care. This applies especially when plants are shut down for days, weeks or months. With the exception of start-ups and stops as has been pointed out above it should be avoided to take bags through dewpoints. After a shut-down, bags should be cleaned until the differential pressure reaches or lies just above that of the new bag installation. This ensures that during porlonged shut-downs reaction between humidity in the air with any residual dust does not occur, and the bags are ready for start-up. Such reactions may be unavoidable where strongly alkaline, acid or oxidative dusts are involved. Case have been recorded where chemical attack continued during plant shut down so that the bags split on re-establishing operating conditions. In such instances bags must be removed, washed, neutralised and dried. They can then be remounted even when a protracted shut-down is envisaged.

2.16 Trends and the Future

If the risk is undertaken to look into the future and forecast trends from the vantage point of the filter medium as the focus of the dust collector, it cannot be done in isolation from political and economic events of the world. Questions of future energy availability, alternative energy sources, environment protective legislation in a time of rising unemployment, continued industrialisation of third world countries with all the risks for ecology and health remain unanswered. The "plundering of the planet" continues. Raw material sources are the precondition for a secure future, not only for the balanced living standards of industrialised nations, but also for the populated developing countries. Raised living standards and protection of the environment, often considered a conflict, is the ideal of many nations. But these concepts must not stand in opposition: Quality of life without a wholesome environment is unimaginable, if it is not to be fiction.

2.16 Trends and the Future

All trends point towards the fact that industry of the "first" and "second" as well as "third" world (whateverer they may be) must respect people and be environmentally considerate, if it is not to destroy mankind. This signifies clean air and clean water.

If environmental protection is still to have a place in times of world-wide shortages of private and public funds, gas cleaning will take on special significance in those areas where greatest efficiencies, lowest cost and minimum resource requirements are vital. Product recycling has increased in importance, because raw materials are too valuable to be blown into the air. "Recycling" will be the catchphrase of the future, not only for raw materials, but also for energy in every form. Destroying energy by cooling will be obviated by the increased application of hot gas filtration.

How does this affect the filter type dust separator and filter medium? Unless entirely new concepts appear on the horizon, for which so far there is little indication, the fabric filter will become more important, because it offers many advantages in relation to investment costs, maintenance, emission control, simultaneous dust collection and gas cleaning, and adaptability to many different process requirements. At the same time demands for better mechanical, thermal, chemical and filtration efficiencies will rise.

Face velocities will be improved to make the filter smaller, more compact and more cost efficient. For this, filter media are required which are more resistant and robust and which can withstand continuously the "cyclones" unleashed on them every time they are cleaned. Jet cleaning will be refined and further developed. Bag lengths for jet filters will increase.

Temperatures in the filters will also increase. Filter materials of fibreglass, PTFE, metal and ceramic fibres will dominate. Gas between 200 and 1400 °C will be handled in appropriately constructed dust chambers. Dust removal under high temperature conditions will be called for from coal fired boilers, initiated by the high price of fossil fuel, the removal of dust from smelting processes, removal of reaction products above condensation or below sublimation points, cleaning hot gases from high pressure coal gasification, or the gas cleaning for gas turbines. The use of heat exchangers with cleaned gases, the heating of driers with hot cleaned gas, remote heating with cleaned gas for high density population areas, the warming of arable land to improve yield and to extend seasons, steam generation with hot gases and other forms of energy generation will give hot gas filtration a wide field.

Chemical absorption, i.e. the binding of detrimental gases to filter dusts or the addtional dosing of particulate absorbers within the dust separator, or in conjunction with upline reactors will be extended. Filter media will have to withstand dynamic chemical reactions on their surface and within their innards.

Often suggested but rarely achieved is the combination of electrostatic precipitation and filter type collection. Metal fibre felts, nonwovens with high metal content, or other conductive filter materials can be positively or negatively charged and can function as a porous electrode. This opens many avenues of separation techniques. Filter efficiency may be improved by the use of electrical forces even in such cases where the medium, as a consequence of high air permeability cannot function effectively. Cleaning may be regulated by electrical means, the differential pressure can be kept constant with constant separation efficiencies. Differential pressures which are too high may be lowered with electrical discharge, when they become too low (by overcleaning) they can be raised by a charge build-up. Higher and higher face velocities will become possible by combining mechanical and electrical separation forces in conductive filter materials. This may more than compensate for the gas expansion at high temperatures.

Is this wishful thinking, imaginative idealism or erroneous fantasy? Maybe, but not only. Much remains to be done in modern research. Empirical developments spurred by competitive pressures leads a more fundamental approach to dust separation. Science has made fundamental and worthwhile contributions, albeit by its very nature in greater depth and at a slower pace. It supports, corrects, adds to what in the field has already hit the market as "state of the art".

There are many uncharted areas, if this were not so, the dust removal field would lose interest. The future will demand that we advance sooner than we want. Both practitioner and scientist remain essential. The test will be whether the fabric filter will continue in its meteoric rise to the end of the century and conquer new territories; or whether it will gain "emminence" in the more conventional arena. The preconditions for a further rise of the filter type separators have been fulfilled. The filter medium will be the crucible for this development.

Literature

[1] *Dietrich, H.:* „Eigenschaften und Eignung von Faserstoffen für Filter", Zement – Kalk – Gips 7 (1978), 349
[2] *Hünlich, R. u. H.:* „Textile Rohstoffkunde", S. 16–24, (1968), Verlag Schick & Schön, Berlin
[3] *Doehner, H., Ruermuth, H.:* „Wollkunde", S. 15, (1964), Verlag Paul Parey, Berlin
[4] *Glanzstoff:* „Chemiefasern", (1960), S. 161 ff., herausgegeben vom Vorstand der Vereinigten Glanzstoffabriken AG, Wuppertal-Elberfeld, 2. Auflage
[5] *Pinnekamp:* „Filterstudie 1979", Forschungsstelle für allgemeine und textile Marktwirtschaft, Universität Münster 1980
[6] *Dietrich, H.:* "New Developments with Fabric Filters – Filter Media especially for Use under difficult chemical and thermal Conditions", Staub – Reinhaltung der Luft 41 (1981)
[7] *Hansmann, J.:* „Glasfaservliese – Eigenschaften, Einsatz, Prüfung und Kontrolle", Spinner + Weber 85 (1967)
[8] *Hansmann, J.:* „Glasvlies als Umhüllungsstoff für die Dränung", Zeitschrift für Kulturtechnik und Flurbereinigung 2 (1968)
[9] *Heintze, E. F.:* „Chemiefasern in porösen bis skelettartigen Artikeln – Spezialprodukte zwischen Kunststoff und Textil",, Werkschrift der Fa. Bayer, Leverkusen, März 1978
[10] VDI 3677, „Filternde Abscheider", Beuth Verlag GmbH, Berlin
[11] *Dietrich, H.:* „Der Nadelfilz und die Filtration", Verfahrenstechnik 2 (1971)
[12] *Schrade, H. C.:* „Anwendung von Gewebeentstaubern bei der Abscheidung problematischer Stäube im Kraftwerksbetrieb", Sonderdruck der Fa. Buckau-Walther, Köln (1978)
[13] *Jörder, H.:* „Kurze Einleitung zur Klassifikation und Terminologie der nichtgewebten textilen Flächengebilde", 12. Weiterbildungsseminar für Ingenieure, RKW Düsseldorf (1970)
[14] *Schmid, W.:* „Der Einfluß von Gewebe und Nadelfilz auf die technische Funktion und Wirtschaftlichkeit filternder Abscheider", Studienarbeit 336, Uni Karlsruhe (1979)
[15] *Blankenburg, G.:* „Ein Beitrag zur Theorie des Filzens", Dissertation, Aachen (1961)
[16] *Sachtleben, R.:* „Die Kunst, Filz zu machen", Kosmos 51 (1955), 282
[17] Ciba Rundschau, Bd. 12, Nr. 139 (1958)
[18] *Dietrich, H.:* „Die Wollfilzherstellung und ihre wichtigsten maschinellen Einrichtungen", Textilpraxis 1 (1958), 3
[19] *Offermann, H.:* „Die Beurteilung des Rohstoffs Wolle im Hinblick auf ihre Verarbeitung in der Filzindustrie", Zeitschrift für die gesamte Textilindustrie 2 (1961)
[20] *Dietrich, H.:* „Fasergekreuzte Wollfilze", Melliand Textilberichte 40 (1959), 591
[21] *Fröhlich, H. G.:* „Überblick über die Herstellung von echten Filzen aus tierischen Haaren", Textil-Praxis 4 (1957), 347
[22] *Jörder, H.:* „Fortschritte auf dem Gebiet der nicht gewebten textilen Flächengebilde", 31. Weiterbildungsseminar für Ingenieure, RKW Düsseldorf (1971)
[23] *Jörder, H.:* „Neue Entwicklungen in der Herstellung und im Einsatz nicht gewebter Textilien", 52. Weiterbildungsseminar für Fach- und Führungskräfte der Textilindustrie, RKW Düsseldorf (1974)
[24] *Newton, A., Ford, J. E.:* "The Production and Properties of non-woven Fabrics", The Textile Institute Manchester, England (1973)
[25] *Lennox-Kerr, P.:* "Nonwovens 71", "Collected and revised papers on 'Non woven Fabrics'", The Textile Trade Presse, Manchester, England (1971)
[26] EDANA Proceedings of "INDEX 78 Congress, Amsterdam (1978)
[27] TA Luft – Weka Verlag, D-8901 Kissing
[28] VDI 3678 Entwurf, „Elektrische Abscheider", Beuth Verlag GmbH, Berlin
[29] VDI 3679 Entwurf, „Naßarbeitende Abscheider", Beuth Verlag GmbH, Berlin
[30] VDI 3676 Entwurf, „Massenkraftabscheider", Beuth Verlag GmbH, Berlin
[31] *Dietrich, H.:* „Die heutige Rolle textiler Filtermedien im Emissionsschutz", Melliand Textilberichte International 4 (1980), 331
[32] Beth-Handbuch, Selbstverlag Maschinenfabrik Beth GmbH, Lübeck (1964)
[33] *Flatt, W.:* „Textilfilter für die Luft- und Gasreinigung in industriellen Produktionsanlagen", Chemische Rundschau 28 (1975), 11

[34] *Dietrich, H.:* „Filternde Entstauber und ihre technische Anwendung", Umwelt 1 (1979), 26
[35] *Dietrich, H.:* "The Profit from sophisticated Filter Felts", Proceedings of the "Third APCA Specialty Conference", Niagara Falls, USA (1978)
[36] *Menden, G.:* „Filternde Abscheider", Aufbereitungstechnik (im Druck) 1982
[37] *Menden, G.:* „Druckluftabgereinigte filternde Abscheider", Aufbereitungstechnik 21 (1980), 1
[38] *Flatt, W.:* „Superkompakte Düsenfilter mit Zyklondimensionen mit optimierter Einzelschlauchspülung setzen neue Maßstäbe", Staub – Reinhaltung der Luft 37 (1977), 412
[39] *Güthner, G.:* „Verminderung von Staubemissionen durch Einsatz von Gewebeabscheidern", Staub – Reinhaltung der Luft 40 (1980), 324
[40] *Löffler, F.:* „Die Abscheidung von Partikeln aus Gasen in Faserfiltern", Chem.-Ing. Tech. 52 (1980), 312
[41] *Meyer zu Riemsloh, H.:* „Abreinigung von Faserstoffiltern durch Druckluft", Zement – Kalk – Gips 7 (1978), 355
[42] "Standard Method of Test for Maximum Pore Diameter and Permeability of rigid porous Filters for Laboratory use", American Society for Testing Materials E 128/61
[43] *Dietrich, H.:* „Möglichkeiten und Grenzen der Herstellung von Filterstoffen", Vortrag „Haus der Technik", Essen, 22.11.1979
[44] *Dietrich, H.:* "Dust Filtration with Needle Felts", Proceedings of Filtration Society Conference, London 28.–30.9.1971
[45] *Lünenschloß, J.; Gupta, V. P.; Berns, K.:* „Untersuchungen der Zusammenhänge zwischen den Merkmalen neuartiger Nadelfilzkonstruktionen und ihren Filtereigenschaften bei der Entstaubung", Forschungsberichte des Landes Nordrhein-Westfalen Nr. 2970, Westdeutscher Verlag GmbH Opladen (1980)
[46] *Samaey, W.:* "Metal fibre media for hot gas filtration", Proceedings of the International Symposium organised by the Royal Society of Flemish Engineers (K. VIV) in Bruges, September 1980
[47] *Dietrich, H.:* „Neuere Erkenntnisse bei der Abscheidung und Rückgewinnung von Pulvern und Feinstäuben mit dreidimensionalen Filtermedien, insbesondere Nadelfilzen", Staub – Reinhaltung der Luft 34 (1974), 176
[48] *Güthner, G.:* „Stand und Entwicklungstendenzen bei Filtern mit Druckstoßreinigung", Hot Gas Symposium der Fa. Du Pont, Düsseldorf (1974)
[49] *Meyer zu Riemsloh, H.; Krause, U.:* „Beschreibung und Beurteilung mechanischer und mittels Druckluft abgereinigter Schlauchsysteme", Aufbereitungstechnik 16 (1975), 245
[50] *Menden, G.; Elberling, N.:* „Druckluftabreinigung bei filternden Abscheidern", Zement – Kalk – Gips 31 (1978), 459
[51] *Rieber, M.:* „Über die Prüfung der Hydrolysebeständigkeit von Textilien", Farbwerke Hoechst
[52] *Dietrich, H.:* "Filtermedia for Dust Collectors", Zement – Kalk – Gips 34 (1981), 535
[53] *Dietrich, H.; Gürtler, H. C.:* „Nadelfilze für die Filtration – ohne und mit Chemikalienausrüstung", Hot Gas Symposium der Fa. Du Pont, Düsseldorf (1974)
[54] *Dietrich, H.:* „Erfahrungen mit der Hydrolyseschutzausrüstung „CS 32" bei Aramid-Filtermedien", Glastechnische Berichte 49 (1976), 131
[55] *Dietrich, H.:* „Eigenschaften und Einsatz von Filtermedien", Erzmetall 31 (1978), 257
[56] *Huber, H.; Jörg, F.:* „Einfluß von Stickstoffoxiden auf Kunststoffe", Staub – Reinhaltung der Luft 35 (1975), 184
[57] *Käppeler, G.:* „Trockene Abscheidung gasförmiger und kondensierbarer Emissionen", Umwelt 1 (1980), 29
[58] *Gürtler, H. C.; Dietrich, H.:* "High Temperature Filtration", Vortrag beim Shirley Institute Manchester, April 1974
[59] *Gürtler, H. C.; de Bruyne, R.:* "The Economics of High Temperature Filtration", Filtration & Separation 6 (1977), 1
[60] *Dietrich, H.:* "High Temperature Filtration under difficult chemical Conditions", Vortrag beim 6th Industrial Air Pollution Control Seminar, Cherry Hill, N. J. (USA), April 1976
[61] *Dornieden, M.:* „Indirekte Heizung und Kühlung (Wärmeaustauscher)", Ullmanns Encyklopädie der Technischen Chemie, 4. Auflage, Bd. 2, Hrsg. von E. Bartholomé, E. Biekert, H. Hellmann und H. Ley, Weinheim/Bergstraße 1972, S. 423/73

Literature

[62] *Davis, C. N.:* "Air Filtration", Academic Press (1973), S. 79

[63] *Pich, J.; Binek, B.:* "Temperature characteristics of fibre filters", Tschechoslowakische Akademie der Wissenschaften, Prag, Aerorols (1965)

[64] *Dietrich, H.:* „Reinigung von Heißgasen mit Hilfe textiler Filtermedien", Stahl und Eisen **10** (1978), 492

[65] *Verplancke, W.:* "Progress in Hot Gas Filtration with Metal Fibre Media", Filtration & Separation **2** (1982), 143

[66] *Dietrich, H.; Gürtler, H. C.:* "A new Textile for High Temperature Filtration of Dust and Gases", Proceedings Filtration Society Conference, London (1973)

[67] *Gürtler, H. C.:* „Nadelfilzfiltermedien aus mineralischen Fasern", Keramische Zeitschrift **8** (1973)

[68] *Verplancke, W.:* "Progress in Hot Gas Filtration with Metal Fibre Media", Proceedings Filtration Society Conference, London (1981)

[69] *Vansteenkiste, P.:* „Edelstahlfasern in textilen Filtermedien als Schutz gegen Explosionen durch elektrostatische Aufladung bei der Trockenfiltration", Staub – Reinhaltung der Luft **41** (1981) Nr. 11, 425

[70] VDI 2263: „Verhütung von Staubbränden und Staubexplosionen", Beuth Verlag GmbH, Berlin 30

[71] *Franke, H.:* „Bestimmungen der Mindestzündenergie von Kohlenstaub/Methan-Luftgemischen (Hybride Gemische), VDI-Berichte Nr. 304 (1978)

[72] *Burgoyne, J. H.:* "The Testing and Assessment of Materials liable to dust explosion or fire", Chemistry and Industry **2** (1978)

[73] *Theimer, O. F.:* "Cause and Prevention of Dust Explosion in Grain Elevators and Flour Mills", Powder Technology **8** (1973)

[74] *Martin, R.:* "Preventing Dust Explosions in Factories", Filtration Society Meeting, London, June 15 (1976)

[75] *Gibson, N.:* "Electrostatic Hazards in Filters", Filtration Society Meeting on "Electrostatics in Gas/Solid Separation", London (1979)

[76] *Raftery, M. M.:* „Untersuchung von industriellen Stäuben auf Explosionsfähigkeit", Staub – Reinhaltung der Luft **4** (1971)

[77] *Lüttgens, G.:* „Elektrostatische Aufladungen – Ursachen und Beseitigung", Technische Akademie Esslingen, Okt. (1979)

[78] *Vansteenkiste, Ph.:* „Bekinox®-Metallfasern als Lösung für antistatische und leitfähige Textilien aus Sicherheitsgründen", Werksmitteilung der Fa. Bekaert S. A., Zwevegem/Belgien, Mai (1980)

[79] *Cocks, R. E.:* "Recognition and Control of Dust Explosion Conditions", Tappi Eng. Conf. Proceedings (1978)

[80] *Einsele, U.:* „Flammhemmende Ausrüstung von Vliesstoffen und Nadelfilzen", Textil-Praxis **11** (1975), 1559

[81] *Einsele, U.:* „Über die Flammfestausrüstung von Textilien", Textil-Praxis **3** (1972), 172

[82] *Einsele, U.:* „Veredelung und Verarbeitung von Vliesstoffen und Nadelfilzen", Reutlinger Kolloquium, Oktober (1974)

[84] DIN 53906: „Bestimmung des Brennverhaltens – Senkrechtmethode", Beuth Vertrieb GmbH, Berlin 30

[84] DIN 54330: „Entwurf des Brennverhaltens von Textilien, Begriffe und Anwendungsgebiete" Beuth Vertrieb GmbH, Berlin 30

[85] DIN 53887: „Bestimmung der Luftdurchlässigkeit von textilen Flächengebilden", Beuth Vertrieb GmbH, Berlin 30

[86] *Hilden, J., Roßbach, V., Zahn, H.:* „Zum thermischen Verhalten von Polyacrylnitrilfasern in Abhängigkeit vom chemischen Aufbau der faserbildenden Acrylnitrilpolymere", Colloid and Polymer Science, Vol **225** (1977)

[87] *Roßbach, V.:* „Schadensanalyse bei technischen Filzen – Untersuchungsmethoden und ihre Anwendungsmöglichkeiten", Organisation für die Filztuchindustrie in Europa, Den Haag (1978)

[88] *Weingarten, W.:* „Filternde Staubabscheider", Seminar Haus der Technik, Essen, 22./23.11. (1979)

[89] *Dekowski, H.:* Werksvortrag der Fa. Felten & Guilleaume, Köln, März (1977)
[90] *Naumann, W.:* „Erfahrungen mit filternden Entstaubern in einem Betrieb der NE-Metallindustrie", Umwelt 3 (1980), 330
[91] *Bettanini, C.:* "Four Years Experience with Fabric Filters in the Ferrosilicon Industry", Filtration & Separation 4 (1977)
[92] *Dietrich, H.:* „Neue Einsatzbereiche für filternde Abscheider und Entwicklungstendenzen", Staub – Reinhaltung der Luft 39 (1979), 314
[93] *Margraf, A.:* „Realisierte Lösungen zur trockenen Fluorwasserstoffabscheidung aus Rauchgasen von Brenn- und Schmelzprozessen", Vortrag im Fachausschuß VI der Deutschen Glastechnischen Gesellschaft am 17.4. (1980)
[94] *Dietrich, H.:* „Entwicklungsstand von Filtermedien für filternde Abscheider bei der Entstaubung von Glaswannen", Glastechnische Berichte 52 (1979), 243
[95] *Scherer, V.:* „Entwicklung eines Entstaubungsverfahrens zur Reinigung bor- und bleihaltiger Abgase aus Glasschmelzanlagen", Glastechnische Berichte 51 (1978)
[96] „Maximale Arbeitsplatzkonzentrationen" Harald Boldt Verlag, 5407 Boppard
[97] The Asbestos Research Council: "The Cleaning of Premises and Plant in Accordance with the Asbestos Regulations", Control and Safety Guide 9 (1973)
[98] The Asbestos Research Council: "The Control of Dust by Exhaust Ventilation when Working with Asbestos", Control and Safety Guide 7 (1973)
[99] *Schütz, A., Coenen, W., Engels, L.-H.:* „Staub am Arbeitsplatz", Rudolf Haufe Verlag, Freiburg i. Br. (1975)
[100] Hauptverband der Gewerblichen Berufsgenossenschaften, Bonn: „Einrichtungen zum Abscheiden gesundheitsgefährlicher Stäube mit Rückführung der Reinluft in die Arbeitsräume", ZH 1/487 (1973)
[101] *Dietrich, H.:* „Abfallentsorgung und Entstaubung nach LV/HV-System", Textil-Praxis 30 (1975), 1131
[102] *Dietrich, H.:* „Vollautomatische Staub- und Faserabscheidung in der Textilindustrie", Intern. Textil-Bulletin 4 (1970)
[103] *Funke, G.:* „Standzeiten von Filterstoffen in Zementwerken", Zement – Kalk – Gips 33 (1980), 190
[104] *Goosens, W.; Grabenhorst, U.:* „Einsatz eines Schlauchfilters hinter einem Schmelzkammerkessel", VGB Kraftwerkstechnik 60 (1980), 12
[105] *Davids, P. et al.:* „Luftreinhaltung bei Kraftwerks- und Industriefeuerungen", Brennstoff – Wärme – Kraft 33 (1981), 170
[106] *Hobson, M.:* "Review of Bag House Systems for Boiler Plants", Journal Air Pollution Control Association 26 (1976), 22
[107] *Leutbrecher, A.:* "Australian Experience, Filtration of Flyash from very low Sulphur Coal", (1977), PB 284 969
[108] *Humphries, W.; Madden, J. J.:* "Fabric Filtration for Coal Fired Boilers: Nature of Fabric Failures in Pulse-Jet Filters", Filtration & Separation 6 (1981), 503
[109] *Viktorson, A.:* "Pilot-testing and Full-scale Experience of Filtering Flue Gases from Coal Fired Boilers", Vortrag vor der Filtration Society, Universität Sheffield, April (1981)
[110] *Makansi, J.:* "Fabric Filters Vie for larger share of Flyash-Collection Duties", Power 1 (1982), 41
[111] *Perkins, R. P.:* "State-of-the-Art of Baghouses for industrial Boilers", Industrial Fuel Conference Purdue University, West Lafayette, Indiana, Oktober (1977)
[112] *van Dooren, K. G. J.:* "Review of Operating and Maintenance Experience with High Temperature Filter Media on Coal-fired Boilers", Vortrag vor der Filtration Society, London, September (1981)
[113] *Bergmann, L.:* "Fibres and Fabrics for SO_2 Dry Scrubbing Baghouse Systems", Filtration & Separation 3 (1981), 222
[114] Standard Havens, Kansas City, Miss.: "Why outside Bag Collection instead of inside Bag Collection on large Gas Volume Jobs?", Werksmitteilungen November (1976)
[115] *Davis, R. A. et al.:* "Dry SO_2-Scrubbing – With special Reference to the Antelope Valley Station Project", Vortrag vor The American Power Conference, April (1979)

Literature

[116] *Gamble, R. L.; McCloy, W. J.:* "Fluidized Bed Combustion", American Power Conference, Chicago, Ill., April (1980)

[117] *Lützke, K.; Wilkes, R.:* „Filternde Abscheider hinter Feuerungsanlagen – Bericht einer USA-Reise", Staub – Reinhaltung der Luft **38** (1978), 443

[118] *Schmid, W.:* „Erarbeitung von Marktdaten über das Filtermedium Nadelfilz sowie Darstellung und Interpretation technisch-wirtschaftlicher Zusammenhänge anhand von fünf ausgeführten filternden Abscheidern", Ing.-Arbeit, Universität Karlsruhe (1980)

[119] VDI 2260: „Technische Gewährleistungen für Abscheideanlagen – Abscheidung fester und flüssiger Luftverunreinigungen", Beuth Vertrieb GmbH, Berlin 30

[120] *Engels, L.-H.:* „Gewährleistungen, Betrieb und Wartung bei Einrichtungen zur innerbetrieblichen Luftreinhaltung", Die Berufsgenossenschaft **9** (1978)

[121] *Engels, L.-H.:* „Entstaubungstechnik – Verfahren, Leistung, Aufwand", TÜ **14** (1973), 224

[122] *Engels, L.-H.:* „Technische Verfahren zur Emissionsverminderung – Anwendungsbeispiele – Erfolge – Verbesserungsmöglichkeiten", VDI-Bildungswerk (1975)

[123] DIN 40 042: „Zuverlässigkeit elektrischer Geräte, Anlagen und Systeme; Begriffe", Beuth Vertrieb GmbH, Berlin 30

[124] *Schmid, W. D.:* Werksinformationen der BWF Offingen

[125] DIN 53 854: „Gewichtsbestimmungen an textilen Flächengebilden mit Ausnahme von Gewirken und Gestricken", Beuth Vertrieb GmbH, Berlin 30

[126] DIN 53 861: „Wölb- und Berstversuch", Beuth Vertrieb GmbH, Berlin 30

[127] DIN 53 857: „Einfacher Streifen-Zugversuch an textilen Flächengebilden", Beuth Vertrieb GmbH, Berlin 30

[128] DIN 53 863: „Scheuerprüfungen von textilen Flächengebilden", Beuth Vertrieb GmbH, Berlin 30

[129] DIN 54 201: „Quantitative Bestimmung der Anteile von Fasermischungen", Beuth Vertrieb GmbH, Berlin 30

[130] DIN 54 345: „Beurteilung des elektrostatischen Verhaltens", Beuth Vertrieb GmbH, Berlin 30

[131] *Preston, J., Economy, J.:* "High Temperature and Flame Resistant Fibres", Applied Polymer Symposia, The American Chemical Society, New York (1972)

[132] *Bakke, E.:* "Optimizing Filtration Parameters", Journal of the APCA **24** (1974), 1150

[133] DIN 53 482: „Bestimmung der elektrischen Widerstandswerte", Beuth Vertrieb GmbH, Berlin 30

[134] *Karsten, H.:* „Moderne Filtermedien – ihre Verwendung in filternden Abscheidern", Chemie-Technik **8** (1979), 51

[135] DIN 54 345 Teil 4: „Beurteilung des elektrostatischen Verhaltens", Beuth Vertrieb GmbH, Berlin

3 Filtration Plants

Werner Flatt

3.1 Introduction

Over the past years it has become more and more accepted that filtration plants can fulfil a useful function in many areas of industry. Environmental protection laws make their use mandatory at times. Heat exchangers or sound absorbing elements often tolerate only cleaned gas. The recovery of products is also a large application area for dust collectors of all types; work places must be kept free of dust and smoke, and dust explosions must be prevented.

This chapter tries to illucidate essential characteristics of different textile filtration systems, equipped with automatic reverse air cleaning. It deals with their application, as well as their constructional, functional, and economic features. The information is intended for the specialist in filtration, whether he be a manufacturer or an operator. It is also addressed to all those that come into contact with a separation problem for the first time. It should enable the purchaser of an installation to make a critical selection using technical and operational features, rather than simply by comparing initial capital outlay. Very few systematic examinations and treatments on this topic appear to have been published, and this book should fill the gap.

The operational data for a dust collector is made up of a large number of parameters; e.g. the type of cleaning system and filter medium, or the face velocity, dust load and temperature of the flue gas, to mention only a few. Because of the limited literature these parameters are often not given sufficient weight by filter operators.

To ease the entry into this material a summary of the history of development, as well as the most commonly used filtering systems is useful. So as not to lose the overall view, no claim is made for completeness; instead I point to the comprehensive literature index.

3.2 Summary

3.2.1 History, Patents and Definitions

Up to the 1950's most of the filtration systems in place separated dust by means of a woven fabric, on the surface of which a base coat was deposited. The regeneration, i.e. cleaning of the bags, was effected compartment by compartment, by sealing them off individually, or in sequence, from the flue gas stream. Bags were then shaken or vibrated. Simultaneously a stream of cleaned gas was blown through the filter bags in the opposite direction to the filtering. Depending on whether the fan is on the clean or flue gas side,

3.2 Summary

such filters are known as suction- or pressure filters, respectively. This basic concept has been applied worldwide in many countries with different constructions. It is still used today for many applications as we shall see later.

The 50's and 60's saw significant development activity by filter manufacturers. New solutions were researched worldwide.

The use of the cleaning gas for regeneration, rather than mechanical shaking or vibration became increasingly important. Also, mounting the filter medium in flat form, e.g. pockets, gained acceptance. New high efficiency needlefelts were developed which further extended the areas of application. The 1980's saw a decline in development activity, so that today's constructions and cleaning systems can be regarded as stable for the time being. Depending on the shape of the filter medium, one differentiates between

- Bag filters and
- Pocket filters (envelopes).

All subsequent comments apply equally to bag as well as pocket filters, since for both the same cleaning principles apply. The way the separated dust is removed from the filter medium is the most important operation within the dust collector. For this reason subsequent headings will be according to the method used for cleaning.

The systems dealt with here are probably the best known from the point of history of development. Those types which are not mentioned here can be related to one or another of the categories. It appears relevant merely to point out that some systems have found particular significance in certain application areas. The developments are ordered chronologically and describe the evolution of the most fundamental patents on which today's cleaning technology is based. All types mentioned are represented on the market.

A grouping into Low Pressure Cleaning – when centrifugal fans are used, Medium Pressure Cleaning – when rotary compressors are used, and High Pressure Cleaning – when compressors or compressed air networks are used seems to suggest itself. There is an unavoidable overlap between these categories, partially because specific trade names have already become well established.

There is no intention to write a "jet filter" book. It will however be evident that the cleaning with compressed air (usually filtered air) is widely practised and will increase in significance in the future.

The year 1886 appears to be of historic significance for textile filters with automatic media regeneration: Two of the first patents, both subsequently put on an industry scale, stem from this period:

- *Shaker Type with Reverse Gas Cleaning* (sect. 3.2.1.1)

    A pocket filter of Nagel and Kaemp (Hamburg), Italian patent no. 157, Vol. XXXIX of 1886, and the bag filter of Beth (Luebeck, Germany) with DRP 38 396 of the same year count among the first textile filters built commercially in large numbers.

    Both systems take single filter elements out of the gas stream in cyclic sequence and remove the dust from the fabric by shaking, i.e. rapping, with a simultaneous reversed flow of cleaned gas. These form the foundation for a series of shaker dust filters with counter current cleaning. In a modern version they are still used for special applications.

The arrangement of the filter media into pocket form (envelope types in the USA) was pioneered in 1939 by W. Rehfus of Stuttgart with German Patent No. 719188. Many pocket filter designs were developed since, some with shaking mechanisms, some with reverse gas cleaning.

- *Blowring Filters* (sect. 3.2.1.2)

These were developed in 1915 by Ernest J. Sweetland of New Jersey, USA, under US Patent No. 1 324 490 as bag filters in which concentric jets moved slowly up and down the length of the bag. These blowjets remove the filter dust from the medium. Such filters became later known as Blowring Types and are still made today by different manufacturers.

Carter-Day of the USA began to build blowring filters in 1948 based on the patents of J. J. Hersey and Day. Also the German patent of H. Junkmann, Muenster, Germany, No. 936 181 of 1955 deserves mention, because reverse blow systems of recent times have become synonymous with the name Junkmann. Diverse manufacturers offer similar systems.

- *Reverse Flow Filters* (sect. 3.2.1.3)

Reverse flow filters where a centrifugal fan forces clean air in reverse through the filter medium, and where a cyclic cleaning of pockets or bag filters takes place became known in 1951 through P. W. Garbo with the US Patent No. 2 526 651. They are still in use today and further developed by various manufacturers.

- *Reverse Jet Filters* (sect. 3.2.1.4)

The year 1951 can also be considered a significant milestone in the history of dust collector development.

H. Church of Baton Rouge, USA, announced in 1951 a filter system with jet stream and venturi tube which was granted a patent (USP No. 2 804 168) on the 27th August 1957. It was taken on by the Micropul Company of Summit, USA, and further developed by T. Reinauer who brought it to its current state of development with German Patent No. 1 228 130 in 1957. Initially this type was built under licence by several firms, and was introduced against strong opposition from traditionalists. Today this system is widely used and has been refined with many variations.

H. Oetiker of St. Gallen in Switzerland established in 1961 that the conversion process of the pressure energy of the cleaning air into kinetic cleaning work within the bag was extremely inefficient, because of the small valve openings and small duct cross-sections. With his patent application in Switzerland on the 18th November, 1963, which became patent no. 413 565 subsequently, achieved decisive improvements in the efficiency of the reverse jet type bag cleaning.

Envelope and leaf type filters also benefitted by the application of the reverse jet principle during the 1960's. The British Patent No. 914 187 of 1962 of A. J. Well should be mentioned in this context. This fundamental concept has also been adopted by many manufacturers.

3.2 Summary

The basic design ideas of the jet type filters, which were all patented between 1951 and 1971, contained a central jet stream that blew into a venturi at close to sonic velocity or into a venturi type slit. Secondary air is drawn in at the clean gas side of the filter; depending on efficiency of the system the ratio is $1:2$ to $1:3$.

- *Ring Jet Type* (sect. 3.2.1.5)

Subsequent developments of the reverse jet filter occurred in the 1970's with the aim of improving the cleaning efficiency and cleaning intensity. Instead of a central blow tube annular jets were developed which enabled the proportion of primary to secondary air of the jet stream to be raised to $1:4$ to $1:8$. Patents on which these improvements are based were notified as follows:

1970 Swiss Patent No. 527 635 of F. J. Espel and E. Ackermann of Uzwil, also 1973 German Patent No. 2 332 031 of H. Meyer zu Riemsloh in Langenberg.

Central and ring type jets with venture injectors increased significantly the total clean gas quantity by utilising the injector principle. There is an adverse side effect in that the pulse of the jet is much reduced.

- *Direct Pulse Filter* (sect. 3.2.1.6)

The Swiss Patent No. 480 084 of E. Eisenegger, Bronschhofen, applied for in 1968, deliberately eliminated the inclusion of secondary cleaning gas. This increases the effectiveness of the primary pulse and raises the pressure built up inside the filter bag during the cleaning cycle.

The cleaning effect is proportional to the acceleration of the filter bag through the initialising pulse. The direct impulse is activated via valves of very large cross section; the expanding gas gives the filter bags an air shock.

It has been demonstrated that the impact of the initialising pulse together with the magnitude of the reverse flow pressure within the bag is principally responsible for the high cleaning ability of the filter medium.

In 1974 B. Axelsson and S. O. Rosby of Vaexjoe in Sweden applied for a patent which also eliminated venturis to enable a direct pulse ("Optipulse" Filter).

H. Geissbuehler, Niederuzwil, Switzerland applied for a patent in 1975 for a plate valve, and in 1977 W. Borst of Uzwil produced a version of the direct pulse filter which added a reverse flow phase to the first pulse ("Superjet" Direct Pulse).

All these developments show that specialists have given high priority to cleaning systems to increase plant performance, i.e. make them more compact and smaller, with lower total resistance, lower investment and operating costs.

Special facilities for maintenance, supervision and localising of defective filter bags will be expanded on later. The accessibility to the filter bags or pockets from the clean gas side is of great importance in maintenance and has only been recognised on a broader basis since the 1970's.

It is surprising that despite the huge flood in patent applications in the sector of cleaning filters, relatively few fundamentally new ideas have surfaced. In the 100 year development history mechanical (shaking, vibrating and pneumatic), blowing (jet types) dominate totally in regenerating the filter medium. This may be indicative that the cleaning systems on the market today have reached a stage of maturity. Today, attention is directed to refining maintenance and to improving operation and to lower explosion risks.

Despite this state of the art in media regeneration there is still uncertainty in regard to arrangement of filter surfaces in relation to the cleaning gas volume. The total filter resistance dominates the layout of the filter; this resistance is expected to stabilize after a period of months or even years of operation. The total filter resistance changes according to different applications (depending on the overall desired efficiency). To reach a stabilized total filter resistance aspects such as pre-filtering, flue gas handling and distribution are of critical importance besides the efficiency of the cleaning system itself.

Operating conditions and properties of a dust or product can vary, even though they may be chemically identical. Thus, the design and performance specifications for a dust collection system must be given exacting and careful attention.

Differing cleaning systems, different flue gas conditions and other constructional factors will lead to very different configurations.

Even the VDI Standard 3677 of July 1980 table 2 gives air to cloth ratios for shaker type filters which vary up to 30%. These air to cloth ratios are based on pressure differentials of 600–2000 Pa; this requires that the fan be fitted with adjustable dampers, because all mathematical models cannot accommodate the variability of dust deposition within the medium, and cannot therefore accurately predict the filter resistance. Thus, field experience is invaluable.

In reverse jet type or air shock filters of the direct pulse type air to cloth ratios of 2 to 3 times the values given in table 2 of the above VDI Standard for shaker type filters are possible.

Because of competitive pressure manufacturers are reluctant to publish basic data for their designs. Values given in the technical literature are incomplete or are biassed towards a particular design. This is understandable in as much as the supply is much larger than the demand in the filter sector, and there is tough competition. Layout and dimensioning will be treated in more detail in sect. 3.4.

3.2.1.1 Shaker Type Filters

This type of filter was introduced in 1887 by Messrs BETH of Luebeck. The market has seen many variants since; only the basic construction is given here. As fig. 3.1 shows, filters are accommodated in several chambers. The flue gas plenum is positioned to the side so that by its flow diversion and delay the larger particles may separate. The cells may be individually sealed from the flue gas by means of dampers on the clean gas side. The bags are made from woven fabrics and dust deposits on the inside of the bag suspended on a frame.

This frame is moved up and down with a simple shaker motion; at the same time cleaning air is either blown or sucked through a valve in reverse direction to filtering. The deposited dust is loosened through the shaking motion and falls, assisted by the cleaning gas flow to the floor, where it is removed from the filter hopper by a screw conveyor. The cleaning air mixes with the flue gas and escapes through the open filter cells.

This cleaning method does not dislodge dust particles that have penetrated deeply into the filter fabric. Some systems have even eliminated shaking, some do without the cleaning air, which further reduces the effectiveness of bag cleaning. The shaker technique has

3.2 Summary

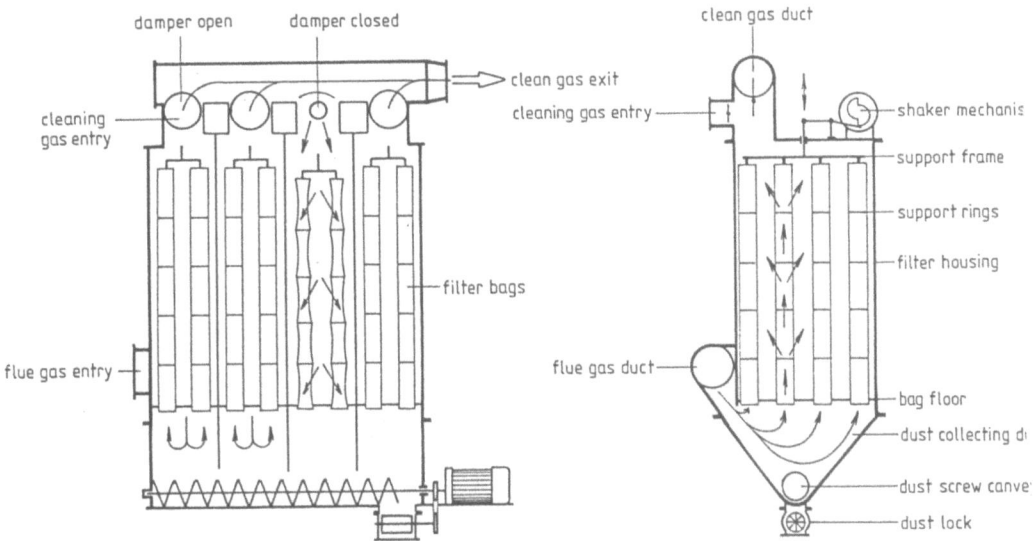

Fig. 3.1 Schematic diagram of a bag shaker filter

seen further refinements. When fibreglass fabrics are used, which require a gentler handling, bags are vibrated with a specially designed mechanism instead of the shaker motion described above. Both horizontal and vertical motions are incorporated. Even a side-lifting and rapping of entire groups of bags is practiced. This avoids bending the bags in the lower region.

Shaker filters are commonly accommodated in rectangular baghouses. They are frequently found in hot gas filtration. When high dust concentrations are involved, they may be offered in circular form with an integrated cyclone separator.

When considering separation efficiency, differential pressure and service life of filter fabric, the face velocities of this type of filter lie below 1.2 m/min. For this reason the ratio of specific volume of the plant to the flue gas volume is relatively large.

The pocket shaker type filter consists of a row of flat rectangular filter elements (pockets) – fig. 3.2. A fan on the clean gas side draws in the flue gas through the pockets. A support cage ensures that air can circulate evenly, and that the pocket surfaces do not touch. The entire filter assembly is moved to and fro with the aid of a cleaning motor which is actuated via an eccentric cam motion and vibrator rod. The induction fan is normally switched off during this cycle, because adequate cleaning can only proceed in the absence of flue gas deposition.

By switching several filters in parallel continuous filtration is possible with increased flue gas capacity. Shaker techniques vary widely, but adopt similar methods to those used for bag filters. For smaller installations the cleaning air pulse is sometimes eliminated.

Their range of use extends from dust extraction from processing machinery, feed shutes, ventilating silos, bunkers or mixers to very large industry installations.

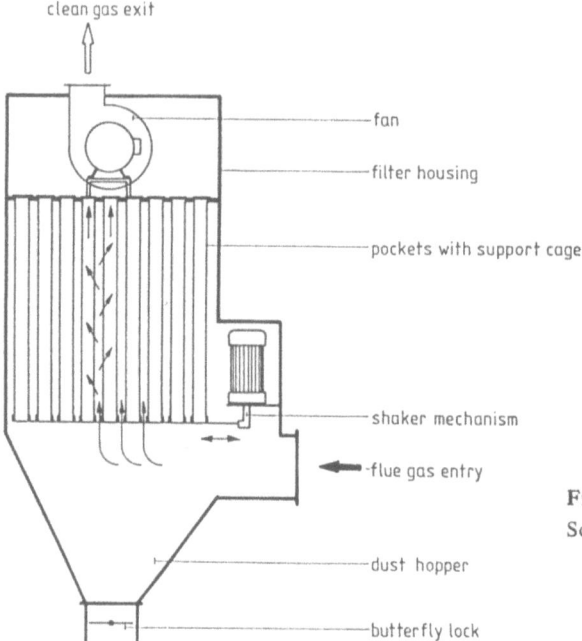

Fig. 3.2
Schematic diagram of pocket shaker filter

3.2.1.2 Back Blowing Type Filters

The method of blowing back with fan operated ring jets is based on a US patent by J. Sweetland. The principle will be exemplified by the DAY "AC" Dust Filter.

An annular jet traverses the entire length of the bag. By blowing from the outside through the bag walls it dislodges the dust which sits on the inside (fig. 3.3). The dust falls by gravity to the base where it is transported by a screw conveyor for removal via a dust trap. The cleaning air for the jets is generated by a separate fan. The jets may be individually guided up and down or joined into a moveable box, as in the standard Junkmann system.

Fabrics are commonly used as filter media. Blowing type filters are relatively rare.

3.2.1.3 Reverse Flow Filters

Bags are reverse cleaned with large quantities of clean gas ducted to a rotatable or moveable distributor (fig. 3.4). To support the cleaning action a pulsator may be superimposed. Compared to cell cleaning single bag cleaning has the advantage that a relatively small filter surface area is withdrawn from service at any one time. When cleaning cell by cell the entire compartment needs to be shut off.

Air fans which may be mounted externally or within the clean gas compartment deliver the cleaning gas. In some instances the cleaning gas is stored in pressure tanks.

3.2 Summary

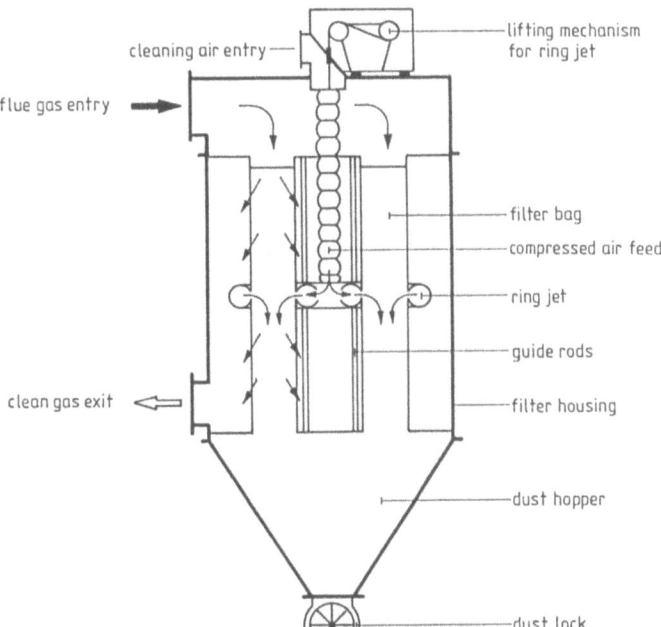

Fig. 3.3
Schematic diagram of blowback filter

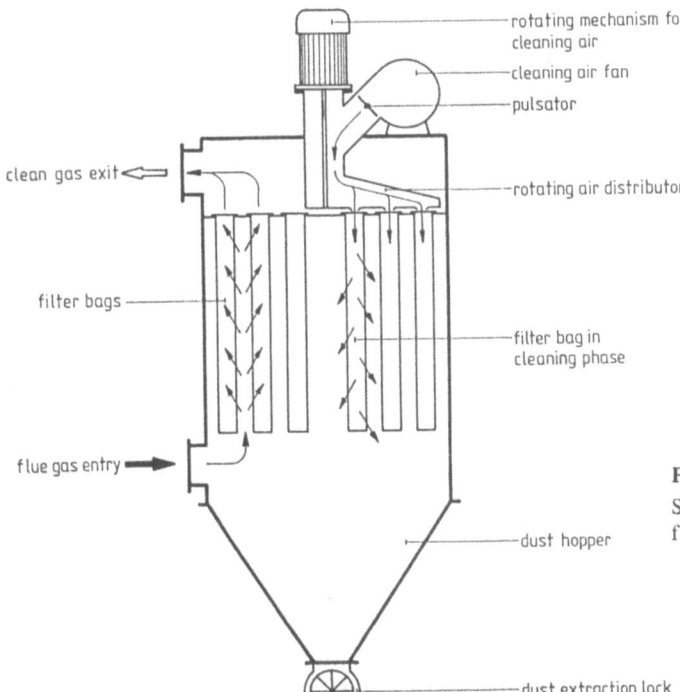

Fig. 3.4
Schematic diagram of a reverse flow bag filter

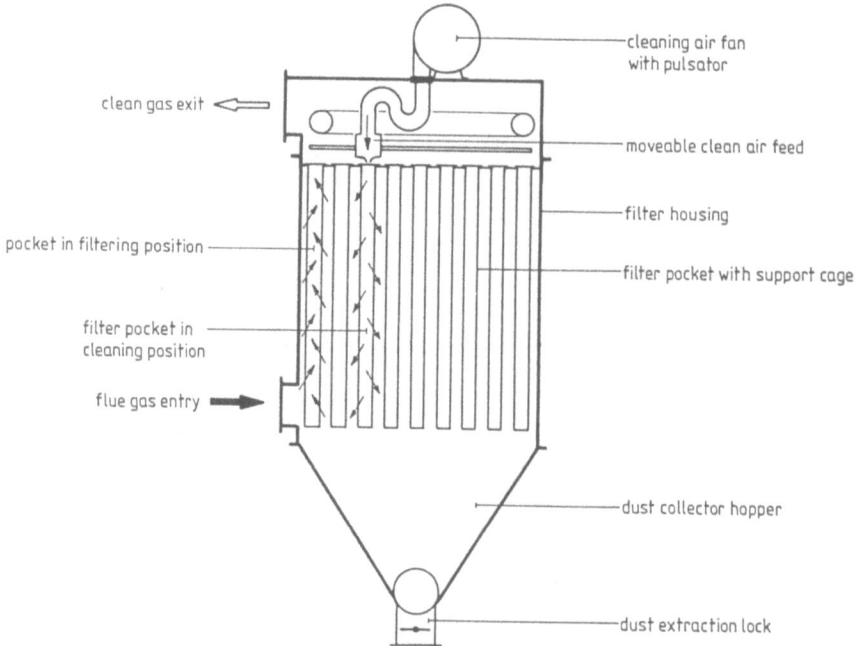

Fig. 3.5 Schematic diagram of a reverse flow pocket filter

Pockets can be used in place of bags (fig. 3.5). They may be suspended from the top or inserted from the side. Depending on whether the fan is on the flue gas or clean gas side, the action is by pressure or suction, respectively. In this case a mobile gas feed moves from one pocket to another.

For smaller installations the cleaning gas is taken from a pressure tank. Pockets are regenerated by fixed cleaning ducts with control valves. Alternatively the whole filter is turned off while all pockets are cleaned at the same time.

3.2.1.4 Reverse Jet Filters

Originally only a few bags, later whole rows, were pulsed with small exhaust jets operating at 6–7 bar gas pressure, and acting from inside to outside (fig. 3.6). By this action the dust, which is deposited on the outside of the bag, is ejected. The pulse lasts approximately 1/10th of a second and is controlled either by electromagnetic valves with electronic control or by pneumatic valves with fluidic control.

With most systems the primary pulse is about 15 dm^3 (standard conditions). The cleaning gas leaves the jet of approx 4 mm diameter at sonic velocity and blows into a venturi which points into the filter bag. Secondary injection air, to the tune of 30 dm^3, is added to this stream and blown counter currently through the bag. The venturi converts part of the velocity energy into pressure energy. The efficiency of this cleaning system is governed by the ratio of jet to venturi; if the venturi is too narrow, it contributes to overall drag

3.2 Summary

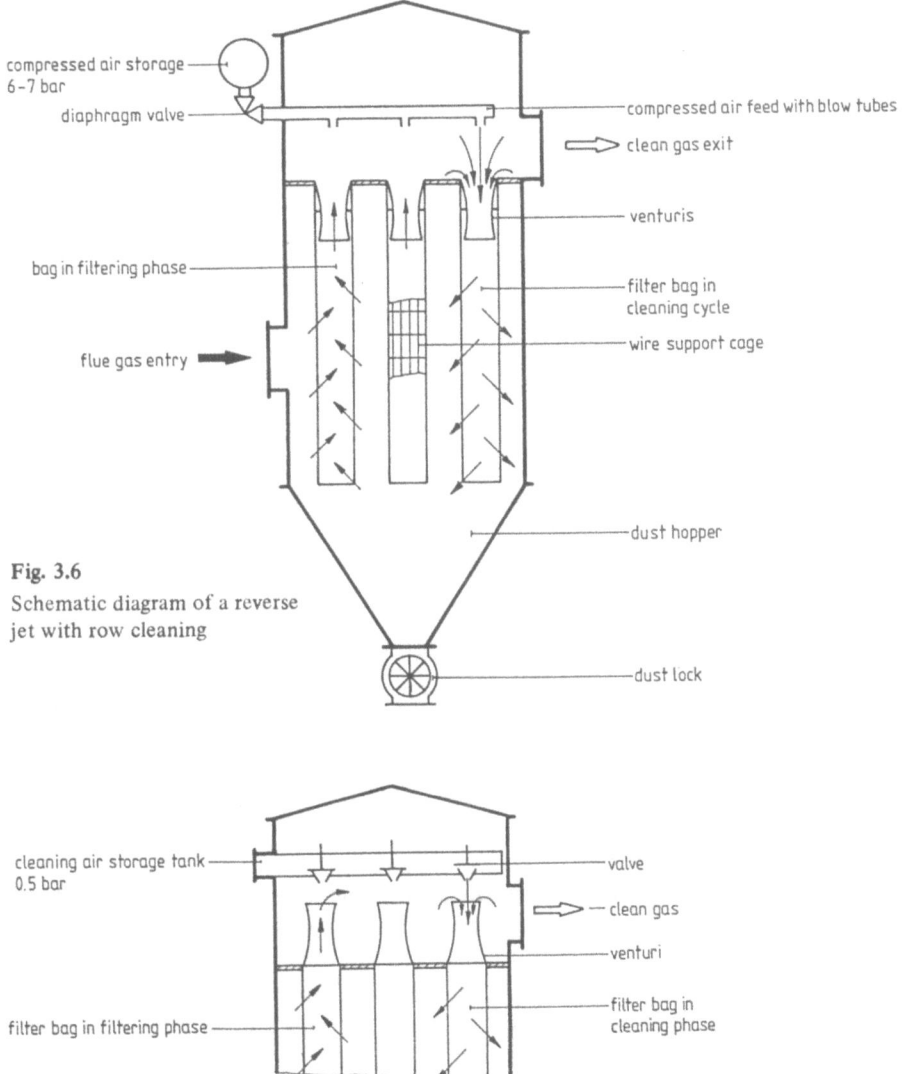

Fig. 3.6 Schematic diagram of a reverse jet with row cleaning

Fig. 3.7 Schematic diagram of reverse jet filter with single bag cleaning

during filtration. Because the venturi reaches into the top of the bag, the upper part is not cleaned effectively or not at all, so that if the dust binds together, bridges can form near the top of the bag.

The food and pharmaceutical industry, or any hygroscopic dust for that matter, demands a dry and oil free cleaning air. Bühler Brothers Ltd. of Uzwil have developed a reverse jet filter with single bag cleaning. The compressed air, which is oil and water vapour free, is delivered by a rotary piston compressor or two stage side channel compressor. The clean air requirements are only 0.5 bar (fig. 3.7). To achieve better cleaning of the bag tops the

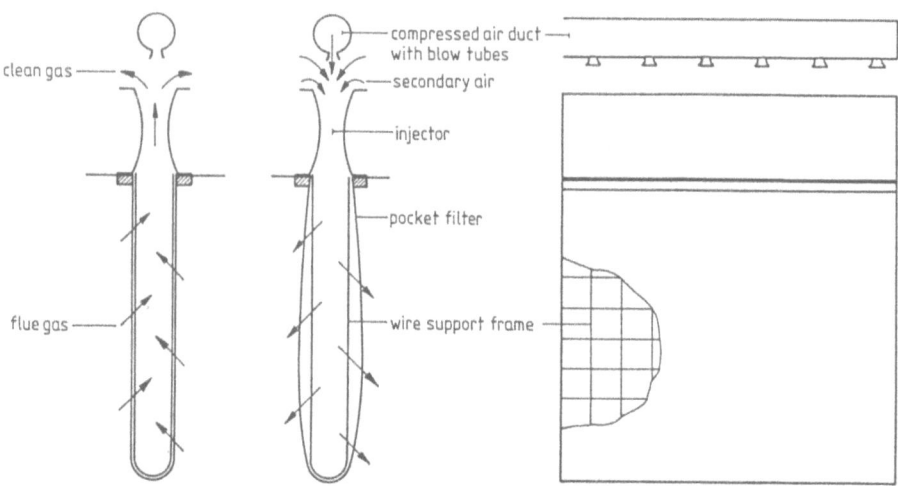

Fig. 3.8 Schematic diagram of a pocket type jet filter – showing a single pocket

venturi is mounted above the attachment collar. This improves the intensity of the cleaning pulse also. In this design for the first time superlarge valves (Oetiker valves) were attached directly to the compressed air storage tank. This is fitted inside the filter housing.

This is a departure from previous cleaning system designs. They operate in the range 6–7 bar cleaning pressure with rectangular diaphragm valves; it achieves an energy conversion with minimal pressure loss.

The reverse jet principle has become worldwide, and is incorporated into the designs of many manufacturers since the patent expired. No other basic design has found such universal acceptance.

Consequently, the system has been adapted to pocket filters (fig. 3.8), primarily by Dallow Lambert – now DCE, which is widely represented in Europe. The system also is incorporated by many plant manufacturers.

Several slit type or circular blow jets, fed from a central trunk, blow cleaning air into the injector, which draws in secondary air. The injector usually has a venturi shape, and the stream spreads over the entire length of the pocket.

3.2.1.5 Ring Jet Filters

All of the above mentioned reverse jet filters (pocket or bag) – are based on a central jet (fig. 3.9). The jet is aimed into venturi or venturi type slits and draws secondary injection air with it. The proportion of primary to secondary air lies in the range 1:2 to 1:3 depending on construction and layout.

The aim of further development has been to improve this ratio, i.e. to increase the secondary air quantity with the same amount of primary air. The quantity of air that is drawn in by the jet stream is determined by its velocity as well as the contact area

3.2 Summary

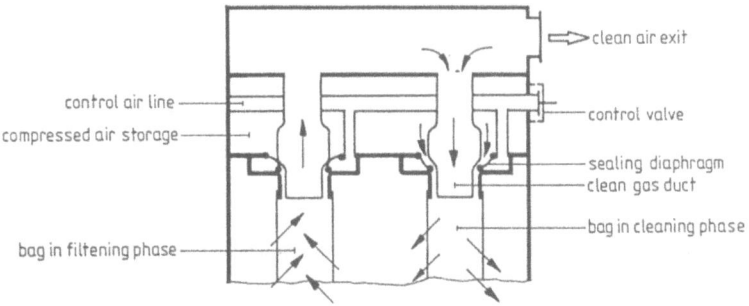

Fig. 3.9 Schematic diagram of a ring jet filter

with the stationary air. There is not much scope to increase this velocity significantly, because all pressure conditions are already at critical levels, and velocity of sound must be taken as approx. constant. Thus, significant improvements could only be achieved by changing the shape of the jet stream. Enlarging the jet diameter is also out of the question, because it requires added primary air. A solution was found in the ring jet which significantly increases secondary air contact for the same primary air quantity. Ratios of 1:4 to 1:8, primary to secondary, are possible with such systems. However, this is achieved at the expense of simplicity; the ring jets are more complicated.

Bühler ring jet filters were the first representatives of these systems. The clean air reservoir at 0.5 bar forms an integral part of the filter head to prevent any losses between storage tank and filter bag. The floor of the tanks serves as anchorage for the bags. Placing the reservoir directly above the bags is now commonly adapted by later designs.

When the pressure increases in the control line, the ring shaped diaphragm valve is pressed against the clean air duct which in turn seals the connection between pressure tank and bag. During the cleaning cycle the control line is vented, the membrane opens to a ring shaped air jet into the bag; secondary air is drawn in from the cleaning air duct.

This annular ring which utilises the Coanda effect (fig. 3.10) was pioneered and distributed by Intensive-Filter in Langenberg. The pressure pulse of 5–6 bar is injected tangentially

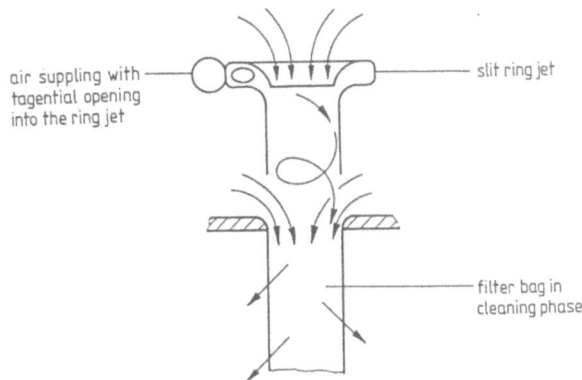

Fig. 3.10
Slit ring jet with coanda effect and double action injector

into an annular slit. This causes a spiral turbulance which increases the quantity of secondary air.

The ring jet principle with twin bags, as built by Airmaster Engineering of Leed, falls into a similar category. The ring shaped configuration conforms to the special double bags, whereby the jet itself is composed of several circularly positioned jets in concentric form.

All such systems attempt to raise the effective air volume by drawing in the largest quantity possible of counter current cleaning air to dislodge the dust with explosive force. The resultant pulse velocity of the jet is dampened by the stationary air which must be accelerated. Thus the cleaning cycle is characterised by a large clean air volume and a rather shallow pressure gradient. A basic approach is to divide the cleaning of bags into 3 phases — each of roughly equal importance.

1. Rapid inflation of the bag by the cleaning air $\frac{dp}{dt}$.
2. Peak of the pressure build-up p_{max}.
3. Cleaning air penetration of the bag $\int_{t_1}^{t_2} \Delta p \, dt$.

To what extent each of the three factors contributes to optimal cleaning is the subject of current investigations, which will be referred to in subsequent sections. Recent findings emphasize the rate of rise of pressure and the peak pressure rather than a large air throughput. This points towards the eventual elimination of the venturi.

3.2.1.6 Direct Pulse Filters

The "Airshock" Filter is the first system to do away with secondary air. By improving all air passage dimensions a method was found which permitted the jets to be attached directly to the bags. This increases the pressure gradient velocity and the total maximum cleaning pressure within the bag.

In this, the first direct pulse filter that appeared on the market, an air distribution system forms an intimate connection between filter bag and pressure tank (fig. 3.11). This

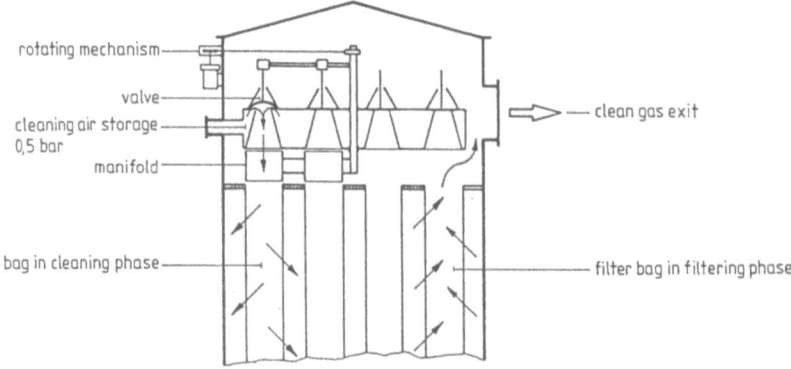

Fig. 3.11 Schematic diagram of the first direct pulse filter with single bag cleaning.

prevents secondary air access, and results in a pure pulse. The distributor manifold, together with the control valve arm, is guided to the bags requiring cleaning. This increases the kinetic effect within the bag by increasing the rate of pressure rise (α) and maximum pressure p_{max}. The effectiveness of such a cleaning system in no way lags behind the venturi injector. This gives support to the contention that the first 1/100th of a second of the pulse impact is decisive, while a large cleaning air volume is secondary. In more recent designs the distributor manifold has been eliminated. The design focuses on the correlation between valve opening size and blow pipe diameter; the pressure in the tank and the distance between the blow tube and bag. Fig. 3.12 shows a modern version of a pulse filter with single bag cleaning.

This system requires only 0.5 bar operational pressure to reach very respectable pressure peaks. Using a PE/PE 501 needlefelt with an air permeability of 210 l/min at 2 mbar filter resistance a gradient of 1.2–6.5 bar/s was recorded with a peak of 38 mbar. 90 % of the cleaning pulse effect is achieved within the first few 1/100th of a second; only a minimal quantity of cleaning air is required for this pneumatic pulse/impact conversion. This pressure build-up plays a significant role in stripping clogged or crusted filter bags; under such conditions pressure gradients of up to 28 bar/s with peak of 270 mbar have been recorded. It should be emphasized that such an effective cleaning presupposes single bag cleaning and is thus used only in super compact filter assemblies of 100 bags (approx. 100 m² surface area). Using face velocities of 4 m³/m² min (for cement) and up to 8 m³/m² min (grain handling) filtration problems with volumes up to 50 000 m³/hr can be handled with minimal space requirements. To gain optimal effectiveness of this cleaning system it is necessary to integrate the blow tube valve into the air tank. To this end special valves were developed with extremely large cross sections that could function at 3–7 bar at elevated temperatures. Large cross section rightangled diaphragm valves are marketed by Goyen Valves in Australia or Mecair of Italy (refer inserts). For higher operating temperatures plate valves made by Bühler Brothers Ltd. (Swiss Patent No. 592 466 of 1975) are more suitable. It has become possible to develop systems which clean whole rows of bags up to 12 in number and 3 metres long simultaneously by optimising cross sections of the blow tube and air distributor.

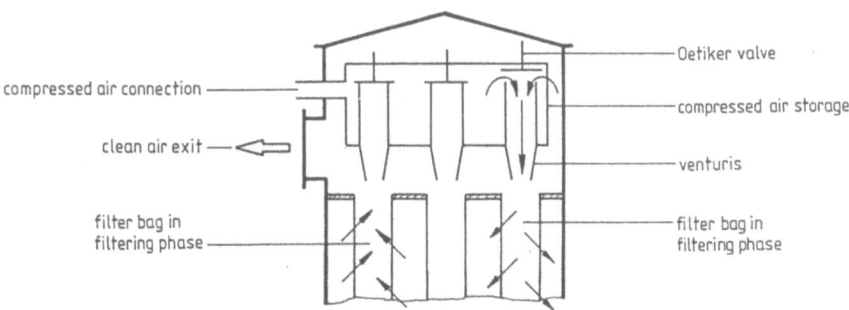

Fig. 3.12 Schematic diagram of a direct pulse filter with single bag cleaning

To assemble a filter system today, nearly all components are readily available "off the shelf", so that filter manufacture is becoming almost an assembly job. However, the responsibility and warranty for an entire installation can only be borne by a manufacturer that is thoroughly familiar with the problems of the user, and that has the experience in the field.

Despite its brevity it is hoped the reader will have appreciated by now the many possibilities at his disposal to clean a textile filter surface. In order to give a balanced view a summary based on information published by various filter manufacturers is given.

It should be emphasized that this tabulation makes no qualitative judgements, it merely puts the systems into a historic framework.

Each and every one of these systems will do justice to its intended application, as long as it is used within the limits of its design specification. Decisive for the user is total cost, which can only be assessed by paying due regard to the specified operating conditions. These designs move along a trend line, incorporating previous experience.

Ease of maintenance of filter surfaces and of moving parts will determine who will succeed in the future. It will be particularly important to localize and replace faulty bags. Direct pulse systems offer special advantages in this regard in view of their basic design concept. This applies also in regard to their explosion safety.

Fig. 3.13a) and b) illustrate two valve designs with large cross sections that are found in modern filters.

3.2.2 Trends

3.2.2.1 Significant Developments since 1970

To make such an analysis relevant it is neccessary to focus on individual developments which, when looked at together, describe a discernible trend. For the sake of clarity details are avoided.

The Coanda ring slit jets of Intensiv Filter with injector cleaning are among the most pertinent developments of the last 10 years, and have led to fundamentally new and successful installations and sales. Equally significant is the direct pulse technique for

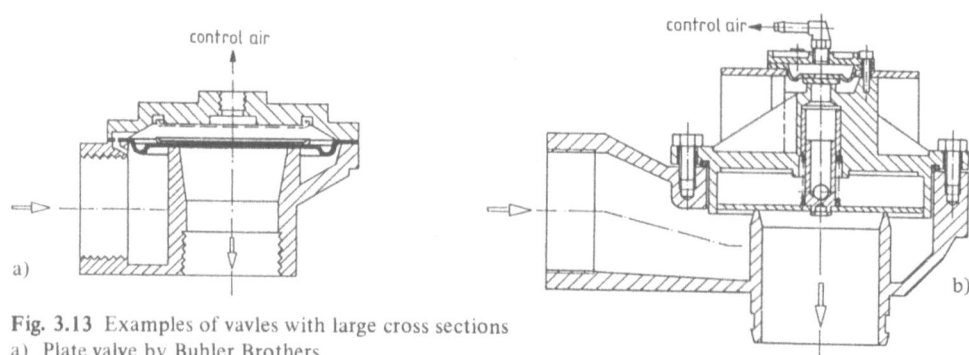

Fig. 3.13 Examples of vavles with large cross sections
a) Plate valve by Buhler Brothers
b) Right angled diaphragm valve by Macair

single bag cleaning developed by Bühler Brothers, as well as the row cleaning system developed by Svenska Flakt.

These techniques pioneered fundamentally new ways, leading to success and setting of new standards.

3.2.2.2 Explosion Protection

The last 10 years have seen a greater sensitivity on the part of filtration manufacturers towards explosion safety of their plants. In this context special mention is made of a publication by Bartknecht of Ciba Geigy AG.

Actual explosion tests with bag and pocket filters paved the way for a design standard laid down under VDI 3673.

This has provided a base for numerous installations to protect man and material with special measures such as explosion proof pressure release and suppression, or shock- and pressure-proof housings.

The potentials for explosions are more easily recognized and preventative action is taken nowadays. The increasing demand for explosion protection is forcing the manufacturers to seek tailormade solutions for a problem situation. One way is to confine explosion or fire damage to the bags, which can be more easily exchanged, and to prevent the damage extending to ancillary parts (melting of aluminium blow tubes and/or alu venturis, spark generation in alu/steel junctions) refer also sect. 3.6. Increased standardisation will be required for pressure release systems and shock resistant filter housings; to this should be added an increased involvement of the engineering skills of both contractor and end user to develop solutions that are both technically sound and affordable.

3.2.2.3 Higher Air to Cloth Ratios

Compared to older types there is no doubt that the improved cleaning systems have enabled higher velocities without sacrificing pressure differential. This fact is not widely recognized, and it is hoped that this book may help to inform − refer sect. 3.3.1.

Design tolerances are narrowing, and greater expertise is being excersised in recognising how these various factors interact. This will result in even more compact filters. This should also give greater cost reductions, better insulation and lower maintenance.

3.2.2.4 Ease of Maintenance

To change filter bags in older type plants the fitter must literally crawl inside the baghouse from the flue gas side, a dangerous and very uncomfortable adventure. Through union pressure in part, new installations are now required, especially when toxic substances are handled, that access to the bags or pockets be solely from the clean gas side. This requirement has gained acceptance in the last 10 years, so that future installations will most likely be solely serviced by this means.

This necessitates appropriate designs of buildings and plant facilities. Increased cooperation and exchange between technical staff, designer and plant operator will further facilitate maintenance.

3.2.2.5 Market Trends

Divo [1] and McIlvaine [2] predict an annual growth rate for filter installations of around 10 % for 1980 to 1985. Against this growth the market for wet filtration and electrostatic precipitators is considered stagnant.

Improved filter media by using more heat resistant fibres, as well as better reverse air cleaning capabilities have made it possible to attack gas cleaning problems with textile filters hitherto the province of wet filters or electrostatic precipitators.

Residual emissions of textile filters are approx 1/10th that of other methods. This has enabled legislation to be tightened, this in turn will give added impetus to the filter market.

The removal of dust in the dry state with the potential of recycling gives a further advantage compared to washers, which bring their own problems of water treatment.

These are cogent reasons for the future importance of gas cleaning with textile filters.

3.2.3 List of Manufacturers of Different Filtertypes

3.2.3.1 Shaker Type with Reverse Gas Cleaning Bags

Air Purification Methods	Air Purification Methods Hopkins, Minnesota 55343, USA, 201 South Third Street
Amer tube	AAF American Air Filter Co. Inc. Louisville/Ky, USA
CEAG	CEAG Verfahrenstechnik GmbH, Postfach 25, 4714 Selm-Bork
LFT	Karl Heinz Frank Luft- und Fördertechnik, Wittkuller-Straße 108, Postfach 190149, 6560 Solingen-Wald
HP	H. Paulus Co., Inh. H. Fenne VDI u. Söhne, In der Graslake 52, 5830 Schwelm i.W.
Intensiv	Intensiv-Filter GmbH Co. KG Langenberg, 5620 Velbert 11–Langenberg
Kastrup	Kastrup Filter, Heinrichstr. 24, 4000 Düsseldorf
Lodge-Cottrell	Lodge-Cottrell Ltd. George Street Parade, Birmingham 3
Mikro Pul	Mikro Pul Gesellschaft für Mahl- u. Staubtechnik mbH, Welserstraße 9–11, 5000 Köln 90
Münster	Robert Münster, Fabrik für Luft- und wärmetechnische Apparate und Anlagen, Hofackerstraße 55, 4132 Muttenz
Schirp	H. Schirp, Bissingstr. 5, 5600 Wuppertal-Vohwinkel
Smico	Smico 500 N MacArthur Blvd., Oklahoma City, OKLA
Sprout-Waldron	Manufacturing Engineers, Muncy, Pennsylvania, USA
Standard	Standard Filterbau, Rösnerstraße 6/8, 4400 Münster
STAEFA	Stäfa Ventilator AG, 8712 Stäfa (ohne Rückspülung)
Venti	Ventilatorenfabrik Oelde GmbH, 4740 Oelde i.W. (ohne Rückspülung)
Wheelabrator	Wheelabrator Tilghman, Entstaubungsgesellschaft m.b.H., Hansaring 49–51, 5000 Köln (ohne Rückspülung)
Handte	Jakob Handte Co. Maschinenfabrik, 7200 Tuttlingen/Württ.
BSH	Büttner-Schilde-Haas AG, Postfach 4 und 6, 4150 Krefeld 11
Flex-Kleen	Research-Cottrel, An der Schanz 2, Colonia-Hochhaus, 5000 Köln 60
EWK	Eisenwerke Kaiserslautern, ASt Zschocke, Barbarossastr. 30, 6750 Kaiserslautern
H. Rühl	H. Rühl KG, Ventilatoren- und Apparatebau, 4900 Herford
KABE	KABE-Werk, Lufttechnik und Entstaubungstechnik GmbH, 6370 Oberursel 5 D

Pockets

Aerob	Gesellschaft für Entstaubungstechnik mbH, Industriestraße 38, 7253 Renningen
Aeronca	Aeronca International Ltd., Penarth Road, Dardiff C F 1 7 UG England
Airfilter	Airfilter Company GmbH, 7101 Oedheim/Württ.
Air Industrie	Air Industrie France, 19/21 Av. 92401 Courbevoie 19/21, A Dubonnet
Armaster	Armaster Engineering Ltd. England, Roundhay Road, Leeds L48 4 BH
Air Pollution	Air Pollution Engineering Ltd. England, 38 Millstrone Lane, Leicester LEI 5JN
Alpha-Mat	Alphastaub Gesellschaft für Staubtechnik GmbH, 6380 Bad Homburg
Arrestal	AAF American Air Filter Co. Inc., Louisville, Ky USA
AZO	Maschinenfabrik Adolf Zimmermann, 6960 Osterburkern
Turbofilter	Turbofilter GmbH, Postfach 1733, 4300 Essen 1
CEAG	CEAG Verfahrenstechnik GmbH, Postfach 25, 4714 Selm-Bork
Humboldt Wedag	KHD Industrieanlagen AG Humboldt Wedag, Postfach 910404, 5000 Köln 91
Jongerius	Jongerius Maschinen AG, Duensstraße, CH-3186 Düdingen
Kiekens BV	Kiekens BV, Kadoelenweg 360, Postbus 3140, 1003 AC Amsterdam
Kost	Edgar Kost Stahlbau, Bavenstedter Straße 67, 3200 Hildesheim
Lühr	Heinrich Lühr Staubtechnik, Enzer Straße 26, 3060 Stadthagen
DCE	Dust Control Equipment Limited, Huberstone Lane, Thurmaston Leicester Le 48 HP, England
Air Industrie tam	Prat Daniel, Av. Dubonnet, F-92401 Courbevoie
Riedel	Riedel Co., Halle i/Westf. 4802 Künsebeck über Bielefeld
Torit	Sogair AG Groupe CBC, Curchod, Baeriswyl Co. SA, CH-1180 Rolle
Turbofilter	Turbofilter GmbH, Rüttenscheider Straße 56, 4300 Essen 1
Wedag	Wedag, Herner Staße 299, 4630 Bochum
Standard	Standard Filterbau, Rösnerstraße 6/8, 4400 Münster
-KABE	KABE-Werk, Lufttechnik und Entstaubungstechnik GmbH & Co., 6370 Oberursel 5

3.2.3.2 Blowring Filters

Amer Jet	AAF American Air Filter Co. Inc., Louisville, Ky, USA
Standard	Standard Filterbau, Rösnerstraße 6/8, 4400 Münster
CEA	CEA Carter Day, Minneapolis, Minnesota 55432, USA

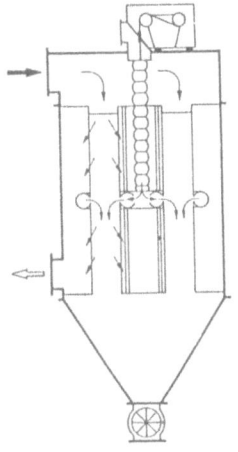

3.2.3.3 Reverse Flow Filters

Bags

Avi Purification Methods	APM Avi purification methods, Inc. USA, 201 South Third Street Hopkins Minnesota 55343
BHS	BHS Entstaubungstechnik, BHS-Werk Peiting, 8123 Peißenberg
Büttner	Büttner-Werke AG, 4150 Krefeld-Uerdingen
CEA	CEA Carter-Day Minneapolis, Minnesota 55432, USA
Amer Term	AAF American Air Filter Co. Inc., Louisville Ky, USA
Dustex	Dustex Corp., P.O. Box 2520, Buffalo 25, N.Y., USA
Lodge-Cottrell	Lodge-Cottrel Limited, George Street Parade, Birmingham 3
Lühr	Heinrich Lühr, Staubtechnik, Enzer Straße 26, 3060 Stadthagen
MAC	MAC Equipment Inc., P.O. Box 205, Sabetha, Kansas 66534
Moldow	Paul Rippert, Am Hanewinkel, 4836 Herzebrock 2
Robinson	Thomas Robinson & Son Limited, Rochdale GB
Semco	Semco, Systems Engineering and Manufacturing Co. Inc., 6330 Washington Ave., Houston, Texas 77007, USA
tam	Air Industrie Prat Daniel Industrieentstaubung, 19–21, Av. Dubonnet, F-92401 Courbevoie
Baumco	Baumco, Gesellschaft für Anlagentechnik, Müller-Breslau-Straße 30 E, 4300 Essen 1
Schirp	Schirp, H., Bissingstraße 5, 5600 Wuppertal-Vohwinkel

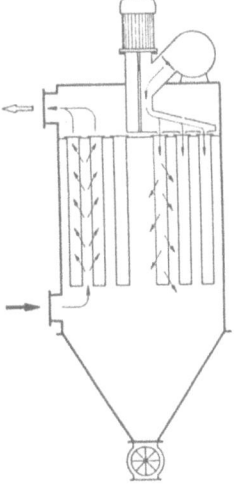

Pockets

Baumgarten	Adolf Baumgarten, Mühlen- und Speicherbau GmbH, 4952 Porta Westfalica
BHS	BHS Entstaubungstechnik, BHS-Werk Peiting, 8123 Peißenberg
CEAG	CEAG-Verfahrenstechnik GmbH, Postfach 25, 4714 Selm-Bork
DCE	Dust Control Equipment Limited, Humberstone Lane, Thurmaston Leicester LE48HP
Lugar	Lugar Gesellschaft für Entstaubungstechnik mbH, Bahnhofstraße 148, 4690 Herne D
Lühr	Heinrich Lühr, Staubtechnik, Enzer Straße 26, 3060 Stadthagen
Standard	Standard Filterbau, Rösnerstraße 6/8, 4400 Münster
Andrew	Andrew Air Conditioning Ltd. Woodbank Works, Stockport
Wiedemann	W.C. Wiedemann Son Inc. 1824 Harrison Street Kansas City, Missouri 64108, USA
Handte	Jakob Handte Co. Maschinenfabrik, 7200 Tuttlingen/Württ.
Krupp	Krupp Industrie- und Stahlbau, Franz-Schubert-Str. 1–3, 4100 Duisburg-Rheinhausen

3.2 Summary

3.2.3.4 Reverse Jet Filters with Venturis
Bags

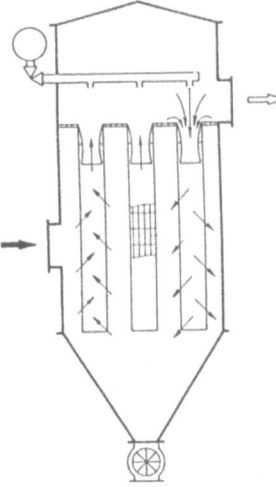

Air Purification Methods	APM Air Purification Methods Inc., USA 201 South Third Street, Hopkins Minnesota 55343
Bühler Düsenfilter	Gebrüder Bühler AG, 9240 Uzwil, Schweiz
Chem-jet	Dust Suppression Limited, Bourne end Mills Hemel, Hempstead, Hertfordshire
Jesma	Hans Jessens Maskinbyggeri A.S., Vejle, DK 7100 Vejle
Vetyfon	Jerslev Maskinfabrik A–S, Stationsvej, DK 4490 Jerslev Sj
Keller „Jet-Air"	Otto Keller AG, Lufttechnik, 9230 Arbon
Kiekens BV	Kiekens BV, Kadoelenweg 360, Postbus 3140, 1003 AC Amsterdam
Loro Parsini	Loro Parsini, Via Savona 129, 20144 Milano
Mikro Pol	Mikro Pol, Gesellschaft für Mahl- und Staubtechnik mbH, Köln, Welserstraße 9–11, 5000 Köln 90
MF	Mohr Fedenhoff AG, 6730 Neustadt/Weinstraße
Pergande	Pergande, Gesellschaft für industrielle Entstaubungstechnik mbH, Muchenstraße 58, 5000 Köln 91
Flex-Kleen	Research-Cottrel, An der Schanz 2, Colonia-Hochhaus, 5000 Köln 60
UOP	UOP Kavag Abt. der Universal Oil Products GmbH, 6467 Hasselroth 3
Standard Havens	Standard Havens Inc., 8800 East, 63rd Street, Kansas City 64133, USA
Beth	Beth GmbH, Postfach 1808, 2400 Lübeck 1
BHS	BHS Entstaubungstechnik BHS-Werk Peiting, 8123 Peißenberg
CEAG	CEAG Verfahrenstechnik GmbH, Postfach 25, 4714 Selm-Bork
Delta	Delta Neu Nederland, Bankrashof 3, 1183 NP Amstelven
BSH	Büttner Schilde-Haas AG, Postfach 4 u. 6, 4150 Krefeld 11
AAF	American Air Filter Direction Commerciale et Bureau de Paris, 37 Av. Pieue ler de Sensie, Paris

Pockets

Aerob	Gesellschaft für Entstaubungstechnik mbH, Industriestraße 38, 7253 Rennigen
Aeronca	Aeronca International Ltd. England, 7UG Penarth Road, Cardiff CF1
Airfilter	Airfilter Company GmbH, 7101 Oedheim/Württ.
Air Industrie	Air Industrie France, 19/21 Av. Dubonnet, F-92401 Courbevoie
Alpha-Jet	Alphastaub Gesellschaft für Staubtechnik GmbH, 6380 Bad Homburg
AZO	Maschinenfabrik Adolf Zimmermann, 6960 Osterburken
BAHCO	BAHCO Ventilation GmbH, Graf-Adolf-Straße 45, 4000 Düsseldorf 1
DCE	Dust Control Equipment Limited, Humberstone Lane, Thrumaston-Leicester EL48 HP
Infastaub	Infastaub, Gesellschaft für Staubtechnik GmbH + Co. KG, Urseler Straße, Niederstedter Weg, 6380 Bad Homburg v.d.H. 1
Jongenius	Jongenius Maschinen AG, Duensstraße, CH-3186 Düdingen
Kost	Edgar Kost Stahlbau, Bavenstedter Straße 67, 3200 Hildesheim
Rüskamp	Lufttechnik Bayreuth Rüskamp GmbH, Pottensteiner Straße 6, 8580 Bayreuth
Meyer	Meyer Mühlenbau AG, Solothurn 2, CH-4500 Soluthurn
Mikro Pol	Mikro Pol Gesellschaft für Mahl- und Staubtechnik mbH Köln, Welserstraße 9–11, 5000 Köln 90
Riedel	Riedel Co, Halle i/Westfl., 4801 Künsebeck über Bielefeld
Robinson	Thomas Robinson & Son Limited, Rochdale GB
Schirp	H. Schirp, Bissinger Straße 5, 5600 Wuppertal-Vohwinkel
Standard	Standard Filterbau, Rösnerstraße 6/8, 4400 Münster
Fläkt	AB Svenska Fläktfabriken, S-35187 Växjö, Schweden

3.2.3.5 Ring Jet Filters

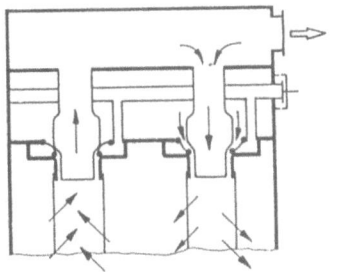

Airmaster	Airmaster Engineering Ltd., England Leeds LS8, 4BH Roundhay Road
Bühler Ring-Jet	Gebrüder Bühler AG, CH-9240 Uzwil, Schweiz
Intensiv	Intensiv-Filter GmbH Co. KG, Langenberg, 5620 Velbert 11–Langenberg

3.2.3.6 Direct Pulse Filters (without Venturis)

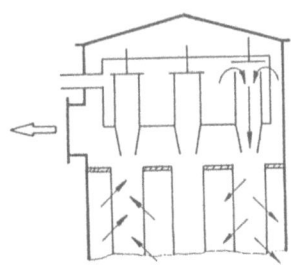

Bühler-Düsenfilter	Gebrüder Bühler AG, CH-9240 Uzwil
CEA	CEA Carter Day, Minneapolis, Minnesota 55432
KKE	KICE Metal Products, 2040 South Mead Ave, Wichita, Kansas 67211, USA
Flex-Kleen	Flex-Kleen Corp., 407 South Dearborn, St. Chicago, Illinois 60605, USA
If	Industrie Filter A/S, Mølledamsvej 10, DK 3460 Birkerød
Lühr	Heinrich Lühr, Staubtechnik, Enzer Straße 26, 3060 Stadthagen
Fläkt	AB Svenska Fläktfabriken, S-35187 Växjö
Young	Young machinery company Inc., Muncy, Pennsylvania 17756, USA

3.3 Filter Configuration and its Effect on Operating Conditions

3.3.1 Method of Measurement and Comparisons

The technical literature contains very little hard core data to compare different filter installations and their operating characteristics. Few opportunities exist in industry to run two plants in parallel under identical conditions, or consecutively. Even in cases where an old plant is replaced by an updated version it is rare to find reliable and extensive records for the old plant to compare with the new one.

Air to cloth ratio and filter resistance directly determine dimensioning of the plant; such design characteristics are often carefully protected by filter manufacturers. Useful and technically reliable comparisons, therefore, remain the domain of larger research institutions with large budgets. This makes an unbiased comparison performed by universities or technical colleges especially valuable. If such work can reach beyond pure academic interest to encompass closer contact with industry − users and manufacturers −, it becomes even more valuable.

Below, examples are given of the characteristics of filter systems that are available on the market today. Systems are compared, therefore, and not manufacturers' models. The magnitude and values of individual diagrams relate to typical configurations and illustrate basics, as well as typical trends.

3.3.1.1 Permanent Test Installations

A comparison of different filter systems under research conditions, e.g. a shaker filter with a jet filter, must be conducted along the following lines − fig. 3.14:

a) Differential pressure as a function of dust concentration and air to cloth ratio,
b) Differential pressure as a function of time and air to cloth ratio,
c) Differential pressure as a function of filtering period and air to cloth ratio.

To take account of variable gas flow, the flow resistance of the filter housing itself must also be considered, as well as the resistance of the filter itself. Temperature and cleaning cycle must be kept constant in comparison a) and b); flue gas loading and temperature must be kept fixed for comparison c).

It goes without saying that one and the same dust should be used for all experiments. Similarly the positioning of the flue gas inlet of the filters to be compared should be as identical as possible. It is very important whether flue gas entry is above or below the bag, or whether it enters tangentially into a circular housing, or flows across in a rectangular box.

Systematic attention must also be given to all comparison tests for filter systems, e.g. shaking, reverse flow, counter current, jet pulse and pocket filters.

Comment on curves ① ② ③ and ④. In all three diagrams curves ③ and ④ represent intermittent operation while curves ① and ② were recorded under stabilised conditions.

3.3.1.2 Pressure Measurement in a Single Bag with Piezoelectric Sensor

To enable the various systems to be compared a technique has been developed using a piezoelectric pressure sensor. This allows a comparison of the various types (reverse jet,

3.3 Filter Configuration and its Effect on Operating Conditions

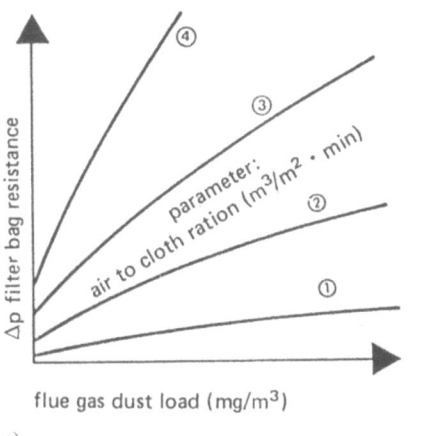

a)

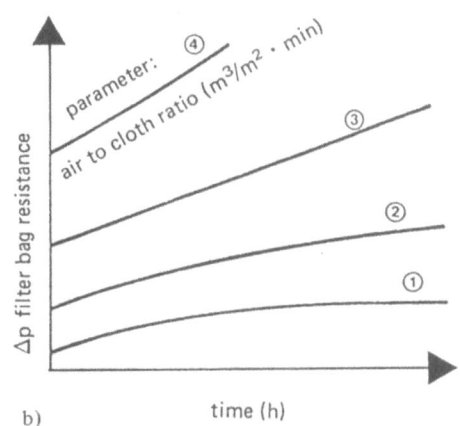

b)

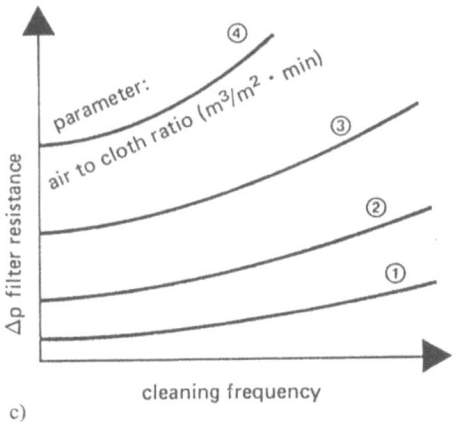

c)

Fig. 3.14
Comparison tests

ring jet, direct pulse). The sensor records the rate of rise of pressure and the peak pressure build-up. The sensor can be moved to different parts along the bag.

It is useful to be able to record the dynamic process to compare the effectiveness of different cleaning modes. Measuring the dust cake distribution along the length of the bag gives additional useful information. Even a pressure measurement in itself allows conclusions to be drawn about the processes occurring within the bag itself. These form the basis for optimising designs for valve cross sections, flow systems, jets, and bag lengths.

The measuring technique is particularly appropriate for optimising reverse jet and direct pulse filters, because it allows a rapid assessment of a large number of variables. Fig. 3.16 shows how the sensors are placed along the length of the bag at intervals of 350 to

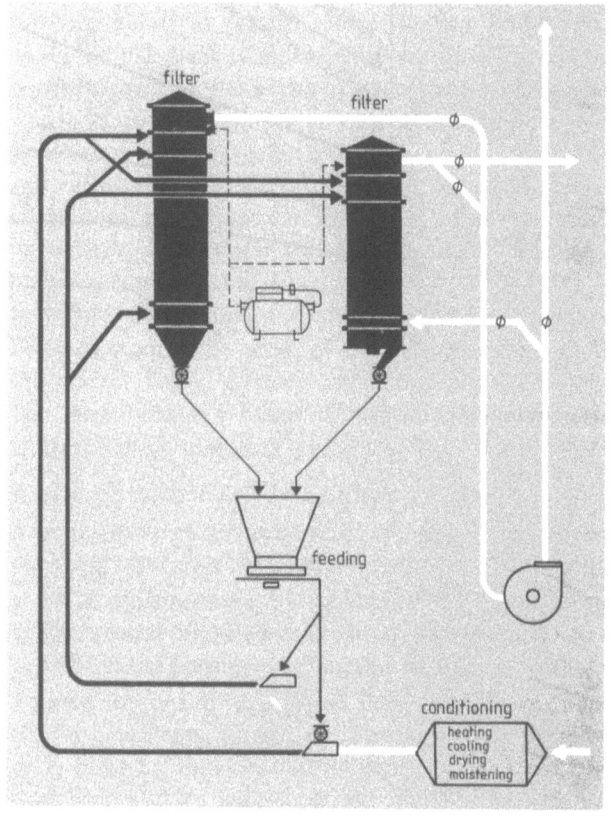

Fig. 3.15
Test arrangement for filter comparisons. The time and material involvement is quite significant. However, the work is useful for customer tests, and to develop new applications.

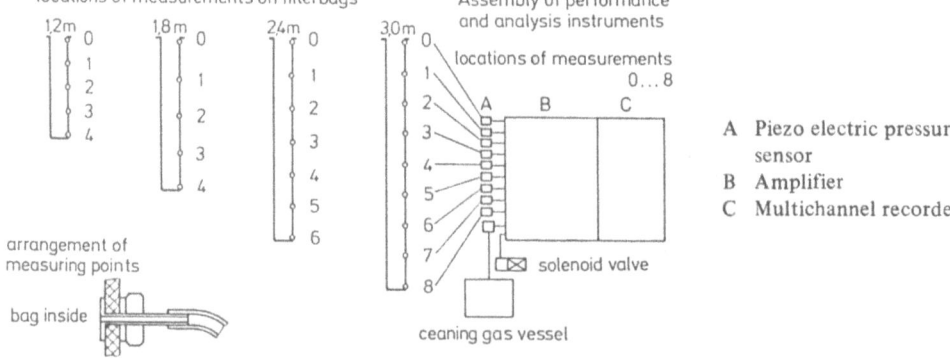

A Piezo electric pressure sensor
B Amplifier
C Multichannel recorder

Fig. 3.16 Pressure measurement on filter bags: principle

3.3 Filter Configuration and its Effect on Operating Conditions

400 mm. These are connected to a pressure recorder. A signal amplifier and a multichannel recorder make the whole sequence visible. Among the variables recorded at the points 0–8 are response time and opening characteristics of valves, the rate of pressure rise α, and peak pressure p_{max}. Finally, $\int \Delta p \cdot dt$ can be calculated, which gives a measure of the total pressure energy transferred to the bag (fig. 3.17).

From fig. 3.18 the values of Δp_{max} and α are derived for all measuring points. Δp_{max} is transferred to fig. 3.19 which then allows conclusions to be drawn about the effectiveness

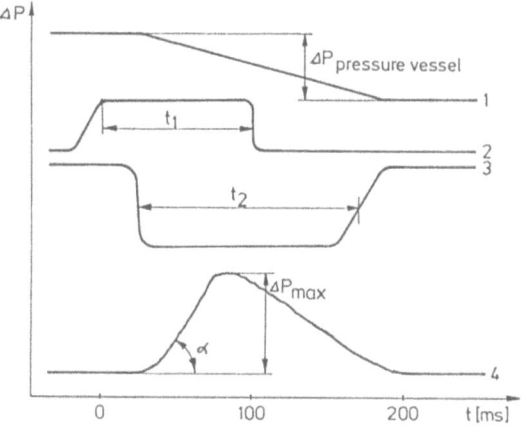

Fig. 3.17
Sample recording
1 Pressure loss in compressed air reservoir
2 Signal pulse for solenoid valve t_1
3 Valve opening period t_2
4 Pressure curve $\Delta p = f(t)$ pressure gradient α

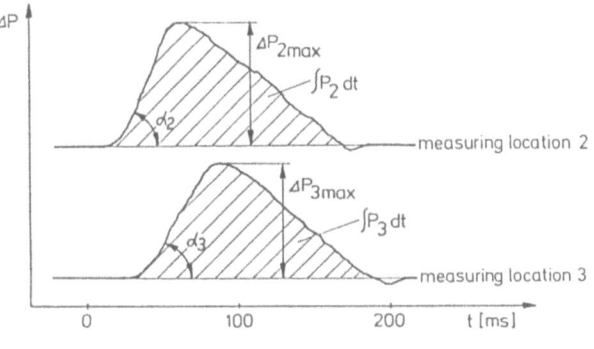

Fig. 3.18
Pressure curve at various points along a test bag

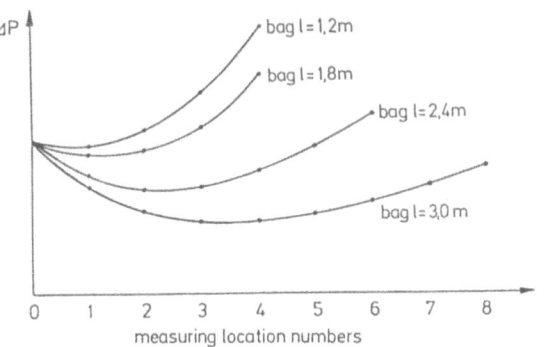

Fig. 3.19
Maximum pulse pressure at various points along test bag

of the jet pulse. To extrapolate these measurements to the limits of such a system, i.e. regenerating a completely blocked filter, the measurements are made not only on dust coated bags, but also on a metal pipe in place of the bag. Such a pipe would represent the limiting conditions of a completely blocked and non-porous filter bag. These measurements show the maximum pressure peak and rate of pressure rise for a system.

3.3.1.3 Determining Area Mass on a Single Bag

At the Technical University in Karlsruhe a laboratory method was developed to test the efficiency of a single bag cleaning system (fig. 3.20). The arrangement to be tested is mounted above the test bag. Dust laden air is fed in at the bottom of the chamber. A radiation gauge records the amount of dust deposited on the bag; the cleaning pulse is initiated after either a finite time, or after a predetermined pressure drop is reached. The entire set of data is recorded and evaluated on automated process recorders with a display and digital plotter. For each trial the following data is recorded at constant filtration velocity:

- Rise of differential pressure with time (Δp/Pa),
- Area weight (w/g/m^2) before and after the cleaning pulse.

Each set of data is collated immediately and the calculations printed out and stored on diskettes.

The values from all detectors are averaged, finally all measurements are integrated to yield a value for the area mass. This method is very involved, but enables experiments

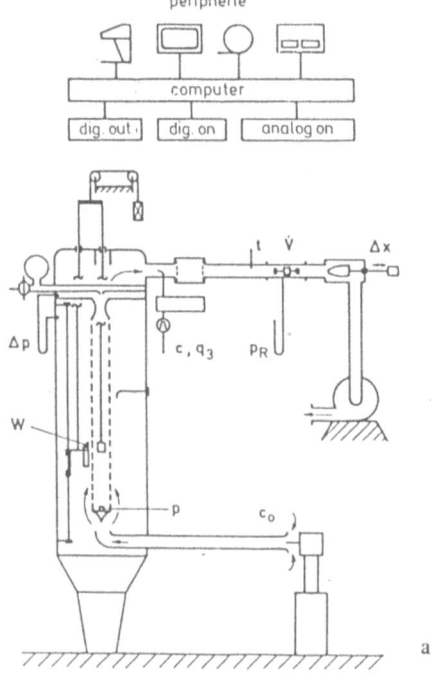
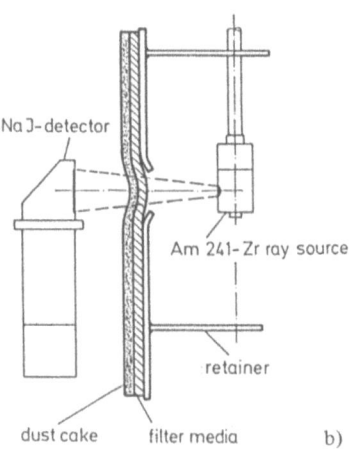

Fig. 3.20

Experimental arrangement using a neucleonic gauge

to be conducted on individual filter bags. Such complex assignments can only be undertaken with the large resources of a government and industry supported academic institution with all its specialist's facilities and staff. However, it is currently the most promising method to analyse the effects of individual phases on the overall effectiveness of the cleaning process.

As with any pilot operation it suffers all the limitations inherent in small scale equipment, e.g. the dust laden air flows from bottom to top within the cell. Also, the flue gas flows parallel to the bag inside the filter housing. The resultant values have comparative significance only, and while they can yield useful knowledge, they should not be taken in their absolute sense when designing full scale installations.

These intensive research efforts have concentrated largely on the filter and filter media. Many questions remain unanswered. The research work continues and will no doubt add to our knowledge. Its results are shedding light on the physical processes that occur on the filter bag, the bag motion, the level and periodicity of the pressure. Much less is known about the build-up and structure of the filter cake, its dependence on deposit rate, flow velocity, particle distribution and flow properties of the dust. Of additional importance is the air velocity distribution along the filter bag and the filtration efficiency. The electrostatic effects require insights, although some results of earlier trials have already been reported.

This list of unsolved or half solved problems is by no means exhaustive; it illustrates that the number of variables are indeed large. It will take more time to gain sufficient know-how so that more design data of general applicability becomes available. Highly desireable in this context is the need to experiment on an industrial scale, so that laboratory trials can be extrapolated. It would be useful to be able to compare filter cleaning systems and configurations under identical test conditions. Modern research has focussed on the jet filter, but shaker and reverse flow systems deserve equal attention. Maybe it would then be possible to formulate general principles for a more affordable and even more efficient dust collector.

3.3.2 Influence of Cleaning

As already mentioned in sect. 3.2.1 there are several ways of cleaning filter bags or pockets. First and foremost it is necessary to detach the dust from the filter medium; then to drive these particles which are lodged in the pore, or that have penetrated into the interior, towards the flue gas side. The success varies with the kind of dust and the cleaning mode adopted. As soon as the dust is mobile it should be quickly removed from the vicinity of the filter medium and extracted from the housing. This speedy removal on regeneration is just as important as the cleaning action itself.

Since the dust must settle to the bottom, the settling velocity of the dust as well as the flue gas flow pattern within the filter housing play an important part. With shaker filters especially it is necessary to take the filter cell out of the gas flow during the cleaning cycle. Some reverse flow and pulse filters also require cell by cell cleaning without flue gas impingement.

This is an inherent design feature in many constructions.

With very light dusts such as active carbon or aerosols closing off the whole filter or individual cells during cleaning is essential.

The magnitude of the differential pressure is a direct measure of the success or otherwise of a filter configuration; this in turn will determine the air to cloth ratio and face velocity. It is therefore well worthwhile to study the processes that operate within the filter medium as well as the flue gas distribution within the filter.

3.3.2.1 Shaker Filters (fig. 3.21)

Constructions differ depending on whether the filter is a pocket or bag type.

Filter pockets are generally vibrated at high frequency. It would appear beneficial to combine this action with cleaning air, yet this is seldom practiced.

Filter bags on the other hand are usually shaken or rapped by a low frequency up and down movement along their length axis. The bags are attached to the floor of the filter housing, thus the downward movement tends to break the cake loose from the bag, while the upward movement tensions the bag and throws the filter cake off.

This method introduces creases at the filter base, so that it is not suitable for fibreglass fabrics. These break with repeated creasing. Fibreglass bags are therefore vibrated at higher frequencies. It is common to support the shaker or vibrator motion with a counter current air stream, just as each cell is isolated from the flue gas flow during the cleaning cycle. The uniform reciprocating movement is dampened by the inertial forces. Thus removal of particles from the interior of the filter medium becomes problematical. Today, this technique is primarily applied to fabrics at low face velocities.

3.3.2.2 Counter Current Cleaning (fig. 3.22)

Single or multiple fans are built into the filter housing. These force fresh air or filtered gas through the filter bags or pockets from the clean gas side.

The efficiency of this system depends on back pressure and the reverse flow air velocity (w_2). The fan must overcome the drag of the filter medium before any cleaning can take place. The flow velocity through the filter should be of the same order of magnitude as that of the flue gas so that the power requirements for a reverse flow fan can be substantial. For added effect some systems pulsate this cleaning air.

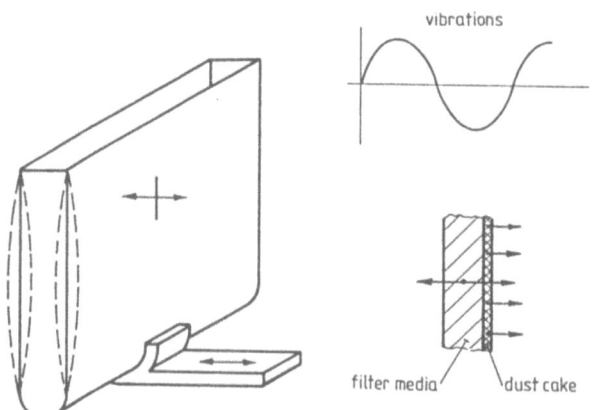

Fig. 3.21
Shaker cleaning of pocket filters

3.3 Filter Configuration and its Effect on Operating Conditions 211

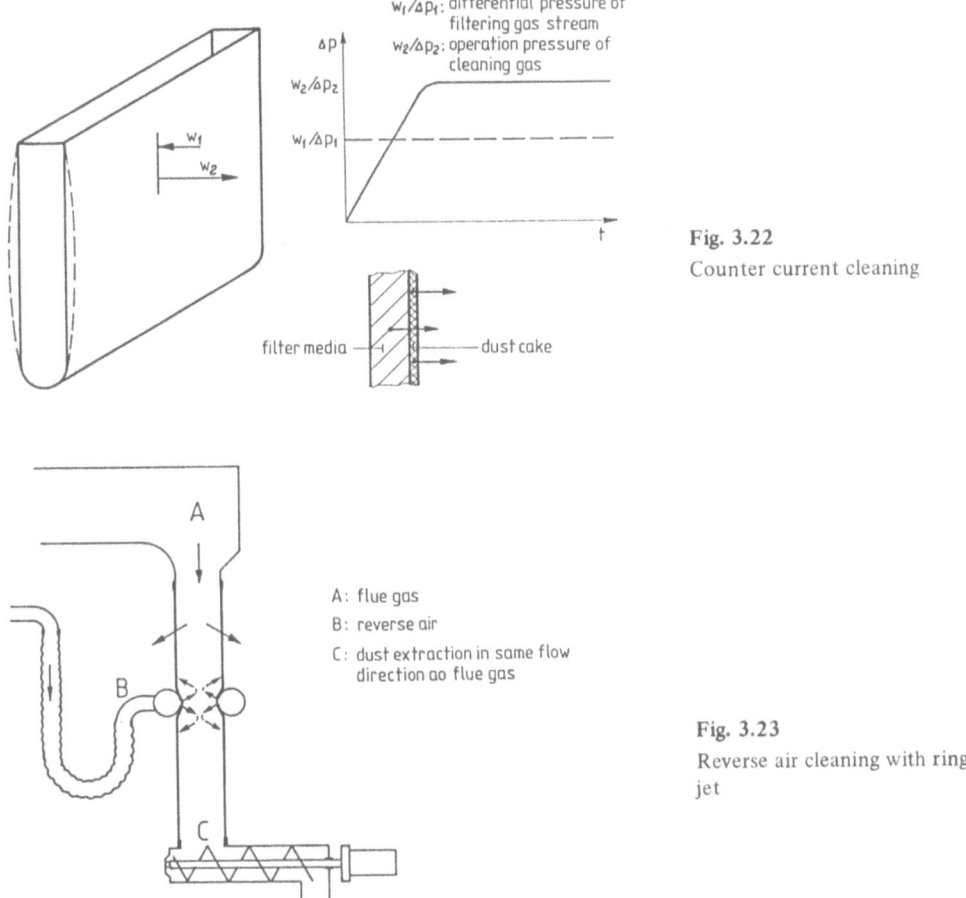

Fig. 3.22
Counter current cleaning

Fig. 3.23
Reverse air cleaning with ring jet

Reverse flow proceeds pocket by pocket or bag by bag within each cell. Since the injected air adds to the total filter load, it must be included in the capacity calculations of the active filter surface. Obviously, the addition load becomes more significant the more bags or pockets are cleaned at any one time. The effect is similar to that of a vacuum cleaner, only in the reverse sense.

In blowback filters where the dust accumulates on the inside of the bag, the cleaning air is localised by the traversing ring blowers (fig. 3.23). This keeps the air requirements to a minimum.

The reverse flow backblow filters were developed as an alternative to the shaker filters at a time when woven filter fabrics were in general use. Maximum air to cloth ratio for these types are similar to the shaker filters.

3.3.2.3 Pulse Cleaning

The rapid expansion of a small volume of cleaning gas produces a pulse which forces the bags or pockets outwards towards the flue gas side. Irrespective of whether jet pulse

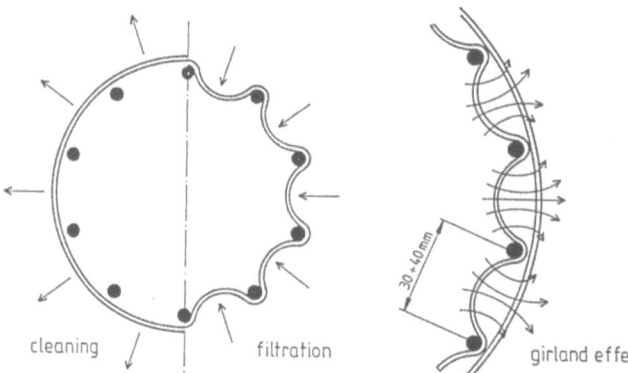

Fig. 3.24
Garland effect with a needlefelt. Support cage with wide wire spacing

filters with venturi, ring jet — with or without venturi, or direct pulse, whether bags are cleaned in series or singly — the basic principle is always the same.

How well each of the cleaning systems perform their function is very much dependent on type and design. The basic function is to convert the energy of the compressed air stored in the cleaning air tank with minimum loss into kinetic energy, which in turn regenerates the filter bags. Number of valves, directional flow changes, diameter of blow tube, shape of the venturi etc. all play a vital role in this conversion process.

The break-up and ejection of the filter cake is affected by the construction of the filter cage, and to what extent it allows the filter medium to move. Bags are usually fitted to the cage slightly oversized. Because of the filter flow resistance, the medium is pushed into the interstices between the cage wires. The pressure pulse blows the bags off their support into a balloon shape (fig. 3.24).

This shatters the filter cake, but also propels it outward during the sudden shape change of the filter medium (fig. 3.25).

The movement is arrested abruptly when the medium reaches the end of its motion, and the inertial forces of the filter cake take over. These forces also assist to bring particles from the interior to the surface. The brief counter current airflow during the pulse further assists the particle movement.

Fibreglass is more brittle and the scalloping effect must be reduced. Reducing the pitch of the wires in the cage and fitting the bags tighter helps (fig. 3.26). Since this reduces the cleaning intensity, it requires a correspondingly lower filtration velocity. The same applies even more to pocket filters, where the envelope fits tightly over the cage, and the garland effect is all but absent (fig. 3.27).

Bag length and pocket shape, distance between elements and flue gas distribution further influence the cleaning action.

Pulse cleaning consists of a complex series of events. An effective design yields high air to cloth ratios with low differential pressure. To summarize, the elements most critical to efficient cleaning are:

3.3 Filter Configuration and its Effect on Operating Conditions

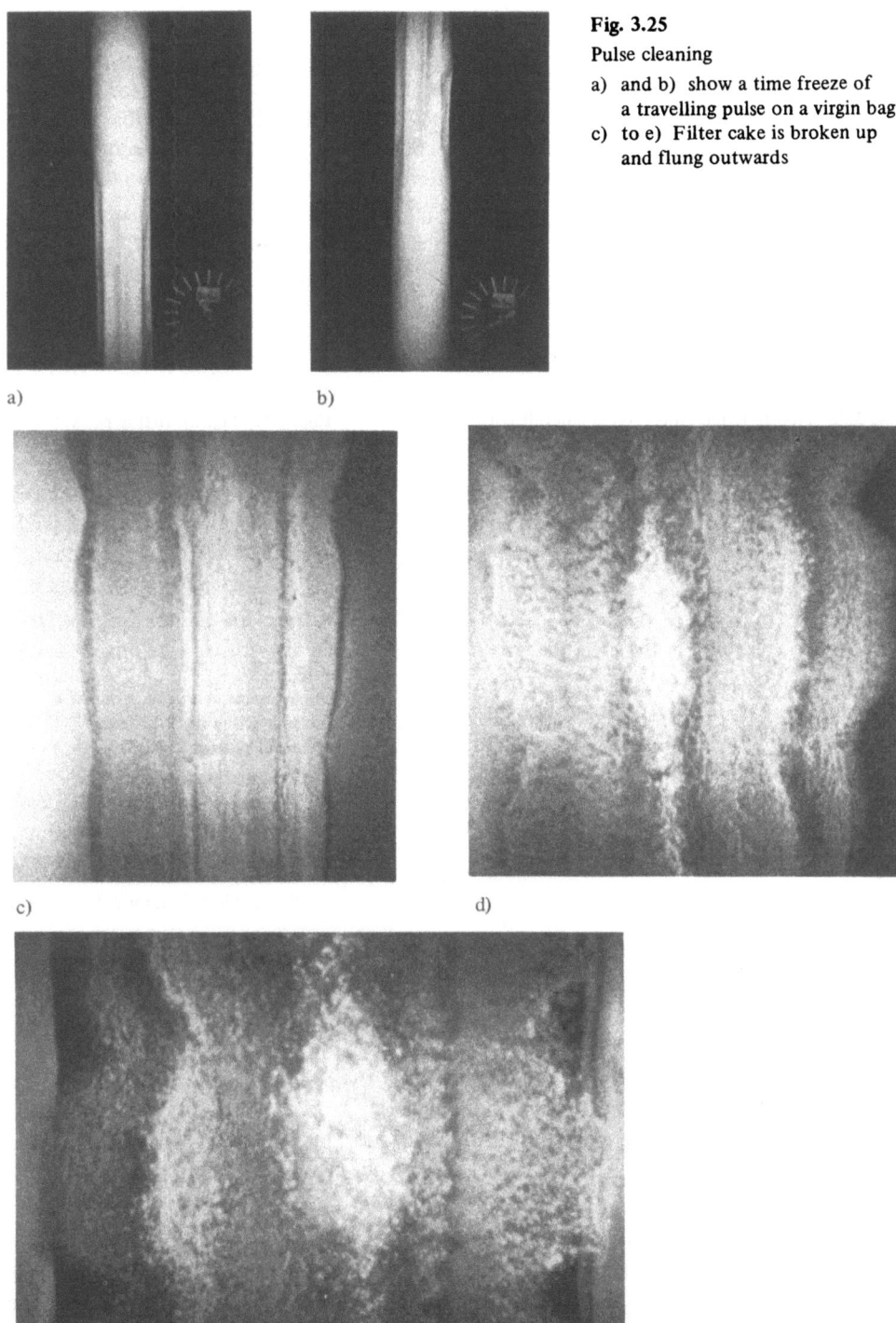

Fig. 3.25
Pulse cleaning
a) and b) show a time freeze of a travelling pulse on a virgin bag
c) to e) Filter cake is broken up and flung outwards

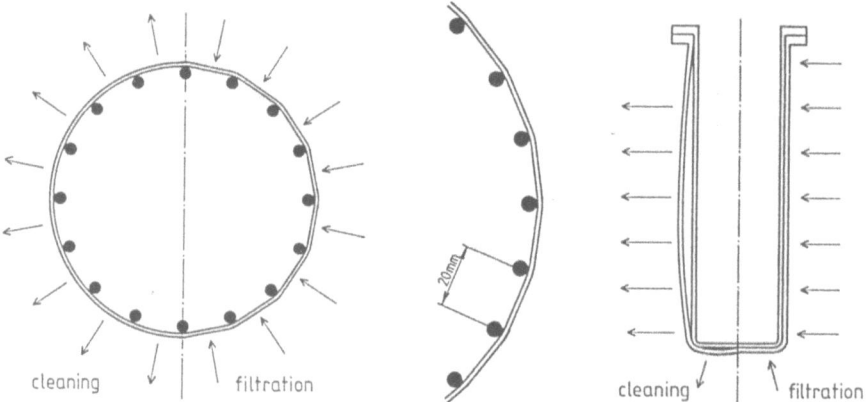

Fig. 3.26 Tightly fitted fibreglass needlefelt with reduced garland effect. Close spacing of support cage wires.

Fig. 3.27 Tightly fitting pocket filter. Large excursion of the surface during cleaning

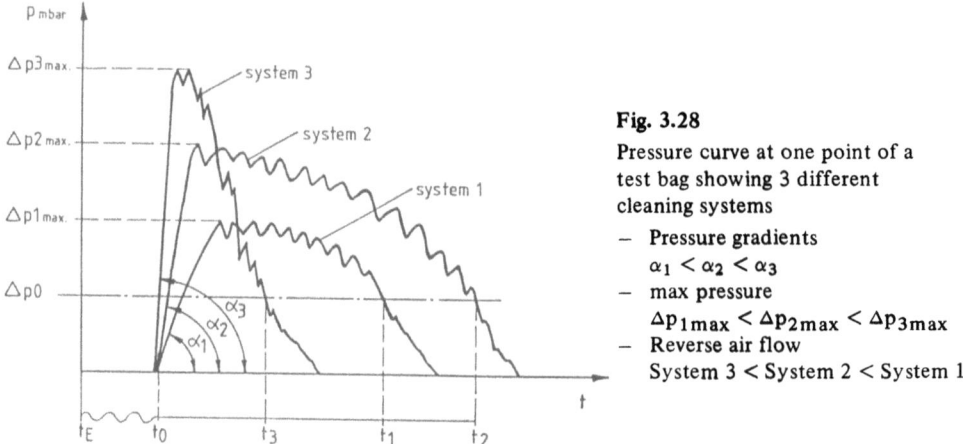

Fig. 3.28
Pressure curve at one point of a test bag showing 3 different cleaning systems
- Pressure gradients
  $\alpha_1 < \alpha_2 < \alpha_3$
- max pressure
  $\Delta p_{1max} < \Delta p_{2max} < \Delta p_{3max}$
- Reverse air flow
  System 3 < System 2 < System 1

- Scalloping effect
- Rate of pressure rise
- Pressure peak inside the bag
- Pulse decay within the bag
- Counter current air stream

Fig. 3.28 illustrates typical pressure curves for three systems at one measurement point on the bag. The systems differ in construction. Depending on the design, higher peak pressures (Δp_{max})- and (α)-values can be achieved with lower cleaning pressures. The lower this cleaning pressure the higher the quantity of air required per cleaning pulse. Flow losses may also be reduced under such conditions.

3.3 Filter Configuration and its Effect on Operating Conditions

The effective cleaning air quantity is not necessarily proportional to the reduction in cleaning pressure. The success of a design depends entirely on how successfully the designer can minimize flow losses between cleaning air tank and filter bag. To this must be added the ratio of primary to secondary cleaning air. In the direct pulse system more emphasis is placed on rate of pressure rise and peak pressure than reverse air flow. Shorter valve opening times conserve cleaning air requirements.

It is reasonable to assume that the scalloping effect is more effective at greater pulse slopes. It is the level of peak pressure (Δp_{max}) that determines the ability of the system to handle blinded bags with very low air permeability.

These limiting values are obtained by replacing the filter bag with an impervious medium such as a plastic foil or a metal tube.

That such extremes show a large difference can be illustrated with measurements performed on different direct pulse systems. Fig. 3.29 shows pressure versus time curves at three different points along the bag. These are placed at the top, middle and floor, respectively. The valve opening pulse together with the point at which the valve is actuated electromagnetically is recorded. So is the pressure within the storage tank. Graphs on the left hand side were recorded with a standard dust caked polyester bag. Graphs on the right hand side indicate significantly higher peak values recorded with an impervious metal tube.

The critical values are pressure pulse slope (α), peak pressure (Δp_{max}), header tank pressure and air quantity per pulse. These values are summarized in table 3.1.

If the cleaning air reservoir is mounted externally to the filter, and if a steering valve and distributor directs the air to rows of bags, a higher initiation pressure is required (3–7 bar). In designs where the header tank is incorporated above the bag floor within the filter air space, and where each bag has its own valve (single bag cleaning), a pressure of 0.5 bar suffices. As can be seen from table 3.1 enormous pressure peaks and pulse slopes are possible at such low level of cleaning air pressure. The ratio between energy input and bag utilisation is calculated from the expression E. This is derived by averaging all values measured along the filter bag. The huge assortment of constructions on the market today can make such a table infinite.

The performance data given here should really be part of the blueprint of the filter and be supplied by the manufacturer to enable the prospective buyer to judge the planned installation for its efficiency. However, the final arbiter of cleaning systems remains the in-use performance.

The bags for systems 4 and 5 were of commercial manufacture. Under conditions of longterm pilot scale use it was expected that both would have the same relationship between flue gas dust load and air to cloth ratio at constant differential pressure. As fig. 3.30 shows, system 5 yields higher air to cloth ratios over the whole range of flue gas dust load for the same differential pressure.

Similar differences appear in longterm tests on commercially produced pocket filters — systems 6 and 7. The curves in fig. 3.31 show a higher ultimate air to cloth ratio for filter bags. In this case the scalloping effect on the bags, in constrast to the more limited movement of the tighter fitting pocket filter, accounts for some of the differences in operating characteristics.

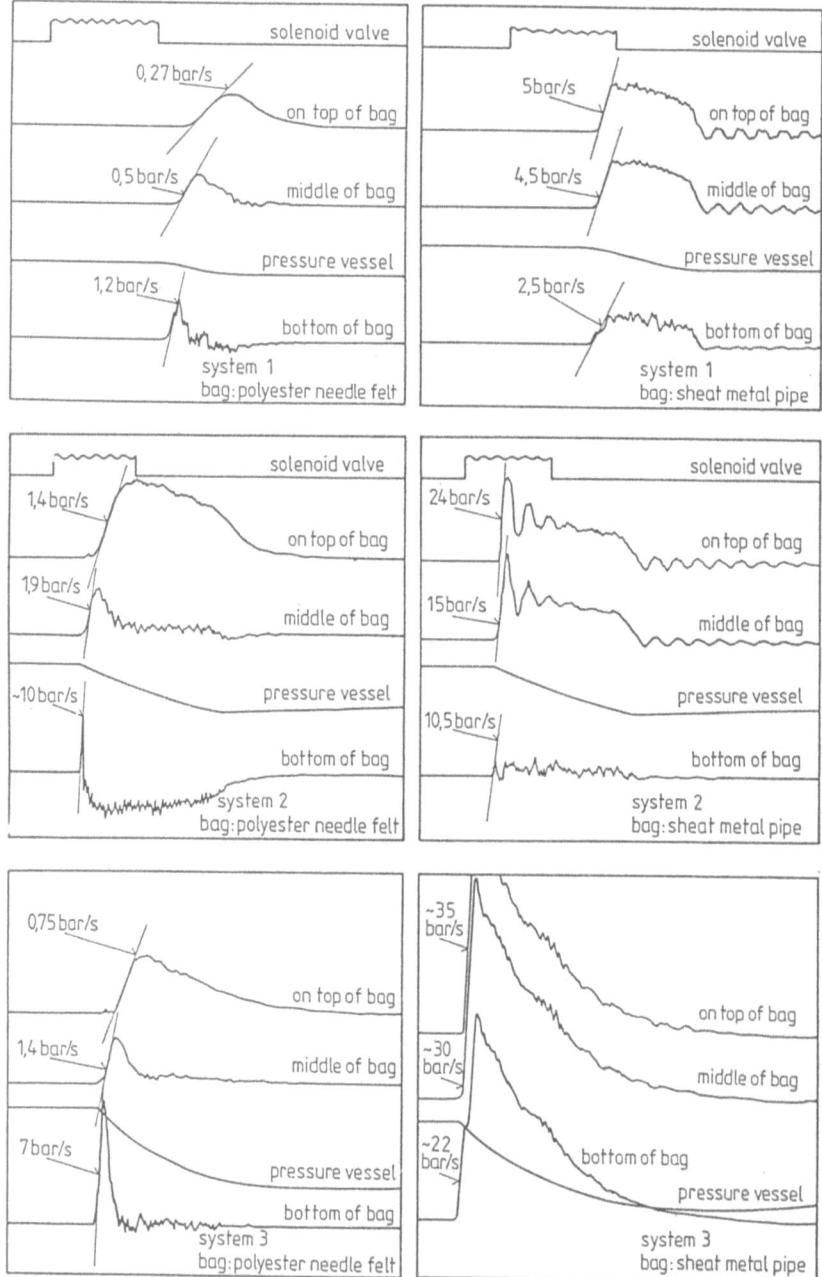

Fig. 3.29 Pulse pressure as a function of time in a filter bag. Measured with a piezo electric sensor. Measurement points: top, middle, bottom. Filter bag: PE/metal tube

a) System 1 Compressed air at 3 bar, 1 valve cleans 10 bags via an air distributor. Compressed air reservoir external to filter

b) System 2 Compressed air at 0.5 bar, 1 valve per bag. Air tank integrated into filter head

c) System 3 Compressed air at 0.5 bar, 1 valve per bag and air distributor. Air tank integrated into filter head

3.3 Filter Configuration and its Effect on Operating Conditions

Table 3.1

Bag length 2.4 metres

	Compressed air pressure in storage tank P/bar	Cleaning air per bag per pulse V(n)/l	Energy required E(E = P·V)	Pressure peak Δp_{max} mbar Bag			Average pressure	Rate of pressure rise α/bar/sec Bag			Average bar/sec
				Top	Middle	Bottom		Top	Middle	Bottom	
System 1 Direct pulse External compressed air supply 1 valve per bag row, each of 10 bags											
Polyester filter bag	3.0	15	45	12	8	8	93	1.2	0.5	0.27	0.65
Limiting value for metal tube	3.0	15	45	45	60	65	56	2.5	4.5	5.0	4.0
System 2 Direct Pulse Integrated compressed air supply 1 valve per bag											
Polyester filter bag	0.5	40	20	14	12	20	153	10	1.9	1.4	4.4
Limiting value for metal tube	0.5	40	20	20	110	110	60	10.5	15	24	16.5
System 3 Direct Pulse Integrated compressed air supply 1 valve per bag with manifold supply											
Polyester filter bag	0.5	65	32.5	32	13	15	200	7	1.4	0.75	3
Limiting value for metal tube	0.5	65	32.5	265	260	260	261	22	30	35	29

218   3 Filtration Plants

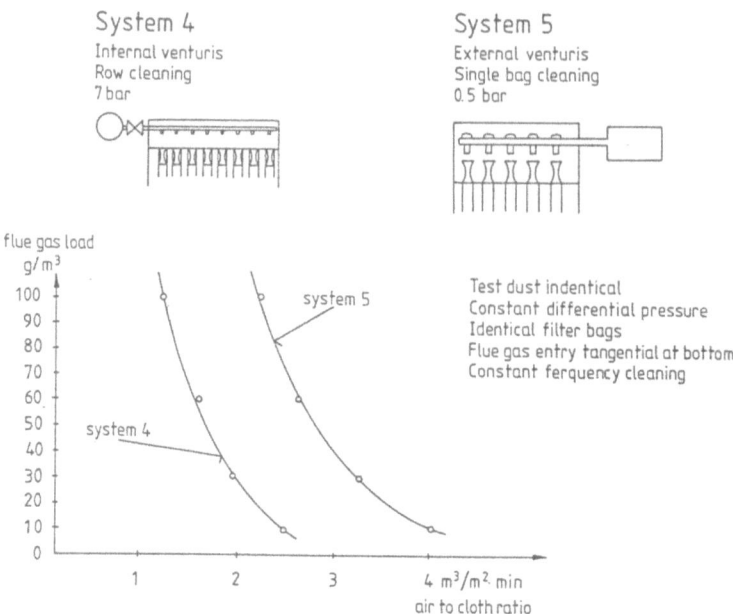

Fig. 3.30 Operating characteristic of two filters differing in cleaning mode. Operating conditions based on extended industrial pilot study.

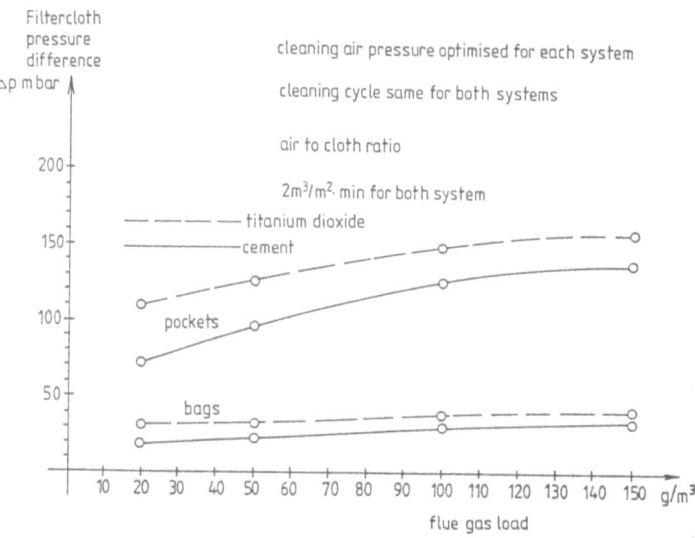

Fig. 3.31 Operating characteristics for one bag and one pocket filter, each with same air to cloth ratio – its dependence on flue gas load. Titanium dioxide and cement. Operating conditions based on extensive industrial pilot studies.

3.3 Filter Configuration and its Effect on Operating Conditions

Such comparisons do not highlight advantages or disadvantages of a system. Only when each system is referenced to operating conditions set down by the user to the manufacturer can a value be assigned to a judgement. To this are added investment and operating cost considerations. However, there are some basic characteristics that need to be considered, if a filter design is to be fit for its intended purpose. It is the purpose of the above examples to show the reasons for this.

3.3.2.4 Bag Length

The limits on how long a bag may be are given by technical considerations and physical limits. During shaking it is the mechanical stress that limits the bag length. As the length increases it becomes more difficult to transmit the shaking movement over the entire length of the bag. There are also limits to the stability of the entire filter assembly when intensifying the shaker movement.

With jet filters the limits are given by the pressure pulse, whose intensity is not infinite. The bag is to be given the maximum quantity of air in the shortest time with the least pressure loss. Valve cross sections must be accommodated within the housing, and opening rates are limited by the inertial mass of the diaphragms and valve plates. Apart from this, the pulse intensity decays as it travels down the bag. This is shown in fig. 3.32 which shows the typical pressure measured along a single bag for different cleaning systems. Each bag has its own valve. These measurements were taken with an unused polyester needlefelt of 500 g/m^2.

Some caution is advised in interpreting these results. The measurements will be significantly affected by both, the air permeability of the filter medium, and by the number of bags serviced from one valve via the air feed distributor. Once a filter cake has been built up the flow resistance rises. This increases the pressure pulse which in turn leads to higher cleaning pressures. The basic shape of the curves as in fig. 3.32, however, always remains the same.

Pressure build-up at the top of the bag varies with different regenerating systems. Systems driven with low pressure have higher pressure gradients, because of their large blow tube and valve cross sections. It is primarily a function of configuration, pressure drop, valve cross section and header tank pressure.

Table 3.2 Pulse Duration 200 ms

System	Valve or blow tube cross section (sq cm)	Storage tank (bar)	Pressure loss in tank (bar)	Air quantity ϱ_N
-·-·-	15	0.5	0.3	~ 74
-··-··-	6.12	0.5	0.3	~ 30
———	0.28	6	2.0	7.0

If the venturi reaches into the throat of the bag, cleaning becomes difficult at that point. A low pressure region or even a suction action occurs at the upper regions of the bag. The same applies to direct pulse systems when the venturis are mounted too close to the

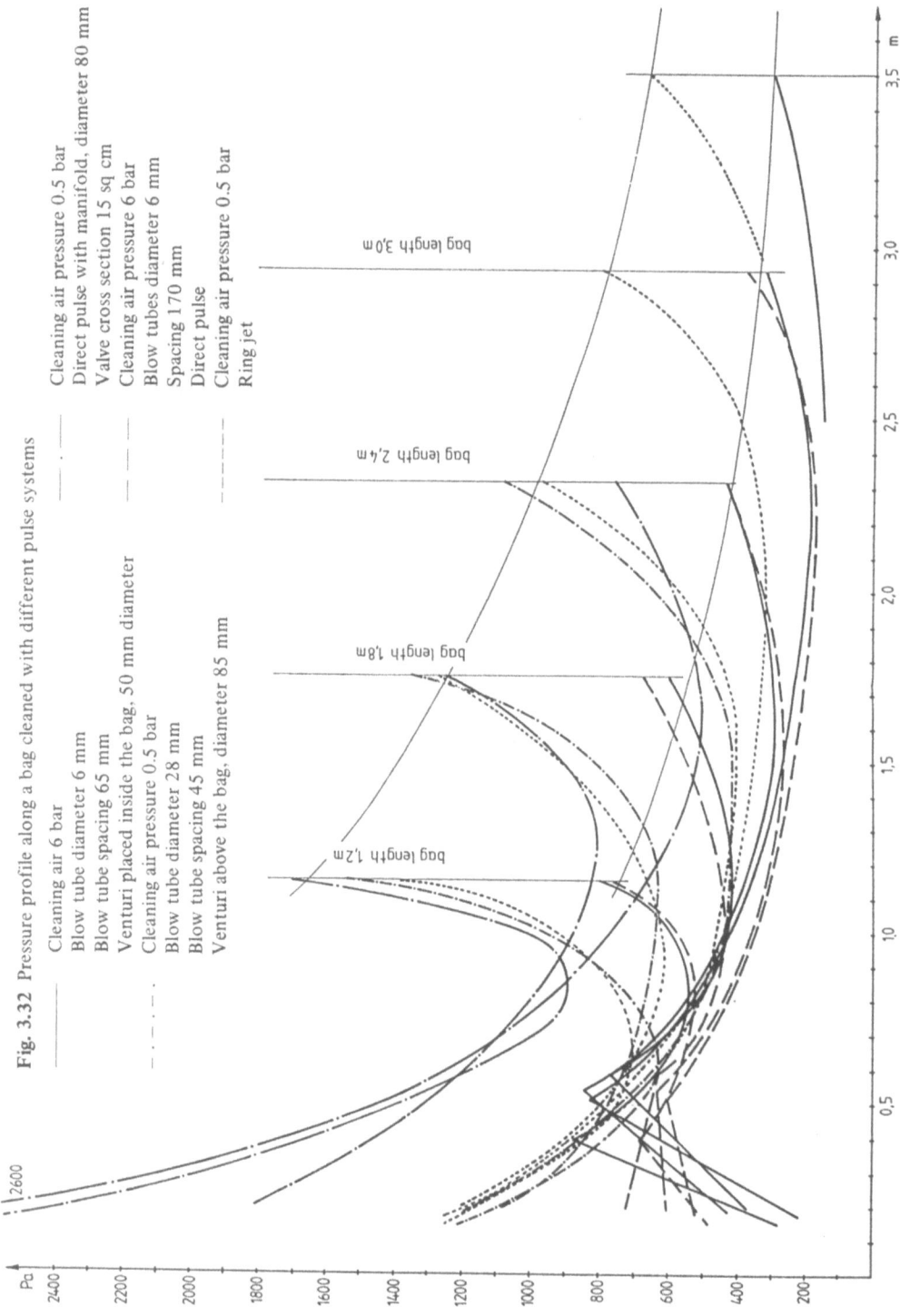

Fig. 3.32 Pressure profile along a bag cleaned with different pulse systems

— Cleaning air 6 bar
Blow tube diameter 6 mm
Blow tube spacing 65 mm
– – – Venturi placed inside the bag, 50 mm diameter
Cleaning air pressure 0.5 bar
Blow tube diameter 28 mm
Blow tube spacing 45 mm
–··– Venturi above the bag, diameter 85 mm

——— Cleaning air pressure 0.5 bar
Direct pulse with manifold, diameter 80 mm
Valve cross section 15 sq cm
– – – Cleaning air pressure 6 bar
Blow tubes diameter 6 mm
Spacing 170 mm
Direct pulse
······ Cleaning air pressure 0.5 bar
Ring jet

3.3 Filter Configuration and its Effect on Operating Conditions

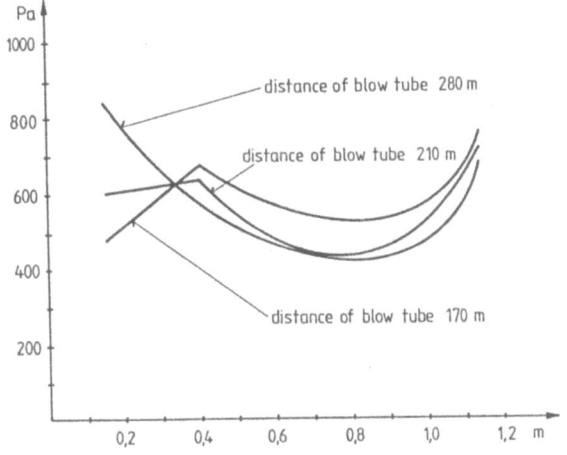

Fig. 3.33
Pressure profile at the head of a filter bag with variable blow tube spacing

bag lip (fig. 3.33). The air stream acts as an injector at this point, because it has not had sufficient time to expand. With cohesive dusts it leads to excessive build-up in this region and sometimes even bridging between bags.

The pressure decays as it travels down the bag, and pressure within the core of the bag drops. The travelling bubble loses energy as it travels along the bag. A portion of the air is pushed through the bag wall. When the pulse can no longer overcome the bag air resistance, any cleaning ceases. For lengths in excess of 3 metres it is recommended to select a lower air to cloth ratio at the design stage, because once a filter has been overloaded the only remedy available thereafter is a reduced filter loading. There are instances where bag lengths can reach 5 metres; space requirements then become critical. Even a 3 metre bag length can be difficult to accommodate.

"Off line" cleaning is usual for bags over 3.5 metres. The consequent reduction in the available active filter surface together with the difficulty in handling bags of this length usually does not justify such a solution. The importance of flue gas distribution will be covered later.

Bags are likely to touch and abrade near the bottom when they exceed 3 metres. This leads to premature wear.

3.3.3 Bag Diameter

Once the air to cloth ratio is fixed and the dimensions of the bags specified, the number of bags that are required can be calculated. This in turn sets the space requirements. This will have a large bearing on cost of the installation. If the bags are designed too narrow, a larger number are needed together with more cages, valves, ducts and other controls. If they are set too wide or are spaced too close together, cleaning becomes less efficient, the drag increases and their service life is reduced.

To optimise the filter surface area in relation to the available space, the following relationship holds (fig. 3.34).

No. of bags along with D: $\quad n = \dfrac{D-a}{2r+a}$

No. of bags along with C: $\quad m = \dfrac{C-a}{2r+a}$

Total no. of bags: $\quad k = m \cdot n = \dfrac{(D-a)(C-a)}{(2r+a)^2}$

If the housing is square, both factors are the same, hence:

$$k = \left(\dfrac{D-a}{2r+a}\right)^2$$

Because the filter surface is proportional to the circumference of the bag, it is sufficient to calculate the total circumference of all the bags, i.e.

$$U_{tot} = k \cdot 2r \cdot \pi = \left(\dfrac{D-a}{2r+a}\right)^2 \cdot 2 \cdot r \cdot \pi.$$

If the distance between bags (a) is considered constant, then radius (r) can be graphed as a function of U_{tot} (fig. 3.35).

It can be shown mathematically that the maximum filter surface area that can be accommodated occurs when the filter bag diameter is equal to the distance between them.

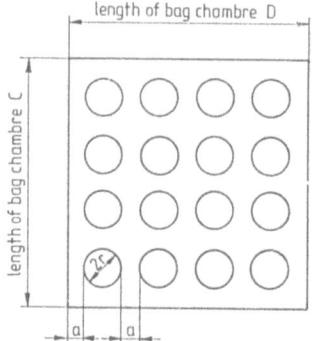

Fig. 3.34 Even spacing of bag positions

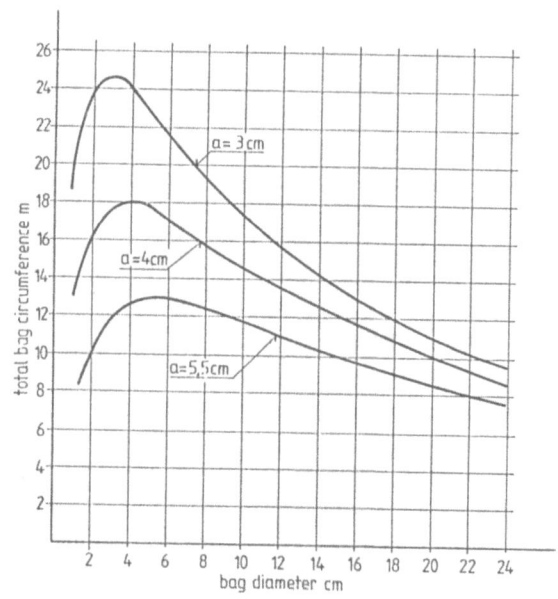

Fig. 3.35
Circumferential sum of bag surface area for a 1 m² floor area

3.3 Filter Configuration and its Effect on Operating Conditions

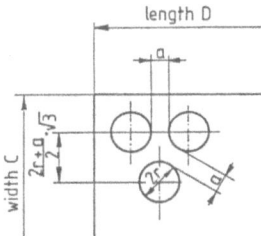

Fig. 3.36 Offset spacing of bag positions

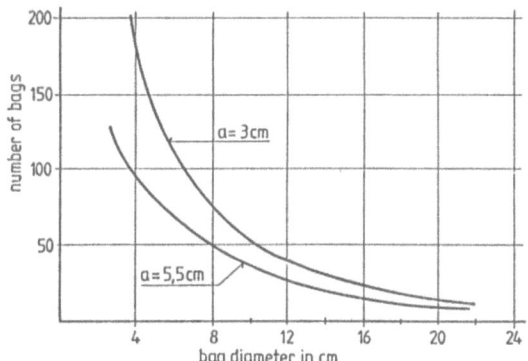

Fig. 3.37 Possible number of bags per square metre floor area

For a staggered bag arrangement the following applies (fig. 3.36):

$$m = \frac{2 \cdot [C - 2(r + a)]^2}{\sqrt{3}(2r + a)} + 1, \qquad n = \frac{D - a}{2r + a}$$

$$U_{tot} = \frac{2\pi r}{\sqrt{3}(2r + a)} \left[\frac{(2D - 2r - 3a)(C - 2r - 2a)}{2r + a} + (D - a)\sqrt{3} \right] \text{ for } m = 3, 5, 7, 9 \ldots$$

$$U_{tot} = \frac{2\pi r}{\sqrt{3}(2r + a)} \left\{ \left[\ldots\ldots\ldots\ldots\ldots\ldots\ldots\ldots \right] - 0{,}5 \right\} \text{ for all } m = 2, 4, 6, 8 \ldots .$$

For filter cells of more than 200 bags the maximum filter surface area can be increased by 10 %. At less than 100 bags the curves sink rapidly below 6 %. The basic shape of the curves remains the same, however. By analogy, the same holds for circular baghouses.

There is a minimum spacing between the bags which should not be encroached. This is to prevent dust that is ejected during the cleaning pulse from being thrown onto adjacent bags. Where cleaning is "off line", the distance can be as little as 3 cm. For jet filters it should be around 5–8 cm.

The number of bags decreases with increasing diameter. However, this is at the expense of space utilisation (fig. 3.37). Final costings will indicate where the balance is to be struck. A large cost factor will be valves, venturis, control equipment and the type of cleaning system considered.

The most common bag diameters today are 120 to 150 mm with 200 mm representing the upper limit. With reverse air cleaning or shaker type bag diameters up to 300 mm are possible. For jet filters 120 to 150 mm seem optimal and appear to fall within the costing structure of most manufacturers.

3.3.4 Flue Gas Distribution

The effect of flue gas distribution must be seen in the context of the required separation efficiency, the quality of cleaning and the filter resistance. In a clean bag condition the flow drag due to the housing varies little with different flue gas distribution.

All potential turbulence that could interfere with the cake build-up is to be avoided. These affect separation efficiency adversely. A uniform flow to the bags is achieved by keeping cross sectional areas as large as possible, and carefully planning any directional changes well away from the bags (figs. 3.38 and 3.39). It is vital to avoid small gas inlets. Increases in cross sectional area should be designed so that they do not accumulate dust. Non uniform flue gas distribution increases total drag and reduces separation efficiency.

Flue gas flow control becomes even more critical, if cells are not taken out of the gas stream during cleaning. An upward movement of the gas prevents the dust particles from settling into the collection hopper. Dust deposition on adjacent bags increases; this in turn requires more frequent cleaning. It is a characteristic of jet filters that a fraction of the dust is redeposited immediately after the cleaning pulse. This fraction becomes less with better flow control. Obviously, cleaning pressure and face velocity also affect this.

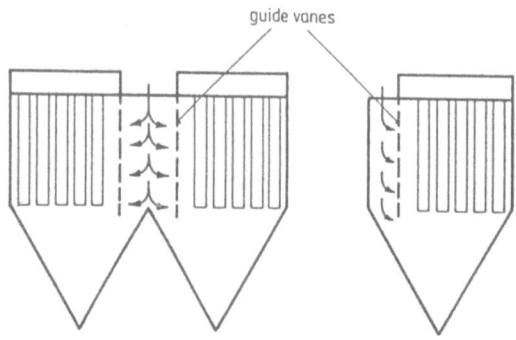

Fig. 3.38
Filter construction with guide vanes

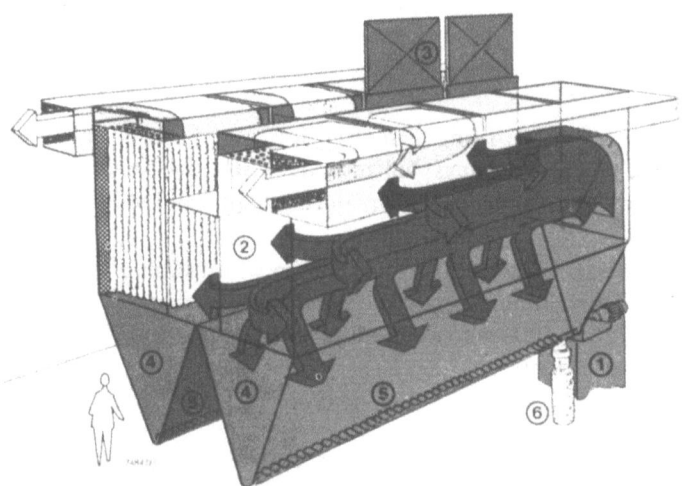

Fig. 3.39 Optimal layout of flue gas flow in a large on-line cleaned filter

3.3 Filter Configuration and its Effect on Operating Conditions

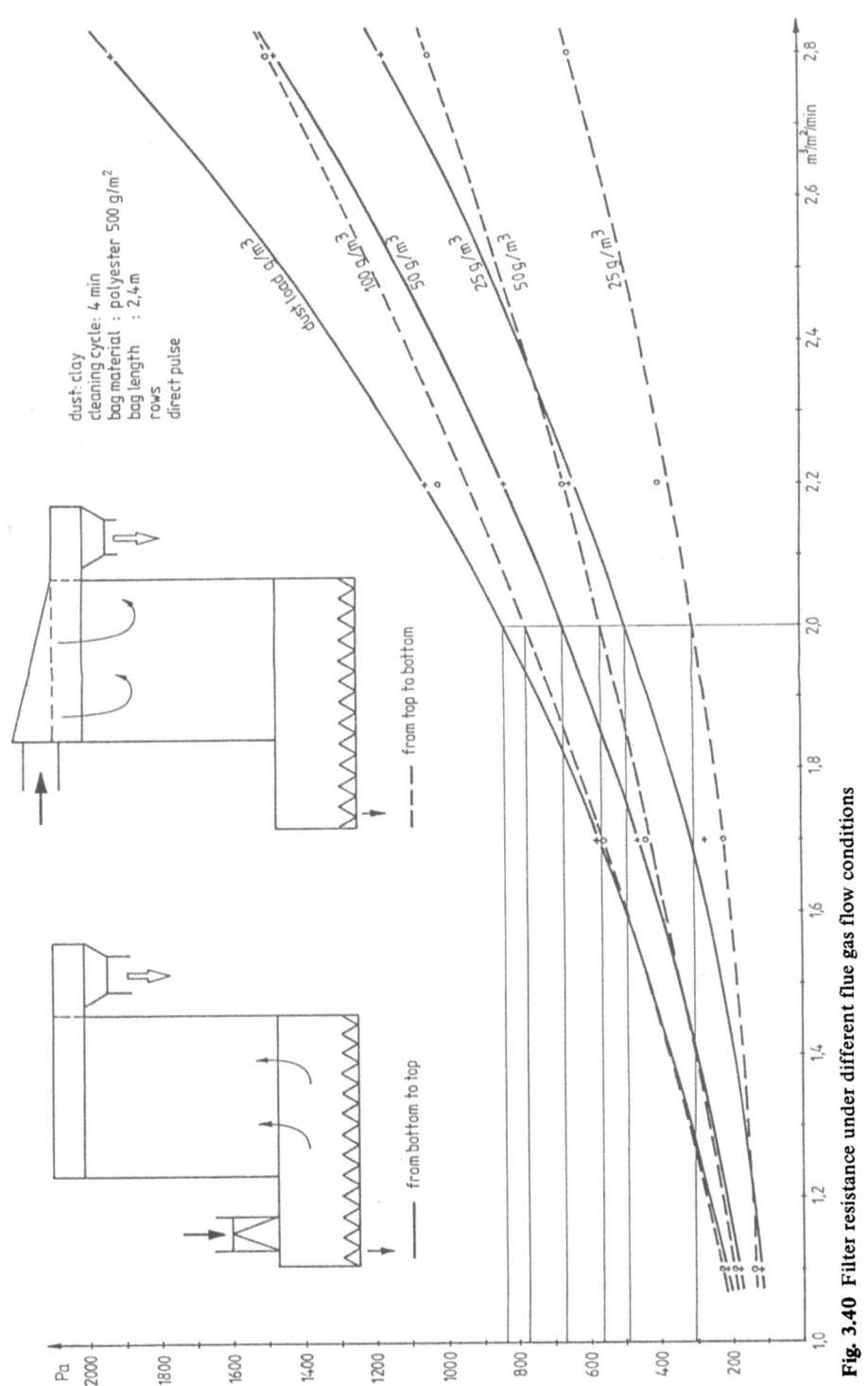

Fig. 3.40 Filter resistance under different flue gas flow conditions

Trials on an industrial scale have shown that with a vertical bag arrangement and low face velocity the overall filter resistance is not sensitive to flow pattern (with due consideration of particle sinking velocity). However, as face velocity rises the dust redeposition, caused by an upward flue gas flow, becomes significant (fig. 3.40). At that stage it becomes essential to reverse the flow to – top to bottom. The extra effort in the layout of the housing is well worthwhile and can yield sizeable reductions in overall differential pressure.

When bags are mounted horizontally the lower bags will accumulate additional dust on their upper surface. This produces a nonuniform cake deposition. Trials have shown differences of up to 5 fold. Leading the flue gas from bottom to top in this case gives improved dust deposition. None the less, even under the most favourable conditions total filter resistance is still 20 % higher than with vertical bag arrangements.

When space becomes limited horizontal spacing may become necessary. If it is essential to retain the same differential pressure with a horizontal bag arrangement, the air to cloth ratio should be reduced accordingly.

Theses arguments apply equally to cirular and square configurations. Thus, a tangential inlet is recommended for aspiration type filters with a top entry and bottom outlet (fig. 3.41).

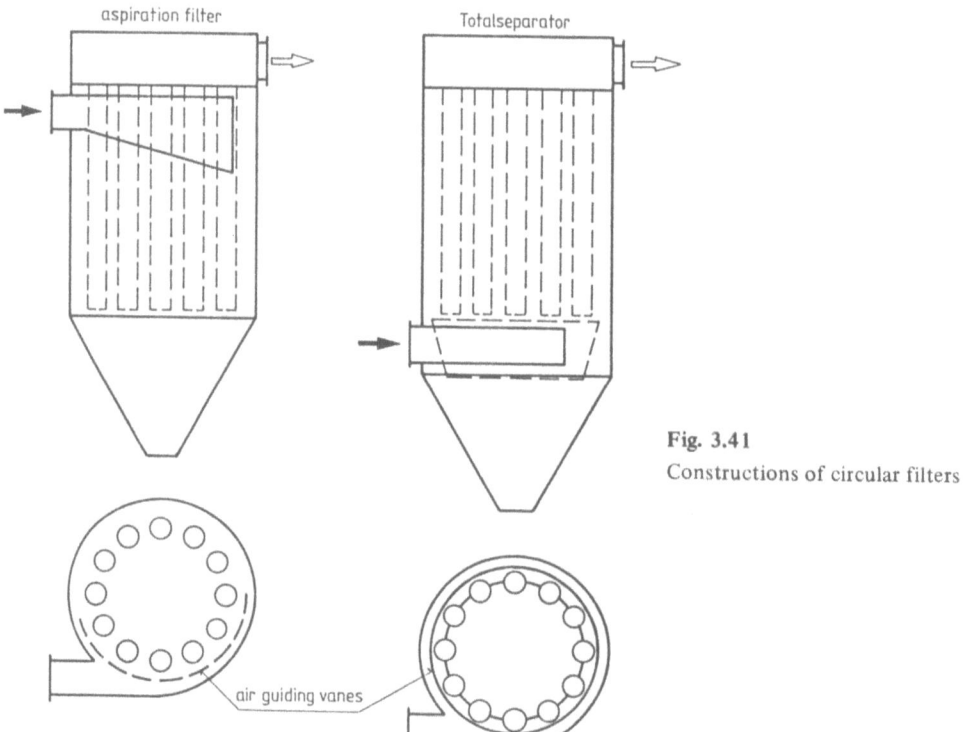

Fig. 3.41
Constructions of circular filters

Circular filters find application in situations of extreme dust concentration, e.g. for dumping at the end of a pneumatic conveyor. To protect the filter bag from overload a cyclone type dust separator is mounted below. To compensate for the resultant adverse flow conditions — bottom to top — extra filter surface area and a larger housing should be considered.

3.3.5 Summary and Conclusions

Sect. 3.3.1 to 3.3.4 covered significant and important design details of dust separators, and how design affects performance.

These performance parameters in turn determine layout and maximum allowable air to cloth ratios at constant differential pressure.

High performance jet systems, such as direct pulse single bag cleaning with compressed air distributor, effectively clean needlefelts at differential pressures of up to 30 mbar. This capability opens hitherto unexplored opportunities for super compact space saving units in the industry. The aspiration ventilator requires high power at high overall differential pressure. This limits its application to relatively low gas volumes, e.g. as part of dumping facilities at the end of a pneumatic conveyor.

Next to the cleaning system characteristic flue gas flow pattern and how the filter medium is arranged (bags, pockets, etc.) will determine the final filter resistance.

A balanced approach needs to be found to accommodate all these factors into the design of a dust collector.

3.4 Designing the Filter Surface Area

3.4.1 General

The parameter that determines the size of a filter installation is the specific air to cloth ratio, or the filtration velocity. This expresses the quantity of gas in cubic metres per hour or minute that flows through the filter surface which is measured in square metres. This value should be chosen so as to achieve a "reasonable" and constant differential pressure, and also to give an adequate service life. The following section of the book will outline what factors determine the air to cloth ratio, and how examples can be used to arrive at this value. An optimum value is obtained by considering the space needed, plant capital cost and operating expenses. If air to cloth ratio is set too low, the filter surface required becomes larger which is more costly. Also, the differential pressure will be low which results in unused filter capacity. If it is set too high, filter efficiency and service life is sacrified. Selecting an appropriate air to cloth ratio is a most difficult task, because it is subject to so many variables.

Fundamental factors that determine specific air to cloth ratio
DUST
Particle Size: The larger the particle size and the greater its density, the greater the available filter loading. Gritty and abrasive particles lead to greater wear and tear and require a lower filter loading.

Flow Properties:	If a product has a tendency to bind together because of temperature, moisture, hygroscopy or chemical reactivity, it lowers filter loading ability. Extremely light or easily fluidized substances leads to lower filter loadings, especially if jet cleaned.
Flue Gas Dust Load:	The higher the dust content per cubic meter of flue gas, the lower the filter loading.
Emission:	The lower the stack emission which has to be achieved the lower the filter loading.

FILTER

Filter Medium:	The type of medium greatly influences filter efficiency, differential pressure and service life, and the resultant potential air to cloth ratio.
Cleaning Frequency:	More frequent cleaning does not always bring the desired result; especially if the medium is already partially blocked, or if the dust forms a crust. At the pilot trial stage it is recommended to set the cleaning frequency to that approximating large scale operating conditions, and to set the air to cloth ratio accordingly.
Filter Type:	A gentle and efficient cleaning allows higher air to cloth ratios. Each cleaning system has its own specific filter loading – in the following general order: shaker – reverse flow – jet pulse – direct pulse (cleaning in rows) – direct pulse (single bag cleaning). When bag filters are used the filter loading can be doubled in comparison to pocket filters.

OPERATION

Differential Pressure:	A low filter resistance assumes a lower surface loading. Depending on filter size and application a target differential pressure should be specified.
Temperature:	Temperature greatly affects service life. Higher gas viscosity causes higher filter resistance. Similarly, fine dusts raise the differential pressure. If there is a risk of operating conditions falling below the dewpoint, preheating of the flue gas may be necessary. In such cases the elevated temperature will increase bag life and allow higher filter loadings.
Time Cycle:	For an around the clock continuous operation the air to cloth ratio should be reduced slightly, if there are no provisions for off-line cleaning. On the other hand, the filtration plant requires no starting up, which can be of great benefit.
Application:	A safety margin should be provided for in the air to cloth ratio in cases where filtration is critical to the production flow, e.g. if the end product is recovered from the filter, or when filtering poisonous substances that are not permitted to be discharged.
Service Life:	The longer the desired service life, the lower must be the air to cloth ratio.

3.4 Designing the Filter Surface Area

It goes without saying that all of the above parameters are interdependent. For additional considerations that affect filter loading refer to sect. 3.3.2.4 and 3.3.3.

Layout Specifications

Operational filter resistances move in the range 6–25 mbar. The higher end applies to jet filters, while the lower end is preferred for mechanically cleaned filters.

When trialling in the laboratory it is important to work with a particle distribution that corresponds to the eventual operation. The more detailed the data on dust and flue gas, the better the layout.

The required filter surface area is calculated from the flue gas flow rate and condition as it arrives at the inlet side of the collector.

FLUE GAS DATA

Temperature	°C
Composition	% volume
Moisture Content	g/m^3
Water and Acid Dewpoints	°C
Density (operating conditions)	kg/m^3
Flow Rate	m^3/h
Gas Pressure at Inlet	Pa, bar
Dust concentration of the cleaned gas	mg/m^3

DUST DATA

Minimum Dust Content	g/m^3	frequency
Average Dust Content	g/m^3	Distribution
Maximum Dust Content	g/m^3	
Particle Distribution (method of measurement)	DIN 66 141	
Bulk Density	g/cm^3	
Chemical Composition (% dry weight)	% weight	
Moisture Content	% weight	
Possible Fatty Content	% weight	
Sensitivity to Fracturing	Y/N	
Hygroscopic Properties	none/little/prone	
Free Flowing	Y/N	
Tackiness	Y/N	
Electrostatic Propensity	Y/N	
Poisonous	Y/N	Class?
Corrosiveness	Y/N	
Abrasiveness	Y/N	
Explosiveness	Y/N	Class?
Basicity	Y/N	
Acidity	Y/N	

GEOGRAPHICAL DATA

Position	height above sea level
Country	
Air Temperature Range	°C
Climate: Moderate, hot, tropical, arctic	
Daily Operating Hours	h
Target Differential Pressure	Pa, bar

3.4.2 Standard Design Guides

There have been many attempts to put order into the methods of calculating an air to cloth ratio. Filter manufacturers rely on the sparse specialist literature or on their own experience, as well as on experimentally derived values. One of the first comprehensive summaries of available data was published in the form of VDI Guide Lines 3677. Valuable basic hints can be found in this publication on textile filters. However, the tabulated filtration velocities for different dusts apply only to mechanically cleaned filters, although reference is made to the fact that higher filtration rates are possible with pulse systems.

Rob. Frey and T. Reinauer [11] in their report deal with venturi type jet filters. Each of the parameters that determine the air to cloth ratio is given a weighting index. This index is then multiplied with a base value that is specific to a dust type. In the present publication an attempt has been made to refine this technique based on the above principle.

W. Flatt [5] differentiates not only between mechanical and pulse type regeneration, but further subdivides the pulse systems into row and single bag cleaning. Mechanically cleaned systems are divided into those with and without reverse air. Load factors for all different systems may differ by up to 6 units.

In a recent publication G. Funke [9] considers the effect of layout, filter media properties and operating conditions on the service life in the cement industry. Further analysis of the differences in load factors between mechanical and pulse cleaning shows a 50–100 % advantage for the latter, without sacrificing service life.

There exist additional publications; these are, however, not sufficiently detailed to allow predictions of filter loadings for different systems.

3.4.3 System Characteristics

The move from a shaker or reverse air filter to compressed air cleaned jet filters has brought about a significant increase in the ultimate air to cloth ratio. The value quoted in VDI 3677 for shaker filters can be doubled for jet filters with dusts that behave "normally". Filter materials have also advanced to allow improved collection efficiency, greater resistance to chemicals and temperature.

It is intended to give each of these systems a place, based on extensive field experience. While not implying exact numeric comparisons, a ranking that covers the most important areas of application and the majority of dusts appears feasible; sufficient at least to allow calculations of filter loadings. Jet filters differ widely in the effectiveness of the cleaning pulse and the extent of the redeposition of dust on neighbouring filter elements.

These effects are less noticeable in shaker and reverse air filters, because the cells are closed off during the regeneration cycle. In this case a pulsating cleaning jet yields far

3.4 Designing the Filter Surface Area

higher service life at the same air to cloth ratio (ref. [9]). Even direct pulse filters show up large differences; especially if single bag cleaning is used. This gives higher pulse pressures within the bag with concomitant higher air to cloth ratio.

Apart from the general configuration of a system constructional details are also significant. These will not be expanded here, because manufacturers are reticent to provide such details. It would also detract from clarity. Thus, in most instances specific air to cloth ratios are quoted as a range (see sect. 3.4.8).

3.4.4 Design Calculations using a Base Value, Indexes and Coefficients

To make calculations easier it is suggested to start from a base value that identifies each dust, and to adjust this value with correction factors that compensate for the various factors. Factor A_1, A_2, ... refer to individual filter systems. Factors B to H take the dust properties, and operating conditions into consideration. Base value A (ref. sect. 4.5) characterizes a bag filter where dust is deposited on the outside with cleaning by a venturi pulse jet. For all other systems this value must be corrected with the appropriate correction factors. The listed base values should be treated as a guide only, since they can change significantly with a more updated filter design.

Some dust cannot be accommodated within this scheme, so that the feasible air to cloth ratio must be derived by experiment.

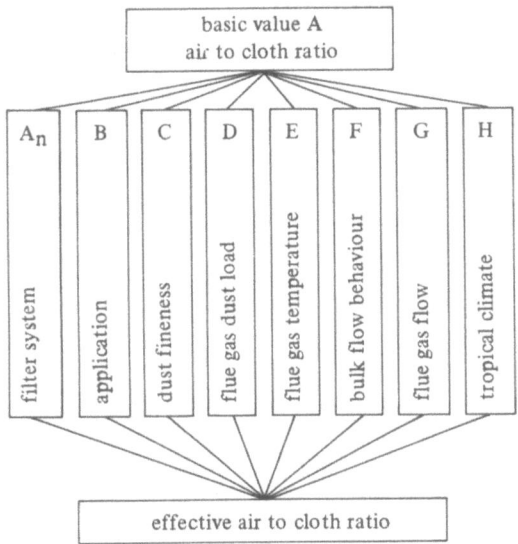

calculating of the effective air to cloth ratio

$$X = A \times A_n \times B \times C \times D \times E \times F \times G \times H$$

Table 3.3 Base values correction for filter system

Shaker and reverse air filter		Jet filter with venturi		Direct pulse filter	
Pockets	Bags	Pockets	Bags	Single bag	Multiple bags
A_4	A_3	A_2	A_1	A_5	A_6
0.45	0.65	0.5	1.0	1.5	1.3

Factor B takes account of different application, i.e. filters are grouped into three sets — simple vacuum cleaning (aspiration filter), filtrate recovery (total separator) and processing filter.

Table 3.4 Factor B Area of application

1.0	Extraction at material transfer points, conveyors, packing and filling stations, sieves, etc.
0.9	Product recovery in grinders, granulators, grading and shaking sieves
0.8	Exhaust gases from spray driers, furnaces, driers, reactors

Factor C takes account of the fineness of the dust by a weighting towards the highest percentage band of particle size (70–80 %).

Table 3.5 Factor C: Dust fineness 70–80 %

above 100 μm	1.2
between 50–100 μm	1.1
between 10– 50 μm	1.0
between 5– 10 μm	0.9
between 2– 5 μm	0.8
below 2 μm	0.7

Factor D compensates for the influence of flue gas dust load (ref. fig. 3.42).

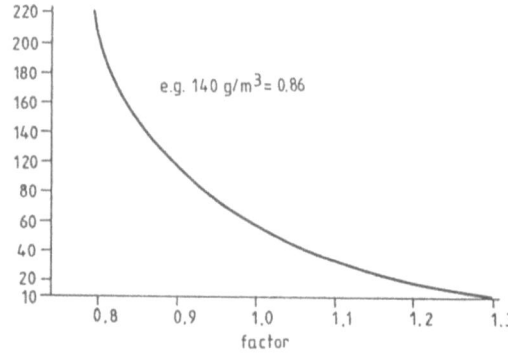

Fig. 3.42
Factor D: Flue gas dust content

3.4 Designing the Filter Surface Area

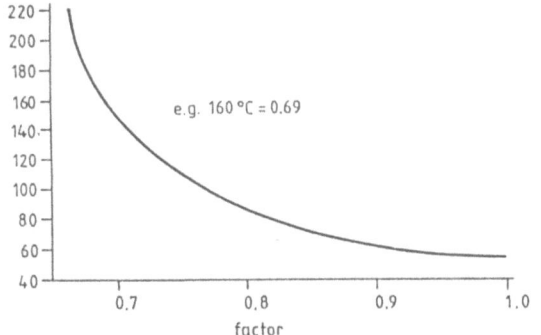

Fig. 3.43 Factor E: Temperature

Factor E grades the effect of flue gas temperature (fig. 3.43). Some dusts are more difficult to separate at higher temperature not withstanding the adverse effect of temperature of service life.

Factor F deals with the problem of cleaning very low density and easily fluidized material. The factor applies only to on-line cleaning systems. With very low densities it is sometimes no longer possible to clean at all, so that even with jet or direct pulse systems off-line cleaning becomes necessary.

Table 3.6 Factor F: Low density easily flowing dusts

Bulk density g/m^3	> 0.6	0.4 ... 0.6	0.20 ... 0.4	< 0.20
Reverse air, pulse, direct pulse, Cleaning – on line	1.0	0.9 ... 1.0	0.8 ... 0.9	0.5 ... 0.8

Factor G treats the flue gas flow control (ref. sect. 3.3.3). It comes into force only when the gas flows upwards past the filter elements, and when cleaning is on-line. There is an element of judgement here, since this air flow is also a function of filtration velocity. Particle density also affects it, though this is compensated for in part by factor F.

Table 3.7 Factor G: Flue gas flow bottom to top

Air to cloth ratio (base value)	≤ 1.4	1.8	2.2	2.6	3.0	3.4	3.8	4.2
Factor G	1.0	0.95	0.90	0.85	0.8	0.75	0.7	0.65

Factor H takes into account tropical climates with high humidity. It is relevant to food production and dusts with hygroscopic tendencies. Tropical factor H = 0.8.

It goes without saying that a filter design can never be too large. Cost/benefit aspects, however, force a compromise. It should be noted that only the essential factors are covered in this scheme. There are special circumstances where these factors are not appropriate for calculating air to cloth ratios. Tacky or hygroscopic dusts, or operating conditions that vary widely are a case in point.

3.4.5 Base Values for Dusts and Fields of Application

Table 3.8

Material	Factor A filter area loading m³/m² min	Material	Factor A filter area loading m³/m² min
Activated carbon	2.5 ... 2.8	Cement:	
Aerosil	2.2 ... 3.0	Limestone, marl	2.5 ... 3.1
Alfalfa meal intake hopper	3.0 ... 3.5	Clinker dust	
Alkali cellulose, dry, flaky	3.0 ... 3.5	(filter with air cooling)	2.0 ... 2.4
Almonds (cleaning)	2.8 ... 3.9	Clinker dust	
Alum	1.8 ... 2.0	Handling, storage,	
Alum earth	2.5 ... 2.9	proportioning	2.2 ... 2.6
Alumina	2.7 ... 3.0	Milling and drying	1.8 ... 2.3
Aluminium	3.0 ... 3.5	Raw material dust	3.0 ... 3.4
Aluminium oxide	1.8 ... 2.2	Raw meal	2.5 ... 3.0
Ammonium phosphate	3.0 ... 3.5	Cement mills	
Ammonium sulfate	3.0 ... 3.5	without milling aids	2.0 ... 2.4
Anhydrite	2.0 ... 2.5	with milling aids	1 7 ... 2.0
Animal meal	1.8 ... 2.4	Cement dust, handling	
Anthracite	2.5 ... 3.5	storage, outloading	2.4 ... 3.0
Argillite	2.5 ... 3.1	Ceramic grinding agent	1 5 ... 2.1
Asbestos	2.3 ... 3.1	Ceramic pigments	2.2 ... 2.9
Ash (fly ash)	2.7 ... 3.0	Chaff	3.6 ... 3.8
Bakelite powder	3.6 ... 4.0	Chalk	2.5 ... 2.8
Barium sulfate	1.5 ... 2.1	Chalk powder	2.0 ... 2.4
Barley meal, hammer mill	2.3 ... 3.0	Chile saltpeter	1.3 ... 1.8
Basic slag	1.8 ... 2.3	Chocolate powder	3.3 ... 3.6
Bauxite	2.3 ... 2.5	Cinnabar	1.4 ... 1.7
Blast furnace slag	1.8 ... 2.3	Clay	2.3 ... 2.8
Blood meal	3.6 ... 4.0	Clay, clay dust	1.8 ... 3.2
Boiler slag	1.8 ... 2.3	Coal:	
Bone meal	1.8 ... 2.3	Anthracite	2.3 ... 2.8
Borax	1.3 ... 1.7	Carbonized lignite	2.4 ... 3.2
Brake linings	3.1 ... 3.6	Charcoal	1.8 ... 2.2
Bran	3.3 ... 4.1	Coke ash	1.5 ... 1.8
Brewer's grains	3.6 ... 4.2	Coke, petrol coke	1.5 ... 1.8
Brick dust	2.5 ... 2.9	Lean coal	2.8 ... 3.6
Brown iron ore	2.3 ... 2.8	Lignite	2.4 ... 3.2
Builder's gypsum	2.5 ... 2.8	Lignite (pit-wet)	3.8 ... 4.1
Cadmium yellow	1.8 ... 2.3	Lump coal	3.6 ... 4.2
Calcium carbide	1.4 ... 1.8	Rich coal	3.1 ... 3.8
Calcium cyanamide	1.3 ... 1.8	Shiny coal	2.1 ... 2.3
Calcium nitrate of ammonium	1.1 ... 1.5	Small coal	2.8 ... 3.6
Carbonate of zinc	2.3 ... 2.5	Cocoa beans	3.9 ... 4.4
Carbon black	2.1 ... 2.5	Cocoa powder (up to 25 %	
Carbon black (pelletized)	1.1 ... 2.1	fat content)	3.1 ... 3.5
Casein	1.4 ... 3.0	Coffee dust (of green beans)	2.1 ... 2.5
Cast iron machining	3.2 ... 3.5	Colophony	1.2 ... 1.4
Cellulose, dry, flaky	3.5 ... 4.0	Compound feed	3.3 ... 3.6

3.4 Designing the Filter Surface

Table 3.8 continued

Material	Factor A filter area loading m^3/m^2 min	Material	Factor A filter area loading m^3/m^2 min
Contact mass (catalyzers)	2.1 ... 2.5	Ground coffee	2.9 ... 2.3
Copper ore	2.3 ... 2.8	Guar meal	2.4 ... 3.0
Copper oxide	1.3 ... 1.8	Gypsum	2.0 ... 2.7
Copra pellets	1.6 ... 2.0	Hazelnuts, cleaning	3.9 ... 4.4
Cork	3.1 ... 3.6	Hemp fibre	4.4 ... 4.6
Cotton	3.0 ... 5.0	Hempseed	3.5 ... 3.7
DDT	3.0 ... 3.5	Heraklith	3.8 ... 4.2
Derivatives	2.7 ... 3.0	Herbicide	1.9 ... 2.2
Detergent	1.5 ... 3.0	Homopolymer HPSO-S	2.8 ... 3.2
Diabase	2.0 ... 2.8	Ink powder	1.8 ... 2.2
Diatomaceous earth	2.0 ... 3.0	Insulating pumice	2.3 ... 2.8
Dicalcium phosphate	4.0 ... 4.5	Iron grinding	2.0 ... 2.8
Dolomite powder	2.5 ... 3.5	Iron ore	2.5 ... 2.8
Dry feed	3.3 ... 3.6	Iron oxide (brown smoke)	1.5 ... 2.0
Dry yeast	2.4 ... 3.0	Kaolin	2.3 ... 2.8
Epoxy resin	3.0 ... 3.4	Lead oxide	2.5 ... 3.0
Epsom salt	1.4 ... 1.7	Lead sulfate	3.0 ... 4.0
Expeller powder (filter residues)	3.1 ... 3.5	Leather dust	2.8 ... 3.6
Extraction meals	3.6 ... 4.2	Lemon peel meal	2.4 ... 3.0
Feed chalk	2.4 ... 3.0	Lignite	2.5 ... 3.2
Feed lime	2.4 ... 3.0	Lime sandstone	2.8 ... 3.3
Feed salt	1.7 ... 2.0	Limestone	2.1 ... 3.1
Feldspar	1.9 ... 2.5	Linen	3.6 ... 3.8
Ferromanganese	1.3 ... 1.8	Linseed cake meal (hammer mill)	2.5 ... 2.9
Ferrosilicon	1.1 ... 1.8	Linseed, crushed	2.4 ... 3.0
Ferrous sulfate	1.3 ... 1.8	Lupolen, gritty	3.0 ... 3.4
Fireclay	1.9 ... 2.4	Magnesia	1.3 ... 1.8
Fish meal	2.3 ... 2.7	Magnesite	2.3 ... 2.8
Flint	1.5 ... 2.1	Magnesium oxide 80 %	2.2 ... 2.6
Fly ash	2.7 ... 3.0	Maize (cleaning)	2.8 ... 3.9
Foundry dust	2.5 ... 2.9	Maize gluten pellets	1.6 ... 2.0
Foundry sand	2.8 ... 3.2	Maize meal (hammer mill)	1.4 ... 2.4
Fruit pomace, dry	3.3 ... 3.9	Maize starch	1.5 ... 3.2
Fruit powder	1.4 ... 1.7	Malt flour	3.6 ... 4.2
Fuller's earth	1.4 ... 1.7	Malt rootlets	1.8 ... 2.2
Galena	2.0 ... 2.5	Manganese ore	1.8 ... 2.3
Galvanizing plants	1.2 ... 2.0	Marble powder	3.1 ... 3.5
Gelatin for film manufacture	1.8 ... 2.2	Marble sand	3.4 ... 3.8
Glass wool	4.4 ... 4.6	Metat meal	1.8 ... 2.7
Gluten (see wheat starch)		Metallurgical sand	1.5 ... 2.1
Gneiss	2.2 ... 2.8	Metal powder	2.4 ... 3.0
Grain dryer	3.9 ... 4.3	Mica	2.5 ... 2.1
Granite	2.2 ... 2.8	Milk powder (depending on fat content)	1.6 ... 2.5
Graphite	1.8 ... 2.2	Millet	3.5 ... 3.7
Grass meal	2.8 ... 3.2	Mirabilit	1.8 ... 2.3
Gravel	2.8 ... 3.6	Moltofil	2.0 ... 2.4
Grey wacke	2.1 ... 2.8		

Table 3.8 continued

Material	Factor A filter area loading m^3/m^2 min	Material	Factor A filter area loading m^3/m^2 min
Moulding sand	2.1 ... 3.2	Soda, light	2.0 ... 3.0
Mullite	1.3 ... 1.8	Sodium bicarbonate	2.7 ... 3.3
Nicotinic acid	1.8 ... 2.2	Sodium perborate	3.1 ... 3.5
Non-ferrous melting dusts	1.8 ... 2.4	Sodium sulfate, water-free	2.0 ... 3.0
Oatmeal	2.5 ... 2.9	Sodium tripolyphosphate	2.5 ... 3.0
Oxide of zinc	1.8 ... 2.9	Soybean dust	2.2 ... 2.8
Paint pigments	2.4 ... 2.9	Soybean meal	1.8 ... 2.2
Paper or corrugated board fibres, paper clippings	3.6 ... 4.0	Soy flour	2.0 ... 2.4
Peat	3.6 ... 3.8	Sphalerite	1.5 ... 1.8
Perlit filler	2.4 ... 3.0	Spices	2.7 ... 3.3
Petrochemicals, dry	3.1 ... 3.5	Spun glass	2.5 ... 2.8
Phosphate fertilizer	1.3 ... 3.4	Stannic oxide	1.4 ... 1.7
Plastic pellets	3.7 ... 4.6	Stearate	2.4 ... 3.0
Plastic powder, polymers	1.6 ... 2.0	Stone dust	2.1 ... 3.2
Porcelain clay	2.3 ... 2.8	Styropor chips	2.8 ... 3.2
Porphyry	1.8 ... 2.3	Sugar	2.8 ... 3.2
Potash	1.4 ... 1.8	Sugar beet chips	2.4 ... 3.0
Potassium bitartrate	1.3 ... 1.8	Sulphur	2.4 ... 3.2
Potassium sulfate chloride	1.1 ... 1.5	Table salt	1.5 ... 1.8
Potato flakes	3.3 ... 4.0	Talc	2.2 ... 3.0
Potato chips	3.6 ... 3.8	Tapioca pellets	1.6 ... 2.0
Potato starch	1.6 ... 2.5	Tea dust	2.5 ... 2.9
Protein (spray drier)	2.6 ... 3.0	Terephtalic acid	2.8 ... 3.2
Pumice sand	2.5 ... 3.2	Titanium oxide	3.0 ... 3.4
Purifier aspiration	3.6 ... 4.8	Tobacco	3.6 ... 3.8
PVC	1.8 ... 3.0	Tobacco powder	2.4 ... 3.0
Pyrite	2.1 ... 2.4	Tomato meal	2.4 ... 3.0
Quartz powder	2.1 ... 2.8	Urea, crystalline	2.4 ... 3.0
Quartz sand	1.8 ... 2.3	Urea, pressed mass	2.1 ... 2.5
Raw phosphate	1.3 ... 1.8	Viscose staple fibre	4.3 ... 4.6
Reduced bones, hammer mill	2.5 ... 2.9	Vulcanized fibre	2.5 ... 2.8
Rice germs	2.4 ... 3.0	Wheat germs	2.2 ... 2.6
River sand	2.5 ... 2.8	Wheat grinding	2.7 ... 3.7
Rock dust (dry)	1.4 ... 1.8	Wheat purifier aspiration	4.2 ... 5.4
Rubber	2.3 ... 2.8	Wheat starch	3.2 ... 3.6
Rubber powder, synth.	4.2 ... 4.8	Whetstone	1.5 ... 2.1
Sand	2.3 ... 2.8	Whey powder	2.2 ... 2.6
Sawdust	3.8 ... 4.2	Wood chips, dry, 1 ... 10 mm	4.1 ... 4.5
Sawdust 0.1 ... 1 mm	3.8 ... 4.2	Wood chips, moist	4.4 ... 4.8
Shell grit	3.5 ... 4.2	Wood dust (grinding dust with proportion of glue)	2.1 ... 2.5
Silica	2.3 ... 2.8	Wood fibres	3.8 ... 4.1
Silica gel	1.3 ... 1.8	Wool	4.3 ... 4.5
Silicates	2.4 ... 3.0	Yeast	3.1 ... 3.5
Skimmed milk powder	3.0 ... 3.6	Zinc chloride	1.4 ... 1.8
Soap	2.3 ... 2.8	Zinc spar	2.3 ... 2.5
Soda, heavy	3.6 ... 4.0	Zinc vitriol	1.3 ... 1.8
		Zinc white	1.5 ... 1.8

3.4 Designing the Filter Surface Area

3.4.6 Creating a Data Bank

It is not possible to incorporate all relevant variables within the above regime. Practical air to cloth ratios depend also on the peculiarities of each make of filter. It would be highly desireable, therefore, to keep a data bank on all accessible operations, dust and gas properties (table 3.9 and 3.10). Only by this means is it possible to refine calculations that enable predictions for similar installations. Giving an unqualified air to cloth ratio for dusts is of no use; only a filter specialist with considerable experience is then able to select the weighing factors. With more comprehensive data it would become easier to accurately transfer values from one installation to another. It would also make it possible for the designer to make valid inferences in a reasonable time on appropriate filter loadings.

3.4.7 Pilot Studies

Potential applications for dust collectors are ever increasing today. New processing techniques raise new problems. Often it is not possible to reproduce the real life processing conditions in the laboratory. The time delay between collecting a dust for trial and using it for filtration experiments may change its properties beyond validating its behaviour in laboratory tests.

A pilot study is the only recourse in such circumstances. If filter loadings are carefully judged together with an analysis of all uncertainties, such an approach can pay dividends, provided the purchaser sees the benefits of such a course. To offset the costs of such a pilot plant the manufacturer of the full scale plant should be granted the opportunity to incorporate the experience gained at this pilot stage into his designs. This opens new markets for textile filters.

3.4.8 Design Fundamentals

To date, an exact mathematical treatment of filter resistance and air to cloth ratio, based on physical models, remains out of reach. Considerable efforts have been expended with many trials only to confirm that the uncertainties arising from time- and system-dependent variables are still too large, so that additional experimental controls are necessary. The design calculations expounded in this book should be seen in this context. The design loads are based on experience gathered in the field. The suggested base values and compensating factors give a guide to determine filter areas for a selected range of systems.

Site experience has shown that this method gives valid comparisons between different operating conditions and different dusts produced by one and the same process. The factors take the most important variables into account, but extreme situations lie outside. There will always be dusts that do not fit into this scheme and that require empirical back-up.

Calculation Examples

1. Fly ash from a coalburning power plant
   ρ_s = 0.38 g/cm^3 , Europe
   Particle Size 3–50 μm (70 %)
   Temperature 130 °C
   Flue Gas Load 17 g/m^3
   Flue Gas Volume 310 000 m^3/h

Table 3.9a

filter test results

product:	asbestos dust	sample no.:		test report no.: PN-516
customer:	Eternit AG, Niederurnen			author: Gämperle date: 27.5.77
manufacturer:				further tests:
				calculation sheet no.:

test data/system data

	test system	required system data		test system	required system data
filter type	MVRS-9/24		recommended filter loading m³/m²·min	3,3	
filter bag grade	PNf-500		raw gas volume m³/min	25.2	
type of gas	air		solids throughput kg/h	0.95	
temp. filter inlet °C	18°		dust loading g/m³	38	
relative humidity %			residual dust content mg/m³	0,27	
dew point °C			pressure at filter inlet mm WG		
duration of test	approx. 50 hrs.		(+ positive pressure, − negative pressure)		
required annual operating time approx.		25 hrs./year	average vertical air velocity in filter m/s		
pressure drop across filter bag mm WG	50				
total pressure drop across filter mm WG	50				
bag cleaning cycle s	9				

material characteristics

air retention capacity s		explosive	no
terminal velocity m/s		toxic	no
velocity whirl point m/s		friable	no
melting point °C		temp.-sensitive	no
moisture content ... % at ___ °C		agglomerative	slightly
fat content %		freeflowing	yes
abrasive	yes	fluidizable	yes
corrosive	no	hygroscopic	no

bulk density t/m³	0,3	el.-stat. chargeable	no
compacted density t/m³		caking	no
particle density g/cm³		caramelizable	no
average particle diam. μm	12	sticky	no
inclination	34°	basic-acid	no
shape	18°		
temperature of material °C	18°C		

remarks, special facts

asbest fibre content after the filter according to report
LA-2496: 0.2 fibers/cm³

PN file card	date: 13.6.77	visa:

3.4 Designing the Filter Surface Area

Table 3.9b

filter loading curves
particle size grid according to DIN 66145

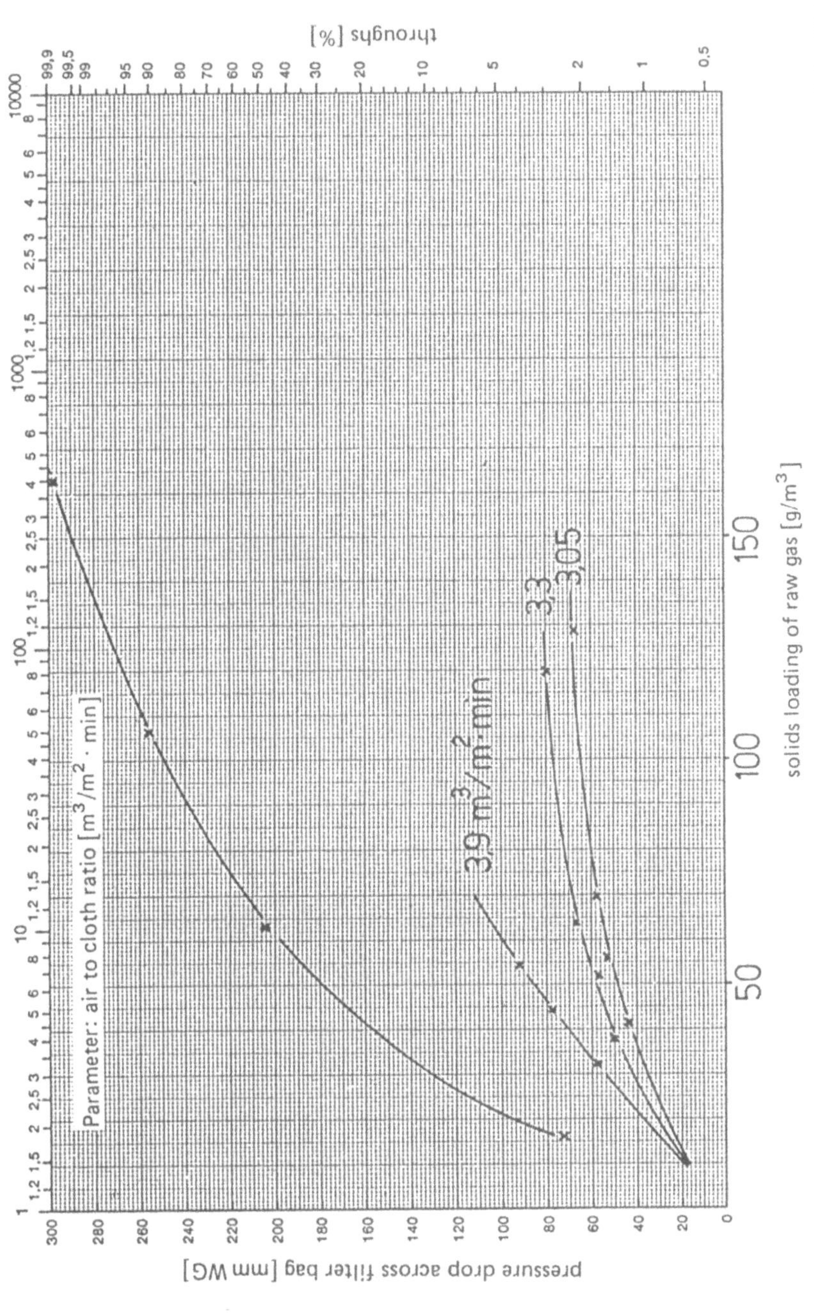

PN file card

Table 3.10a

Operating results of filter system, total separator/exhaust system

customer: Eternit Colombiana SA	any test card?	execution no.
location: Bogotá	test report no. PN-516	start-up report no. 8/81
country: Kolumbien altitude above sea level 2650 m	material sample no.	author: SM date 1.10.81

	stated	measured		stated	measured
					eternit dust/chips
filter type	PGFG-300/30	PGFG-300/30	material to be separated		
filter bag grade	PNf-500	PNf-500	bulk density t/m³	0.35	0.5–0.8
			material temp. filter inlet °C		18–21
gas to be filtered	Air	Air	moisture content %	<10	11
raw gas volume m³/min	899	1136	material throughput kg/h	210	
gas temp. filter inlet °C	0–30	18–21	filter loading m³/m²·min	2.63	3.32
relative humidity %		~51*	residual dust content mg/m³	<10**	
dew point °C	lower than gas	~14.5	pressure drop across filter bag mmWG	max. 110	60
	temperature		pressure drop across total filter mmWG	~150	60
filter cleaning pressure bar		2.5	pressure at filter inlet mmWG	−250	−370***
cleaning air volume Nm³/min			(+ positive pressure, − negative pressure)		
cleanoff cycle s		20–25	average vertical air velocity in filter m/s	1.36	1.72

other observations

dew point exceeded   no

material build-ups in filter housing etc.

other materials

separated in the same system		bag resistance	remarks
material	material throughout	mmWG	*) φ varies from 50–100 % (mist)
	kg/h		**) residual fibre content 2 fibres/cm³
			***) high pressure loss of 176 mmWG at air intake of remotest machine line A.
			Will be modified approx. in 1982

PN file card	date	stated: UT-3.4/SM	visa: 5.6.79
		measured: UT-3.4/SM	visa: 1.10.81

3.4 Designing the Filter Surface Area

Table 3.10b

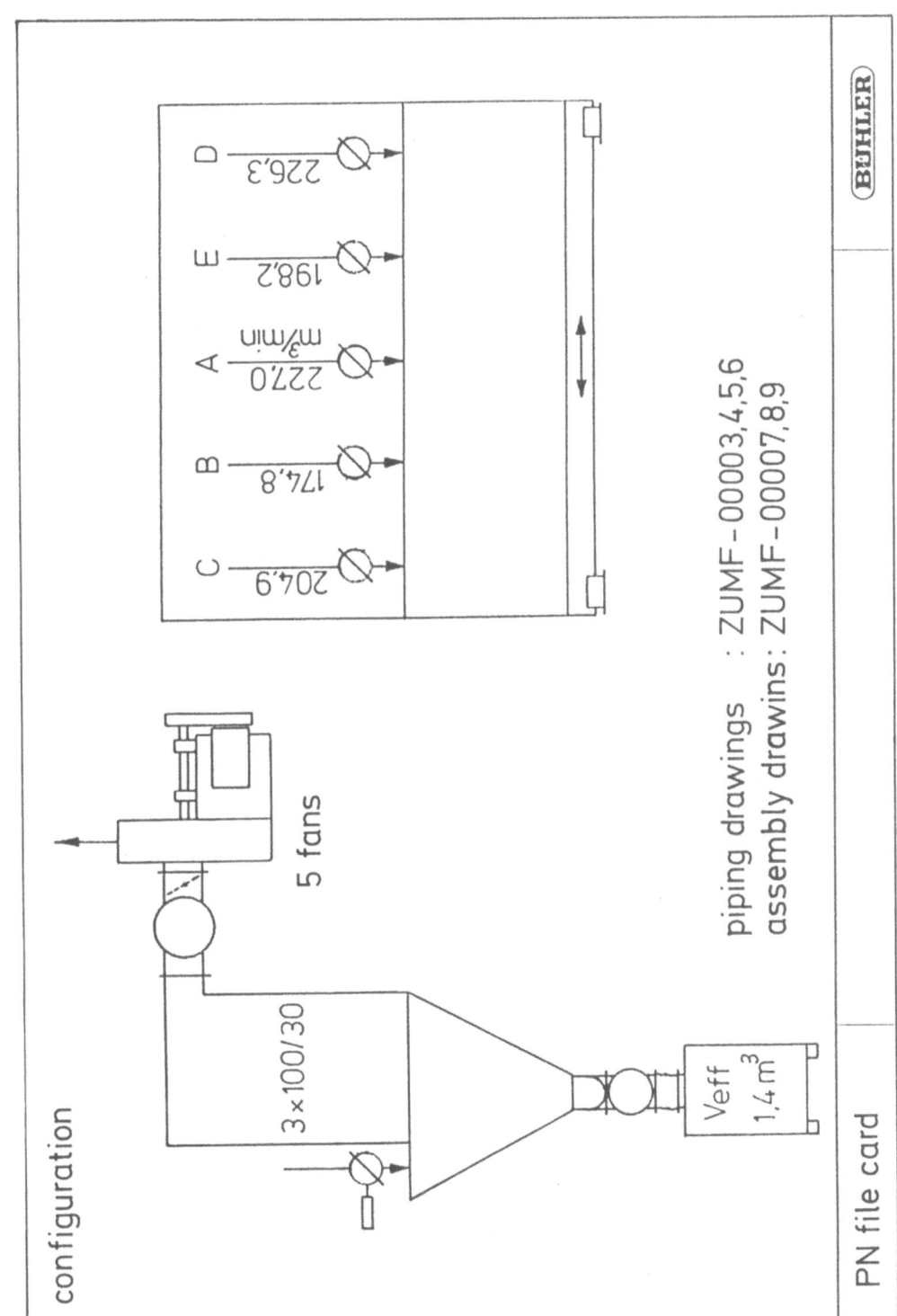

Jet Filter with Venturi (bags)

A = 2.7
A_1 = 1.0
B = 0.8
C = 0.9   X = 2.7 · 1.0 · 0.8 · 0.9 · 1.2 · 0.71 · 0.9 · 1.0 · 1.0 = 1.49 m³/m² min
D = 1.2
E = 0.71
F = 0.9   For a flue gas volume of 310 000 m³/h this requires
G = 1.0   3466 m² filter area
H = 1.0

If another filter system is used, the following air to cloth ratio is obtained:

Shaker filter with bags:

2.7 · 0.65 · 0.8 · 0.9 · 1.2 · 0.71 · 0.9 · 1.0 · 1.0 = 0.97 m³/m² min
310 000/0.97 · 60 = 5326 m² active filter surface.

If this area is distributed over 12 cells, one of which is cleaned off-line at any one time.

Total Filter Area = 5326 · (12/11) = 5811 m²

Direct Pulse Jet Filter with Row Cleaning:

2.7 · 1.3 · 0.8 · 0.9 · 1.2 · 0.71 · 0.9 · 1.0 · 1.0 = 1.937 m³/m² min
310 000/(1.74 · 60) = 2667 m² Filter Area

2. *Kiln Dust Extraction at Conveyor and Screen Shaker*
   Bulk Density = 0.9 g/cm³, Europe
   Particle Size 2–50 μm (80 %)
   Temperature 23 °C
   Flue Gas Load 7.9 g/m³
   Flue Gas Volume 15 100 m³/h, Gas flow bottom to top

   Jet Filter with Venturi (bags):

   A = 2.6
   A_1 = 1.0
   B = 1.0
   C = 0.9
   D = 1.3   X = 2.6 · 1.0 · 1.8 · 0.9 · 1.3 · 1.0 · 1.0 · 0.85 · 1.0 = 2.6 m³/m² min
   E = 1.0
   F = 1.0
   G = 0.85   For a gas volume of 15 100 m³ this requires 97 m²
   H = 1.0   Filter Area

3. *Active Carbon in a Drying Chamber*
   ρ_s = 0.25 g/m³, Europe
   Particle size 10–50 μm
   Flue Gas Temperature 60 °C

3.5 Ease of Maintenance and Intervals

Flue Gas Load 165 g/m³
Direct impulse Jet Cleaning in Rows (bags)
Gas Volume 4860 m³/h, Flow — bottom to top

Direct Pulse Single Bag Cleaning

A = 2.8
A_1 = 1.5
B = 0.8
C = 1.0   X = 2.8 · 1.5 · 0.8 · 1.0 · 0.82 · 0.91 · 0.825 · 0.95 · 1.0 = 1.97 m³/m² min
D = 0.82
E = 0.91
F = 0.825   A gas volume of 4860 m³/h requires 41.2 m²
G = 0.95   Filter Area
H = 1.0

4. *Wheat Flour — Vacuum Filter, Europe*

Particle Size 40–100 μm (70 %)
Temperature 30 °C
Flue Gas Load 5 g/m³

Direct Pulse Single Bag Cleaning

A = 3.5
A_1 = 1.5
B = 1.0
C = 1.1
D = 1.2   X = 3.5 · 1.5 · 1.0 · 1.1 · 1.3 · 1.0 · 1.0 · 1.0 = 7.5 m³/m² min
E = 1.0
F = 1.0
G = 1.0
H = 1.0

3.5 Ease of Maintenance and Intervals

3.5.1 Cleaning and Control Systems

In contrast to the production plant, filters are treated as a necessary evil by the plant operator. It is an expense item. He therefore seeks an installation that requires little or no attention once installed. Thus, safety, low maintenance, and ease of servicing have first priority. Today's high labour costs demand short downtimes and simple maintenance. The fewer the moving parts the less the maintenance. Regeneration systems that are incorporated into the filter head or integrated into the side access doors give easy entry to the filter elements for changing.

If lubrication points cannot be avoided, they should be actuated by automatic means from a central point.

Valves for jet filters must perform at high switching rates and frequencies, and diaphragms should be made of ozone resistant rubber. Ozone resistance and low temperature flexibility become specially important where filters are in the open and where winter temperatures are likely to be low. Ethylenepropylene rubber combines these two properties with high mechanical strength. Unfortunately, it is sensitive to oil. Thus, oil free cleaning air must be given high priority.

With jet systems, if temperatures exceed 100 °C and if inert gases are used for regeneration, there is no diaphragm material to date with adequate service life to withstand these high switching frequencies. In this case valves with stainless steel plates in special casings and steering mechanisms have prooved reliable and durable.

Where electronic control circuits are used, provisions should be made for easy exchange of printed boards, if they fail. Jet filters require provisions to adjust valve opening times within the range 20–200 ms.

Pulse intervals lie between 3 to 60 seconds. As a rule of thumb, the differential pressure stabilises for jet filters when each filter elements is cleaned in 3–6 min intervals. When flue gas loads become very high, such as in pneumatic conveyor systems, this cleaning interval must be shortened.

Systems in which the cleaning cycle commences at a predetermined differential pressure are only successful, if flue gas loadings are kept low, and the dust behaves normally. Such filters also must have a generous filtration capacity.

In any case, the control system should be so adjusted that it continues the reverse cleaning for approximately one hour after the flue gas supply is closed down. This gives additional cleaning to the bags before plant shut down. This is a precautionary measure against crust formation, if the filter cake remains on the bag; it also prevents condensate forming. The filter can also be accessed then for maintenance.

With larger installations it is important to have a damper control upstream of the main fan which regulates the flue gas flow as the differential pressure rises. This prevents filter overload during the critical phase of running in new elements when filter resistance is very low.

3.5.2 Filter Elements

For ease of maintenance defective pockets or bags must be localized rapidly. This presupposes access through the clean gas side and adequate lighting. Provisions must also be made to vacuum clean the area after a defective element has been exchanged.

A faulty bag can be detected by telltale local dust deposits, and they can be repaired during the preventive maintenance cycle (e.g. 6 monthly). When toxic materials are handled, the system must comply with strict safety procedures; it then becomes essential that the filter elements be accessible from the clean air side solely.

Filter constructions are available today where the elements can be rapidly detached and removed. The simpler and more robust this construction, the more does the design approach the needs of the industry. Great attention must be given to properly seal the elements against the floor or partition wall between flue and clean gas.

Filter elements last longer when properly assembled. Sewing threads should exhibit the same or better chemical resistance than the filter medium itself. Triple seams have been found useful for jet filters. This is particularly important for the floor seams.

Filter elements of 3 metres or longer, which usually hang from the top, may abrade at the loose bottom end. It has been found useful, therefore, to reinforce the bottom

3.5 Ease of Maintenance and Intervals   245

10 cm of the bag. Filter elements for jet systems must have a plate insert at the floor of the bag to reflect the pressure wave upwards.

In the most recent filter designs the filter support cage only needs to be unlocked and withdrawn at the clean gas side. The old disused bag drops into the flue gas compartment (obviously with the dust conveyor stopped), and the new bag and cage is inserted from the clean gas side.

The dirty filter bags are recovered through an access hatch at the hopper base, and disposed of in plastic containers. This procedure makes bag changes almost a "white collar" job.

For a 1000 bag filter house such a system allows 2 men to exchange a defective bag within 1/2 h downtime.

3.5.3 Wearing Parts, Spares and Consumables

Wear and tear arises from mechanical and chemical effects, ageing of rubber diaphragms and seals, etc.; that depends on construction of the system.

If very abrasive dusts are handled, parts that wear are elbows in the ductwork, deflector vanes in the flue gas manifold, and gas entry flanges in cyclone type separators. Wear effects can also appear on the clean gas side when handling such substances as cement, clay, fly ash or similar.

During the costing stage of a dust collector it is prudent to include a list of essential spare parts that may be required over a 5 year operating period.

Spare parts for the cleaning system should be available at short notice. Such parts should be stocked by the plant user to enable a speedy return to service.

Stocks of spares should be kept to a minimum, because they constitute idle capital. A selection of exchange bags or pockets must always be at hand. With today's short delivery times it is possible to order whole replacement sets at short notice.

The latest jet and super jet filters require little attention. As a rule, very few spare valves, diaphragms or circuit boards for the controls are necessary. When compressed air is used for cleaning it is important to replenish the absorber in the chemical drier.

3.5.4 Special Maintenance Facilities

The chemical industry manufactures several products on the same filter line; this requires special facilities. Filter types have been developed where whole filter sections can be replaced in one operation. This has led to the use of filter cartridges for each product (ref. fig. 3.44 and 3.45).

An inbuilt washing and drying system can achieve this (clean in place). For extremely toxic substances the "Top lift clean gas system" has been found satisfactory. Spray jets are incorporated into the lid of the filter on the clean gas side; these bind the dust and rinse it down (fig. 3.46). The bags are detached on the clean gas side and drop into a special hopper in place of the usual extraction lock.

In large filtration units, such as in spray drying, an overhead crane removes and inserts filter cartridges of 200–300 bags in a very short time. This technique is found in powdered milk production (fig. 3.47) where health standards require cleaning with minimum downtime (fig. 3.48).

Fig. 3.44 Changing the filter element from the clean gas side

Fig. 3.45 Micropul, Cologne

Fig. 3.46 Cleaning system integrated into the filter head with spray heads (CIP System)

Fig. 3.47 Changing a filter cartridge in a powdered milk spray drier

3.6 Explosion Protection

When these maintenance factors become of primary concern; it becomes important for both supplier and user to discuss how much effort should be allocated to plant layout for minimum maintenance and downtime. Improved efficiency, improved quality of product and protection of personnel are some of the gains that can be realized by greater investment in this area.

Fig. 3.48 Fold-back lid enables a quick change of a filter cartridge

3.6 Explosion Protection

Fig. 3.49 Pioneer work in laying the foundation to VDI 3673 Design Guide

3.6.1 Regulations and Definitions

The VDI Guideline 3673 "Pressure Relief in Dust Explosions" is a basic document that has found recognition in many countries. Bartknecht [15] gives many valuable hints for filter construction. Further concepts and definitions may be referenced in the above literature.

In this chapter the criteria will be dealt with that relate to the protection of filter plants – underlined by typical examples.

Explosion proofing of filter installations is defined as protecting man within the vicinity of a dust explosion, whether it occurs within the collector housing or within the associated ductwork. An additional requirement is that the filter itself remains serviceable. Inevitable damage is to be kept to a minimum and to those parts that can be easily exchanged, so that the plant can return to operation after only a few hours of downtime.

3.6.2 Construction and Tests

Explosion protection involves a number of basic constructional considerations which, it is sad to say, are still lacking in many plants offered to the market today. Thus, low melting aluminium parts, especially diecast components, may not be used either within the flue or clean gas area, close to the bag floor, nor in the form of valves, blow tubes or venturis. The heat generated affects these vital parts during an explosion or fire to such a degree that rehabilitation is not feasible.

The same applies to vibrator and shaker mechanisms with associated bearings, oscillating motions and lubricating points.

For this special requirement direct pulse systems were developed in which these sensitive parts are not within the filter housing. Air distributors, jets and support cages should be made of steel, so that only the bags are destroyed – provided the explosion pressure can be safely handled.

The energy of the explosion is dissipated into the open air through a pressure release hatch. The pressure release fire ducts must have no bends and must be maximum 5 metres long. A flash back into the breathing ducts with consequent propagation to other parts of the plant is prevented by specially designed and proven non-return valves, trap doors, and escape chambers. The mounting and support member must be capable of absorbing the thrust of a gas explosion. A plant equipped with these safety measures is capable of restarting within a few hours of such an event, because the bags can be exchanged rapidly from the clean gas side.

Carbon dioxide gas flooding of the baghouse or water sprays (depending on dust type) bring a fire within quick control.

Commonly available pressure release hatches, non-return valves, trap doors and burst chambers must be certified by a recognized testing authority. A separate calculation chart within VDI 3673 sets out design load for pressure release hatches. Fig. 3.50 shows design parameters for a powerful ignition potential. More details are given in the VDI 3673 Guidelines, as well as in Bartknecht [15].

To arrive at the chamber volume the aggregate volume of bags or pockets is to be deducted from the housing volume.

3.6 Explosion Protection

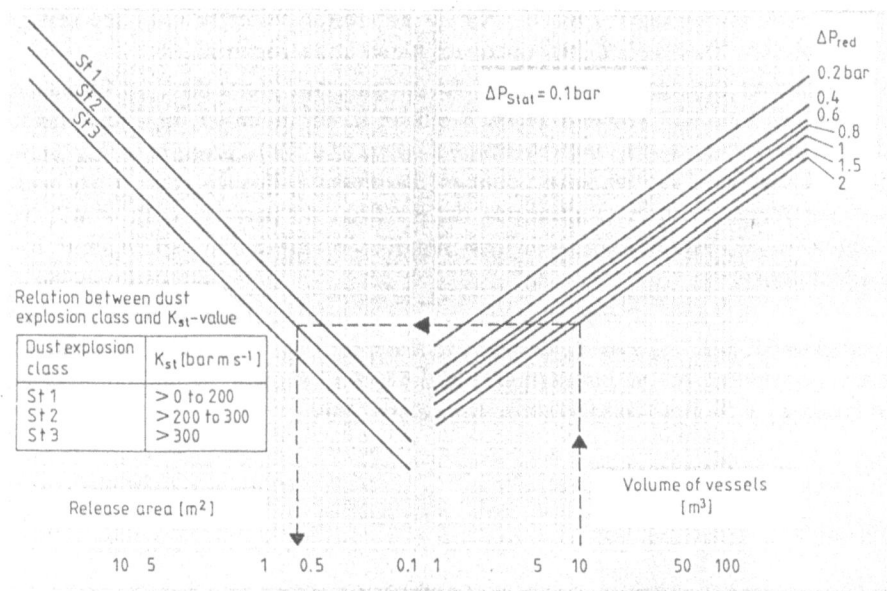

Fig. 3.50 Nomogram

Dusts are classified into risk categories 1, 2 and 3 on the basis of "Explosion Potential Tests". A rough guide is given by the "Hartmannrohr" test method; specialized test services must be consulted for more accurate determinations. Because dusts can vary widely in their explosion behaviour, such values can serve as a guide only. The relevant data must be obtained on a case by case basis.

Fig. 3.51 shows the construction of a pressure release plate which is held in place by a rubber sealing strip. Once opened during a discharge of an explosion these hatches will

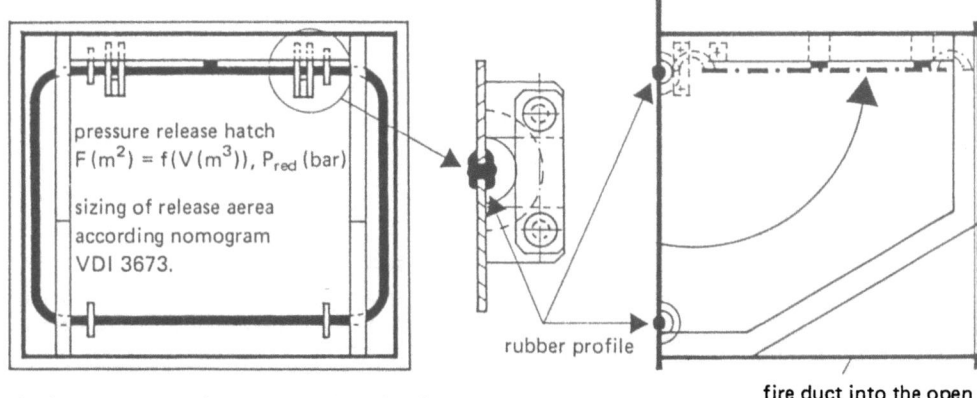

Fig. 3.51 Example of a pressure release hatch

fall closed by their own weight. Even if they do not reseal perfectly until the seal is replaced, they prevent the entry of combustion air for secondary ignition.

To certify triggering settings and to ensure functioning and durability, performance tests are essential. Fig. 3.52 shows a pressure release hatch mounted on a test stand. Access doors (fig. 3.53) are no longer permitted where the filter pressure safety limit exeeds 5 bar. These seals can no longer contain the pressure. In such cases it has been found that the toggle closures (Rathmann) on the filter lid perform well, as will be further demonstrated. Fig. 3.54 shows a test stand to establish the safety limits for chamber access doors.

To ensure the safe discharge of explosion gas and the associated flame jet both, internal layouts, and external facilities must be designed to allow for this. This necessitates careful study of the positioning of the installation. Fig. 3.55–3.57 show a layout and arrangement that complies with these considerations for a class 2 dust.

Fig. 3.52 Pressure release hatch on a test stand at the Central Safety Office of Ciba Geigy CH 4000 Basle

Fig. 3.53 Reinforced access door of a gas chamber for a pressure proof filter

Fig. 3.54
Test stand for door closures at the Institute for Explosion Safety in Dortmund-Derne, Westfalia West Germany

3.6 Explosion Protection

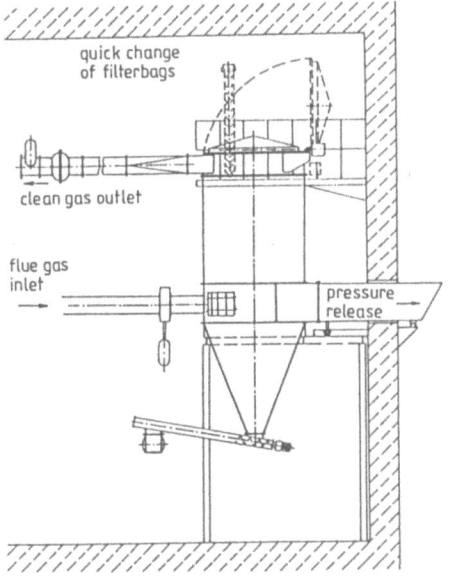

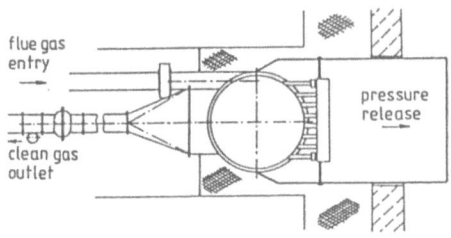

Fig. 3.55
Pressure release vent into the open

Fig. 3.56
Pressure safe filter installation with safety release.
P (red) = 1.3 bar

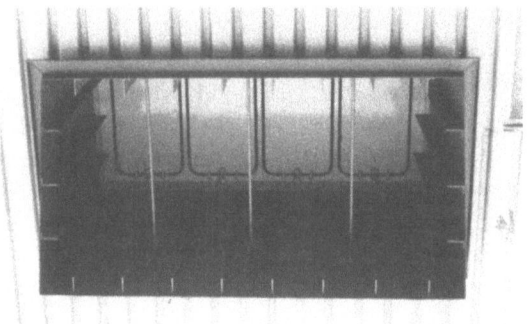

Fig. 3.57
Release hatches with escape duct

3.6.3 Planning Based on Field Experience

Filter plants that are equipped with:
- explosion pressure release
- pressure safe and impact resistant filters
- explosion suppression systems

illustrate how basic research, in cooperation with the previously mentioned institutes, can be applied. A decision on whether a plant under consideration must incorporate provisions for pressure release, pressure resistance, or impact resistance is largely determined by functional characteristics and cost constraints. However, safety should not be compromised for greater efficiency.

There are no fundamental criteria for decision making, but the following points may be helpful:
- pressure release techniques are only appropriate, if the explosion energy can be dissipated safely, either into the open or into fire traps,
- if this is not feasilbe, or if other parts of the plant besides the filter are endangered, then a pressure resistant system is appropriate,
- if explosions are likely to be frequent, then they should be allowed for using design specifications for explosion proof pressure vessels (AD Notification, applicable to Germany and Europe). The design must withstand the maximum expected explosion pressure within the elastic stress limits of the material.

An appropriate solution, in any case, must be sought in discussion between operator and contractor.

Example 1: Pressure Release (fig. 3.58–3.60)

Good relations between customer and contractor enabled the installation of a filter with 240 bags that was cost effective and easy to maintain.

An explosion rating of 1 was determined for a dust dispersed in, and to be recovered from, an air stream. The Ciba-Geigy Safety Office tested and certified the plant as safe. The pressure release system (ref. fig. 3.59) consisted of hatches sealed with special rubber profiles. Possible explosion gases were vented through a short fire duct in the roof. The

3.6 Explosion Protection

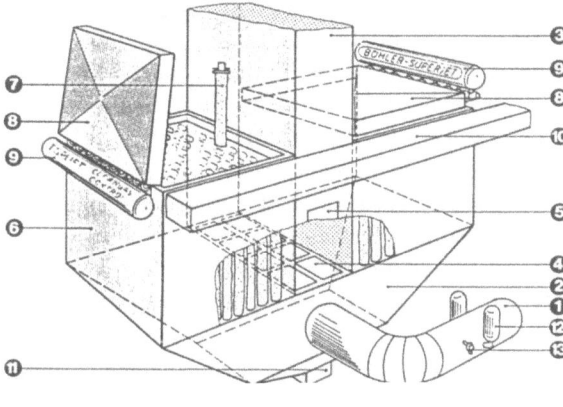

Fig. 3.58 Perspective drawing: Explosion suppression as per DIN 3673. Escape hatches are located at the centre with a vertical release vent. Filter elements are placed on either side.

Capacity 2400 Bm³/h. Filter area 216 m². Total filter resistance 8–15 mbar. Flue gas dust load 20 g/m³. Fineness 100 % 100. Final emission 10 mg/Nm³.

1. Flue gas entry
2. Dust collection hopper
3. Central escape duct to the roof
4. Pressure release hatch
5. Access door to escape duct
6. 2 × PGFG 120 filter elements
7. Filter bas
8. Filter head with integrated cleaning system
9. Cleaning system
10. Clean gas duct
11. Dust extraction
12. Fire extinguishers
13. Detectors

Fig. 3.59 View through the access door into the escape duct. Note the hinges, rubber seals, door stops and vacuum protection

Fig. 3.60 Air duct tested to 1 bar impact pressure with extinguishers on either side and explosion detectors

hatches were designed to comply with VDI Guideline 3673. For easy replacement of the pressure release seals an access door was provided in the venting duct. The up to date design provided for maintenance and bag changes to be performed from the clean gas side. Torsion springs hold the filter lid in balance; this contains the cleaning system within it, so that the filter head can be serviced by one person within minutes.

In the unlikely event of an explosion travelling at sonic velocity into the filter cell through the breathing ducts to cause a secondary detonation, the ducts were equipped with an additional explosion protection system. Detectors and electronic release mechanisms blow quenching substances into the ductwork within fractions of a second. This flame barrier also acts in reverse to prevent the spread of a primary explosion from the filter into the breathing ducts.

All systems, including the filter, were earthed, to prevent static build-up. Such earthing is necessary irrespective of whether the plant is pressure resistant or explosion proof.

Example 2: Pressure Resistant (fig. 3.61–3.64)

Pressure tested jet filters certified to 10.5 bar with pressure resistant breathing ducts, fans and dust extraction locks. A safety certificate was issued after an explosion test.

The venting of highly explosive dust from several production machines and conveyor points within a factory could only be solved by accommodating and dissipating the frequently occurring explosions. Interruptions to production of not more than 24 hours could be tolerated. That implies: "live with the explosion!"

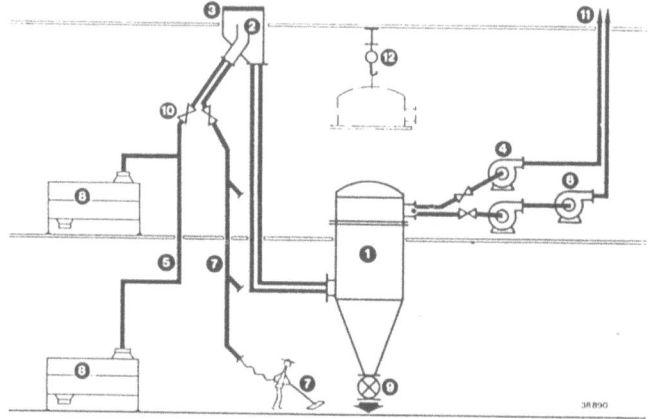

Fig. 3.61

Schematic drawing – protection of man and machine as a total concept.

Explosion protection with pressure tested ducts DIN 3673, tested to 10.5 bar

1. Filter approved to 10.5 bar
2. Burst chamber
3. Bursting foil (red)
4. Low pressure fan for venting
5. Venting ducts (yellow)
6. High pressure two stage fan for dust collection
7. Ducts for vacuum cleaning
8. Process machinery
9. Dust lock with fire protection
10. Damper
11. Clean gas exit above roof
12. Maintenance crane

3.6 Explosion Protection

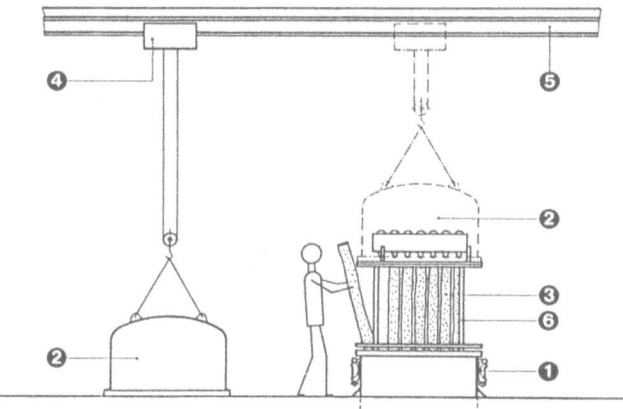

Fig. 3.62
Schematic drawing of rapid bag changing with overhead crane and mezzanine floor

1. Rathmann toggles
2. Filter head
3. Filter bags
4. Overhead crane
5. Rail
6. Support frame

Fig. 3.63 Arrangement for pressure tested MVRS Filters with high pressure fans and air ducts. A gate valve acts as a fire barrier downstream.

Fig. 3.64 Rathmann toggles and clamps facilitate a rapid bag change with easy repair of damage

The maximum explosion pressure for these dusts were measured at 10.5 bar and assigned class 3. Despite the frequent explosions the plant mechanical stress must remain within the elastic limits of the material. For these reasons this installation was designed to operate within a pressure environment.

After an explosion the gases require venting. This is achieved by pressure resistant fans through exhaust ducts, as well as via a recoil chamber with a bursting panel. This pressure release chamber is also a barrier between filter and dust source. Should an explosion occur upline of the filter, the shock will be absorbed by the pressure release chamber. This routes the explosion and flash via the burst panel into the roof, so that it cannot reach the filter. By the same method a primary explosion within the filter is prevented from reaching the intake ducts. All inlet ducts were laid out with design pressure ND 16. The pressure resistant fire traps are activated immediately on an explosion and prevent any spread to the downstream piper or conveyor facilities.

Above 5 bar explosion pressure filter doors and seals have not been found satisfactory. Thus, the bag floor is attached to the filter lid by means of "Rathmann" flanges to enable rapid access to the filter bags. After an explosion the entire bag assembly, together with filter head, can be lifted clear by a service crane to change the bags in minimum time. The system has performed flawlessly in operation.

To summarize, it can be said that effective explosion protection is a reality today with comparatively small effort in relation to damage potential. It should be the aim of all parties to continue the development of safety devices and methods in order to make the workplace safer and to eliminate costly interruptions to production.

3.7 Heat and Sound Insulation

Heat insulation prevents condensation and protects the housing from corrosion. In exceptional circumstance preheating is practiced sometimes.

Depending on the the chemical composition of the gas, acid dewpoints which arise from such substances as sulphur dioxide, hydrogen chloride, hydrogen fluoride may lie above 100 °C. The operating temperature of a filter plant should therefore lie about 20 °C higher.

If there is any risk of condensation whatsoever, insulation becomes essential to protect the filter housing from corrosion and to maintain the operation of the filtering system. Condensation products can combine with the dust cake on the filter medium to form a crust that cannot be removed with normal cleaning. The bags or pockets become impervious. This effect can be prevalent in smoke collection from coal firing or garbage incineration; also, in spray drying plants. Crust formation by the accumulation of water vapour must not be confused with sublimation, i.e. when a gaseous substance solidifies within the pores of the filter medium to cause blinding. Sublimation cannot be prevented by insulation. It must be solved by changing the nature of the chemical reactions or by extending the reaction time within the filter. Hygroscopic products require not only good insulation, but also care to prevent moisture leakage in the form of convection currents during closedown of the plant, e.g. weekends or holidays.

Non-return dampers are required that close automatically at the flue gas and clean gas side as soon as the extraction fan is shut down. The amount of moisture within the closed off filter is easily absorbed by the dust cake that sits on the filter medium. Moisture condensation within the dust hopper collector during idling or shut down should be prevented at all costs. In plants for smoke cleaning, heating of the hopper collector to 70 °C is sometimes practiced. Even more important is the regular removal of the settled dust to prevent crusting. This is also the most effective way to prevent afterglow caused by latent oxidation of the dust (e.g. residual carbon in flyash).

Spontaneous combustion can occur, if the dust is allowed to accumulate for long periods. To prevent running the plant below the dewpoint during start-up of heavy oil burning or spray drier plants it is adviseable to raise the internal temperature within the filter housing with a separate circulation system. Part of the clean gas is passed through a heat exchanger and returned to the flue gas side while the filter is shut off. The amount of preheating depends on available time and minimum operating temperature requirements. Sometimes the bags are coated with chalk at the same time. Obviously, all ductwork associated with the preheating should be insulated.

There are times when heat insulation serves the simple purpose of protecting employees from excessive heat, such as when combustion plants are confined to enclosed buildings. In most cases, however, heat insulation has a functional or operational purpose. This can result in overall improvements in efficiency of the filter plant.

Sound proofing deserves the same attention as heat insulation. Even with well insulated ducts, filter housings and sound proofed blowers and motors the sound radiation at close range may still exceed prescribed limits in densely populated areas. This may require additional sound proofing. Fitting sound absorbers, weatherproofing, thermal insulation and architectural aspects results in a truly integrated design. A separate enclosure, as against separate housings for individual components for a filter installation can have many advantages for operation and maintenance.

When all aspects of a separate building for the entire installation are weighed up against component housings, the decision often falls to the former. This is not surprising when it is considered that one third of the cost is contributed by materials and two thirds is contributed by labour.

Insulation consists of 10–12 cm rock wool; galvanized iron is a better sound insulator than aluminium cladding.

Separate building have simple, flat components that are easier to assemble than the complicated shapes of machinery, with an added benefit of weather protection. A water logged insulation is of no value.

A decision in favour of a building complex for a filter plant can be influenced by a labour unemployment problem in the building industry.

It should be evident from the above comments that a separate building to house the installation may be only fractionally more expensive, but is to be preferred under any circumstance.

There is no universal solution. Today, more than simple thermal protection is called for, and aspects of sound attenuation, long term weather protection, maintenance and operation conditions, building codes and architectural appearance are receiving increased attention.

3.8 The Cleaning Air

3.8.1 Properties and Function

All systems that clean by means of reverse air or pulse require compressed air, the quality of which deserves particular attention. Condensed water, oil vapour and other impurities can contaminate the dust cake. In jet filters, where the stored energy is rapidly converted

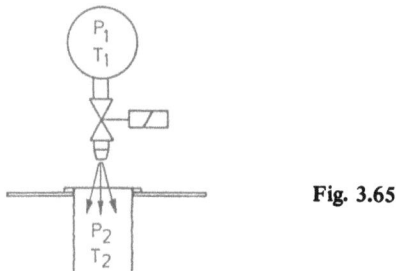

Fig. 3.65

to kinetic energy during expansion of the air, it is possible to reach the water condensation point.

Essentially the process is adiabatic due to the very rapid expansion (approx. 0.1 second) (fig. 3.65).

Water can condense at the exit of the blow tube, if the air is not sufficiently dry. Mixing the cooled compressed air with the clean gas can lead to further moisture condensation. Should this lead to condensation within the filter medium it will interfere with the filtering process. Dust and filter can cake together and block the filter.

The following situations place special demands on the cleaning air:

- Placing the plant in the open, in cold or wintry climates.
- The flue gas is already close to its dewpoint (drying process or fuel combustion, etc.).
- Hygroscopic products such as cement, instant coffee, washing powder, instant chocolate, ammonium zinc chloride, etc. also place demands on the cleaning air.

Under these circumstances it may be necessary to dry the air – as will be described later. As well, the pressure gradient in the blow tube should be kept as low as possible. This keeps the associated temperature drop to a minimum and so minimises the dewpoint depression. As outlined in sect. 3.3.2 the cleaning efficiency must, however, not be adversely affected.

3.8.2 Cleaning Air Production

Different systems require different types of compressors. Ensuring that the air supply is clean extends the life of the compressor. Air requirements vary with cleaning systems and fall into the following groups:

- Low pressure cleaning (up to 0.2 bar) using centrifugal fans. Atmospheric air is used for this purpose.
- Medium pressure cleaning (0.5–1.0 bar) with compressors. Centrifugal or rotary compressors deliver oil and water free air. They require effective inlet filters of paper or cloth.
- High pressure cleaning (3–8 bar) with positive displacement compressors. Piston or screw type compressors supply the cleaning network. Inlet filtering should be ensured.

3.8.2.1 Quality of Cleaning Air

Untreated compressed air carries impurities from several sources:

- from the inlet air (atmospheric dust, moisture and vapours),

3.8 The Cleaning Air

- from the compressor (oil, particles of wear),
- from the distribution system (rust, sealing remnants, welding slag).

These impurities should be totally removed or at least removed to a degree were they will no longer interfere with or damage proper functioning.

3.8.2.2 Line Filters

Line filters are used to extract dust, scale, metal filings and carbonized oil from the compressor. Line filters with a 50 μm pore size are adequate.

Where the compressor working surfaces are lubricated with oil, oil vapour inevitably finds its way into the air supply. The majority of this oil and water vapour will fall out at the intermediate and final condensor stage. Even traces of oil are undesireable for the cleaning air, because they affect the life of the filter medium. Micropore filters can eliminate these traces to deliver compressed air that is technically oil free.

Line filters also remove condensed water. However, only water that is present in droplet form will be removed. Vapourized moisture can only be eliminated by drying. The water collected in line filters is drained either by remote action or manually. From experience automatic systems tend to clog after a while, if they are not given regular attention.

As a general rule, whenever textile dust collectors are housed in enclosed buildings and the dust or product is non-hygroscopic, simple line filters are sufficient.

It is understood that all plant, ancillary equipment and pipes for the compressed air are properly installed, e.g. inlet air temperature as low as possible and condensor, storage tank and piping with downward gravity in the flow direction.

Proper functioning and maintenance of line filters and associated equipments is equally important.

3.8.2.3 Air Driers (fig. 3.66)

Water and oil vapour are gaseous impurities in compressed air. Other extraneous matter is rarely found in vapour form. Oil vapour, i.e. hydrocarbons, cannot be eliminated by mechanical means. Absorbers such as active carbon are required. Water vapour can be removed by drying. Three basic methods are employed:

- freeze drying
- adsorption
- absorption

Each method has characteristics that must be respected to give satisfactory results. Such driers must be located in a cool environment and not subject to freezing temperatures. Freeze drying reduces the dewpoint of the compressed air to just over 2 °C. This method is the most economic for general use. Its efficiency is inhibited, if the drier surfaces are contaminated with oil or other impurities. A prefilter is therefore essential.

Absorption drying can reach condensation points below 0 °C; the drier inlet temperature must not exceed 30 °C. Sodium chloride is the absorber commonly used. It is expended as it takes up water and requires regular replenishment. At 7 bar and 25 °C the nominal usage is 80 grams per 1000 m^3 air. The consumption of the drying agent rises rapidly, if the air contains water in liquid form. It is important, therefore, to keep the inlet temperature low and to use an efficient water separator. A dust filter in the dry air line is

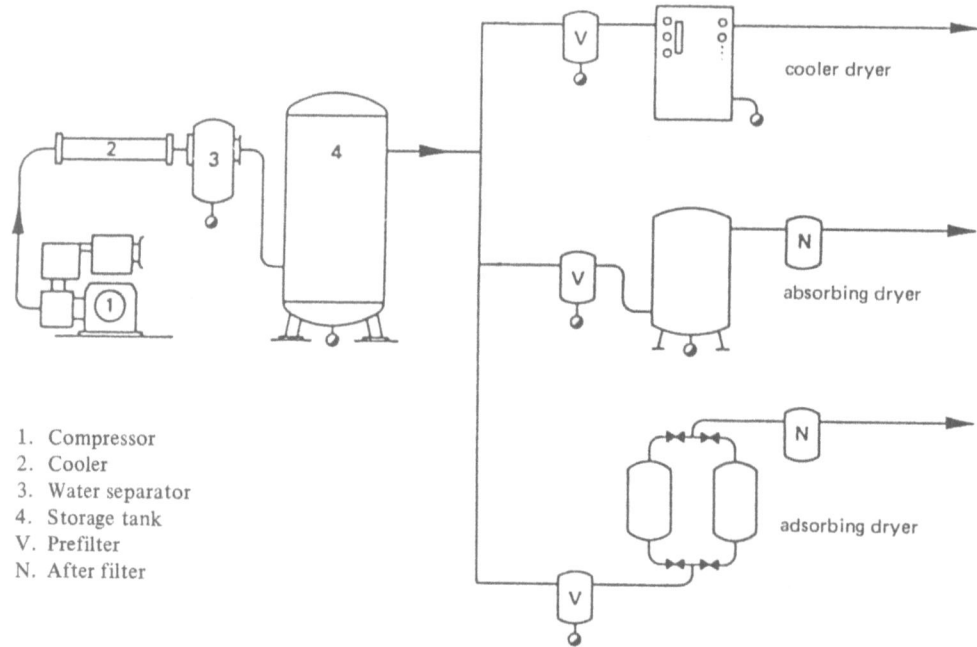

1. Compressor
2. Cooler
3. Water separator
4. Storage tank
V. Prefilter
N. After filter

Fig. 3.66 Compressor with choice of air drying systems

useful to hold back any absorber dust. The corrosive condensate should be drained daily; automated industrial systems have been found unreliable and prone to blocking.

Adsorbers can reach condensation temperatures down to $-90\,°C$. Silica gel, activated aluminium oxide and molecular sieves are among the techniques used. The air stream tends to carry the absorber with it, so that a microfilter must be placed downstream. Because of the adsorber loss and contamination with oil and other substances, it requires changing every two to three years.

3.8.3 Pressure Lines

Cross sections depend on:
- designed velocities
- designed pressure loss
- operating pressure
- number of reducing points
- length of lines

Air velocity should lie between 5–15 m/sec. The pressure loss should not exceed 5 % of the operating pressure. Instruments, elbows and branches act as reducing points. To calculate cross sections such reducing points are expressed in equivalent lengths of pipes

Table 3.11 Equivalent pipe lengths of air line fittings

Fitting	Inches / mm	1/2 / 15	3/4 / 20	1 / 25	1 1/4 / 32	1 1/2 / 40	2 / 50	2 1/2 / 65	3 / 80	4 / 100	5 / 125	6 / 150
Seat valve		4	5	6	8	10	15	20	25	30	50	60
Streamlined valve		2	2.5	3	4	5	7	8	10	15	20	25
Sluice valve		0.2	0.25	0.3	0.4	0.5	0.7	0.8	1	1.5	2	2.5
Pipe angle		0.9	1.2	1.5	2	2.5	3.5	4	5	7	10	15
Pipe elbow [GB: Pipe bend] r = 0		0.5	0.7	1	1.5	2	2.5	3	4	6	7.5	10
Pipe elbow [GB: Pipe bend] r = d		0.2	0.25	0.3	0.4	0.5	0.6	0.8	1	1.5	2	2.5
Pipe elbow [GB: Pipe bend] r = 2d		0.1	0.12	0.15	0.2	0.25	0.3	0.4	0.5	0.8	1	1.5
Hose coupling or T-pipe		1	1.5	2	2.5	3	4	5.5	7	10	15	20
Reducer		0.3	0.4	0.5	0.6	0.7	1	1.5	2	2.5	3.5	4

(table 3.11); these lengths are added to the total lengths. Fig. 3.67 is a computing chart to read off the required pipe cross sections given the known values.

Air lines should be installed so that leaks can be detected and repaired. Horizontal lines should have a fall of 1 % in the flow direction. Downward sloping trunk lines should not lead into an end line, but lead beyond the branch line to prevent condensation entering. Collection points for condensate are at the lowest points in the network.

A pressure switch is useful to monitor the pressure lines and to supervise the operation of compressed air supply.

3.9 Certification

3.9.1 Filter Resistance

3.9.1.1 Basics

One should differentiate between the absolute differential pressure of the filter medium (filter bag resistance) and the total resistance measured between flue gas inlet and clean gas outlet. This value includes losses within the filter housing. For standardised designs only the pressure differential of the filter medium is generally quoted. To calculate the

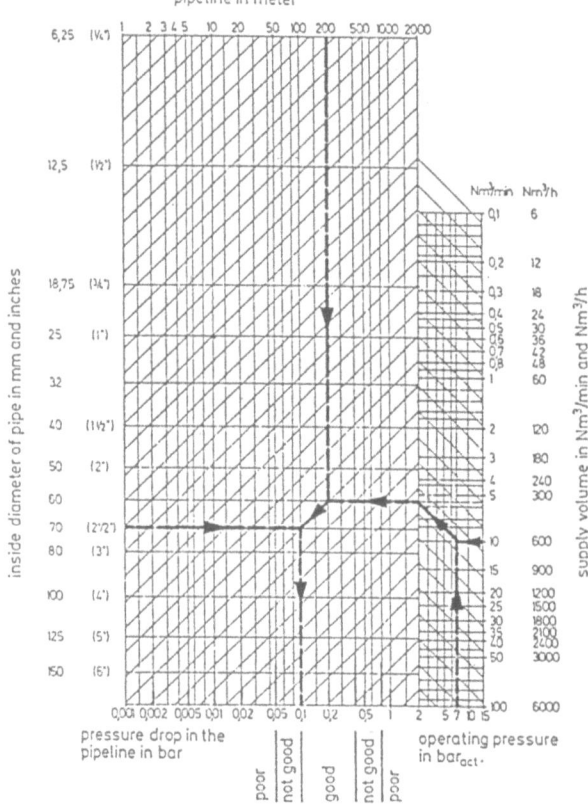

Fig. 3.67
Nomograph for the calculation of pneumatic pressure lines

capacity of the main fan, the total resistance must be considered; its size should be related to the overall filter resistance and be part of any certification.

Mathematical and physical modelling cannot accurately predict the filter resistance. Too many parameters influence dust collection. Even in applications where there is a wide field of experience it is necessary for the maker to set down all relevant factors and to allow for some latitude, if a guarantee for the differential pressure is to be part of the contract. Ref. also to sect. 3.4.

As an example, it is not possible to transfer design data from one galvanising plant to another by simple extrapolation of filter surface area and gas volume. Excessive capacity on the other hand is inefficient in both space and cost.

Certification in regard to filter resistance is only sensible in as far as it is based on all influencing factors. Exact specifications of the dust to be filtered are also part of this. It is further adviseable to include contingency plans, if guaranteed parameters cannot be achieved. A judicious allocation of the business risks and functional risks that are to be borne by the various parties will contribute to a harmonious solution.

3.9 Certification

3.9.1.2 Measurement Techniques

The required fan capacity is a product of differential pressure and gas volume. It is useful, therefore, to monitor this pressure. This is accomplished very simply by a differential pressure gauge. Such an instrument should be calibrated during commissioning of the plant, since commercially availalbe instruments can show considerable variations.

Similarly, in large installations recording the clean gas volume is desireable. Simple Prandtl probe or Pitot tube suffices, but must be maintained regularly.

Anemometers with electronic readout, multichannel recorders and remote sensors are only required in exceptionally large plants, e.g. large coal or waste combustion plants.

Filters equipped with needlefelts and efficient cleaning systems surpass all other dust collectors in collection efficiency. With suitable filter media residual dust below 25 mg/m^3 is easily reached. Values below 10 mg/m^3 are even possible, if the physical, chemical and mechanical properties of the dust cooperate. If the air to cloth ratio is kept below 2 m^3/m^2 min, collection efficiency can be raised to 5 mg/m^3 where dust agglomerates.

If certification below 5 mg/m^3 is required, the performance limits of single stage filters are reached. Such requirments can arise with toxic substances. In this case a second dry filter in series lowers the residuals to less than 0.5 mg/m^3. This second filter, often termed "police filter", is only cleaned if the differential pressure exceeds a predetermined value. Emision values from these systems border on the limits of detectibility (extreme cases are not considered here).

Such filtered air can stand up well to a mountain resort after a thunderstorm.

3.9.2 Filter Efficiency and Emission Limits

3.9.2.1 Basics

A 100 % collection is still out of reach even with the most modern filter technology and most up to date filter medium. Collection efficiency is a function of technique and system, and last but not least available financial resources.

Two examples to illustrate this:

- For PVC spray drying a 100 g/m^3 particle concentration is usual. For a residual emission of 10 mg/m^3 the collection efficiency is 99.999 %.
- In a galvanizing plant the flue gas dust concentration from the zinc bath venting is around 100 mg/m^3. If the filtered air contains also 10 mg/m^3 dust, the collection efficiency is 90 %.

In both cases the separation factors are excellent by today's standards, when compared to electrostatic precipitators, wet scrubbers or cyclones. Since the percentage collection efficiency of a filter is confusing, it is usual to express the residual dust content in absolute terms of mg/m^3, rather than a relative percentage. This has become the basic unit of measurement. Furthermore it relates directly to the manner in which regulations for residual dust content is expressed. It is usual in the filter market to express collection efficiency in mg/m^3.

Worldwide and in Europe no uniform standardisation exists in setting emission limits. "Dust at the Workplace", a publication by Schütz, Coenen and Engels [14] gives a good exposition of the subject. It has applications outside Germany as well.

The "Technical Regulations for Clean Air (TA Luft 1974)" is a basic document within Germany. The latest in this field can be found in a revision of the Regulations of the Federal Emission Protection Legislation (1986). This contains a VDI Manual "Clean Air", which has become mandatory (gazetted Nov. 1983).

3.9.2.2 Measurement of Collection Efficiency

Residual dust and gas measurements are performed during commissioning, as well as for regulatory approvals. They should be repeated at 6 monthly intervals.

VDI Guideline 2066 Part 2 of October 1975, and Part 1 of June 1981, as well as VDI 2579 of August 1980 for the gravimetric determination of dust content applies.

A continual monitoring and recording of dust residue or toxic emissions is intricate and costly. In large installations, however, they assist in overall control and serve as evidence to the control bodies. It would go beyond the scope of this book to enumerate the different methods. Reference should be made to the specialized literature of the subject.

A field example is given based on the standards quoted above.

Measurement Technique (fig. 3.68)

Residual emissions of a dust collector can be measured gravimetrically by instruments such as "Millipore". Chemical absorption methods such as the "Impinger" are used to determine the gaseous composition.

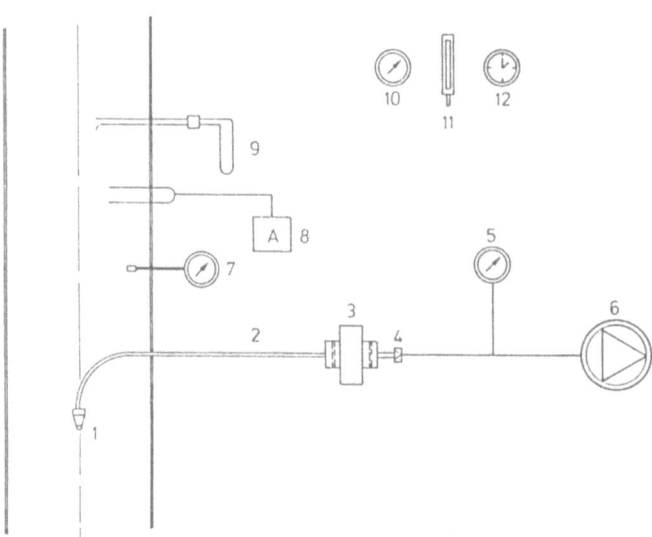

Fig. 3.68 Instrumentation for measuring dust content

1. Sampling probe (VDI 2066 Sect. 3.3.1.2)
2. Sampling line (may require heating)
3. Membrane filter (may require heating)
4. Calibrated orifice
5. Manometer
6. Vacuum pump
7. Thermometer
8. Gas analyser (moisture, composition)
9. Pitot tube for measuring gas velocity
10. Barometer
11. External temperature
12. Clock

3.10 Investment Analysis and Operating Cost Comparisons 265

The gas stream is sampled by a probe. The sampling stream is passed through a sampling device containing two tared membranes in series. Dust particles in the stream are retained by the first membrane; this measures the residual dust. The second filter paper determines any changes in behaviour of the first as a result of temperature or humidity. A calibrated orifice controls the quantity of air that is withdrawn.

The choice of the measuring point is important. A straight section in the clean gas duct should be chosen, whose length in either direction is 6 to 8 times the duct diameter. At this sampling point a profile of gas velocity across the cross section of the duct is taken. The sampling gas velocity at that point should be as close as possible to the clean gas velocity. Any deviation from this produces a bias, because of differential particle movement.

3.10 Investment Analysis and Operating Cost Comparisons

3.10.1 Criteria

Once the requirements for a dust collector have been established, the question arises on what basis are comparisons to be made. The user will generally invite several tenderers. Should he select the lowest capital cost offer, he may have the cheapest, but that is not to say the most economic. A financially sound choice must take both capital cost and operating costs, which stretch over the entire depreciation period of the plant, into consideration. Wage costs, spare parts, energy, chemical and water consumption can influence the choice substantially. The more detailed these cost factors are laid out, the clearer will be the ultimate choice. It avoids later unpleasant surprises. It would be highly desirable that the contractor for the collector provide not only capital cost estimates, but also detailed estimates of running costs for his offer. A suggested layout of such a cost comparison is presented in section 3.10.2.

3.10.2 Cost Evaluation

Supplier:

pos.		Specification	Capital Cost	Running Expenses	Weighting
1.	*General* Purpose Type (pocket, bag, inertial, scrubbing) System (electrostatic, wet, dry) Operating Hours per annum Idle Time per annum Electricity Cost per kWh Water Cost per m^3 Chemical Cost per kg Wage Costs employees per h Wage Costs contractor per h	 ---------- ---------- ---------- ---------- ---------- ---------- ----------			
2.	*Filter Type* Surface Area gross/net (m^2) Filter Medium (water, fabric, felt) Space requirements L · W · H Corrosion Protection (quality, thickness) Insulation (quality, thickness, enclosure) Sound Attenuation (dBA at 10 m)	---------- ---------- ---------- ---------- ---------- ---------- ----------	---------- ---------- ----------		
3.	*Filter* Flue Gas Volume (m^3/min) Air to Cloth Ratio (m^3/m^2 min) Differential Pressure (bar) Certified Efficiency (mg/m^3)	 ---------- ---------- ---------- ----------			
4.	*Fan* Supplier ---------- Type ---------- Total Pressure (bar) Motor Power (kW) Ratings full load (kWh) Idling (kWh)	---------- ---------- ---------- ----------	----------	 ---------- ----------	

5.	*Filter Regeneration*
	 a) Wet Scrubber
	    Cost of Ancillaries
	    Water consumption m^3
	    Chemical Consumption kg/h
	    Pump Rating kW
	    Consumption kWh
	 b) Dry Filtration
	    Cost of Ancillary Equipment
	    Cleaning Air Capacity m^3/h
	    Cleaning Air Pressure bar
	    Motor Rating kW
	    Power Requirements kWh
		
		---------- 	
		---------- 	
		---------- 	
		---------- 	
		
		---------- 	
		---------- 	
		---------- 	
		---------- 	
			
			---------- 	
			
			---------- 	
				
				---------- 	
				---------- 	
				---------- 	
				---------- 	
				
				---------- 		

3.10 Investment Analysis and Operating Cost Comparisons

Continued

pos.		Specification	Capital Cost	Running Expenes	Weighting
6.	*Filtered Products* Quantity Use Disposal Cost	---------- ----------		----------	

7.	*Spare Parts*
	 a) Wet Scrubbers
	   Pumps
	   Jets
	   Valves
	   General
	 b) Dry Filtration
	   Spare Set of Bags/Pockets
	   Service Life
	   Total Replacement Cost
		---------- 	
		 ---------- 	
				---------- 	
				 ---------- 	
				 ---------- 	
				 ---------- 	
				 ---------- 		

8.	*Maintenance* Man hours a) Labour Cost Employees man hours/annum b) Contract Labour man hours/annum	---------- ----------		---------- ----------	

9.	*Costs*
	 Building
	 Erection
	 Commissioning
	 Total Costs
	   Depreciation · no of years
	   Running Costs per year
	   Cost per 1000 m³ of Clean Air
		----------	----------
			 ---------- 	
			 ---------- 	
			 ---------- 	
			 ---------- 	
				---------- 	
				 ---------- 		

3.10.3 Comparison Table for Difference Offers

	filter "A"	filter "B"	filter "C"	filter "D"	filter "E"
System	bag	bag	pocket	wat	venturi
Gas volume (m^3/min)	533	500	580	525	566
Filter area (m^2)	342	180	?	–	–
Air to cloth ratio (m^3/m^2 min)	1.55	2.77	?	–	–
Fan pressure (Pa)	2750	3630	1670	3480	9010
Capital cost	181650.–	125065.–	164011.–	128940.–	162730.–
Annual running costs	28510.–	40280.–	36435.–	51240.–	113640.–
Capital cost per 1000 m^3 Gas (6 yr depreciation)	0.284	0.208	0.236	0.206	0.239
Running expenses per 1000 m^3 gas	0.222	0.335	0.260	0.413	0.835
Total running cost per 1000 m^3 gas	<u>0.506</u>	<u>0.543</u>	<u>0.496</u>	<u>0.619</u>	<u>1.074</u>
Ratio of running/capital cost	0.78	1.61	1.10	2.01	3.49

3.10.4 Discussion

Once the data, as outlined above, is assembled the following becomes apparent: Using the standard depreciation period of 5 years specific capital cost can be calculated. If this is compared to the running expenses, it will become obvious that nearly all systems consume more than they cost over this period. In the above practical example the system with the highest capital outlay has the lowest operating cost, and is thus the most cost efficient. This design is more mature with resultant lower maintenance effort. Once the plant is written off, only running expenses remain. The running costs, therefore, should be given great significance. A detailed cost study will always pay dividends. Only by calculating the specific total cost is a true comparison of systems possible.

As a final point it should be mentioned that an "out of service" estimate for the plant should be included, since this can interrupt production with costly consequences.

Literature

[1] *Divo, I.:* Gesellschaft für Marktforschung Hahnstraße 40, Frankfurt-Niederrad. „Der Markt für Umweltschutz in Gegenwart und Zukunft; Teilbereich Luftreinhaltung" Gemeinschaftsuntersuchung September 1972

[2] *The McIlvaine Company:* Marketing and Financial Analysis 2970 Maria Avenue, Northbrook, Illinois, "Air Pollution Control in West Germany" Jan. 1980, "Air Pollution Control in Italy" Febr. 1980, "Air Pollution Control in France" Jan. 1980, "World Air Pollution Control".

[3] *Flatt, W.:* Staub-Reinhaltung Luft **37** (1977), Nr. 11, November, 412…416 „Superkompakte Düsenfilter in Zyklondimensionen mit optimierter Einzelschlauchspülung setzen neue Maßstäbe"

[4] *Flatt, W.:* Chemische Rundschau **28** (1975) Nr. 16, „Textilfilter für die Luft- und Gasreinigung in industriellen Produktionsanlagen", SIA Fachgruppe für Verfahrenstechnik Arbeitsgruppe ‚Mechanische Verfahrenstechnik' Tagung vom 10. April 1975 an der ETH in Zürich

Literature

- [5] *Flatt, W.:* VDI-Bericht Nr. 304, 1978. Sichere Handhabung brennbarer Stäube, „Druckentlastung von Filtergehäuse, Einfluß von Einbauten"
- [6] *Flatt, W.* und *Strittmatter, W.:* Bühler-Bühler-Miag Nachrichten Nr. 214, 1979; „Explosionsschutz nach VDI-3673 durch Bühler-Bühler-Miag Düsenfilter"
- [7] VDI-3677: Filternde Abscheider, Juli 1980
- [8] VDI-3673: Druckentlastung von Staubexplosionen, Juni 1979
- [9] *Funke G.:* Zement-Kalk-Gips, 33. Jahrgang, Nr. 4/1980, „Standzeit von Filterstoffen in Zementwerken"
- [10] *Bakke, E.:* Journal of the Air Pollution Control Association, Dec. 74, Vol. 24, No. 12 „Optimizing Filtration Parameters"
- [11] *Frey, R.* and *T. Reinauer:* Air Engineering 4/64, "New Filter Rate Guide"
- [12] *Humphries, W.* and *J. J. Madden:* Filtration and Separation 18 (1981), No. 6, S. 503/505, 539, "Fabric Filtration for Coal-fired Boilers: Nature of Fabric Failures in Pulse-Jet Filters"
- [13] Berufsgenossenschaftliches Institut für Arbeitssicherheit, Lindenstraße 60, D-5205 St. Augustin
- [14] *Schütz, A., W. Coenen* and *L. H. Engels:* Staub am Arbeitsplatz", 1974, Rudolf Haufe Verlag
- [15] *Bartknecht, W.:* Explosionen, Ablauf und Schutzmaßnahmen, Springer-Verlag (Berlin)

Index

Abrasion 116
- Resistance 104
Absorption Drying 259, 260
Acrylobutastyrene Dust 118
Acid Dewpoint 128
Acrylic Fibres 146
Additional Quality Parameters 104
Adhesion Efficiency 31
- Forces 33
Aerodynamic Diameter 5
Afterglow 144
Agglomerates 55, 115
Aging 127
Air Driers 259
- Permeability 101, 145
- to Cloth Ratio 88, 118, 124, 184, 195
Aliphatic Polyamides 63
Alkali Content 171
Annular Jet 186
Antistatic Treatment 137
Approximation for Collision Efficiency 30
- Function 12
Aramides 64, 129
Area Mass of Single Bag 208
Area Weight 44, 101
Arithmetic Mean 11
Aromatic Polyamides 64, 129
Aspiration Type Filter 226
Assembly 165
Atomised Droplets 21
Average Degradation 127

Back Pressure 210
Bag Bottom 165
- Change 170
- Diameter 221
- Filters 41
Barbs 90
Base Coat 180
Base Value 231
Behaviour of Particles 2
Blinding 169
Blowback Filters 211
Blow-off Velocity 31
Blowring Type Filters 182, 186
Blow Time 48
Boltzmann Constant 25
Breathing Ducts 254
Bridging 189, 221
Brownian Motion 132

Brush Discharge Effect 141
Bubble Test 111
Burning Behaviour 143
Burst Strength 103

Calculating Total Collection Efficiency 21
Calculation Examples 237
Candle Filters 166
Cellulose 56, 125
- Nitrate 58
Cement Industry 155
Charge Parameter 28
Chemical Absorption 131, 151, 173
- Damage 125
- Fibres 58
Chicken Effect 115
Clean Gas Dust Content 39
Cleaned Filter 1, 19, 36
Cleaning Air 257
- - Production 258
- - Volume 193
- Intensity 39, 43, 46
- Pressure 123, 214
Coal Grinding and Drying 154
Coanda Effect 191
Coefficient of Restitution 33
Collection Efficiency of Fibre Assemblies 23
Compressed Air Requirement 45
- - Pressure 40
- - Reservoir Pressure 47
Concentration Peak 39
Conex® 64
Continuous Filament 82
- Service Temperature 56
Control Line 191
Conversion of Distributions 14
- - Population Means 11
Copolymers 61
Copper Braid 138
Core Resistance 138
Corona Current 141
- Effect 140
Coulomb Forces 28
Counter Current 210
Counting Method 8
Creasing 125, 127
- Failure 112
Critical Velocity 33
Cumulative Frequency 7
Cunningham Correction Factor 6

Index

Darcy's Law 42
Decontamination Factor 18
Deep Bed Filter 19, 35, 36
– – Filtration 24
Degree of Compaction 101
Delay Line 130
Denier 59
Density Distribution 8
– Zone 110
Depth Filtration 114
– of Insertion 90
Design Calculations 231
– Fundamantals 237
Designing the Filter Surface 227
Desulphuring of Flue Gas 158
Deterministic Particle Transport 26
– Motion 25
Dewpoint 256
Diagramatic Representation of Particle Distribution 7
Differential Pressure 41, 204
Diffusion Coefficient 25
– Effect 25
Dimensional Accuracy 167
– Analysis 27
– Stability 100
Direct Heat Exchange 132
– Pulse 183
– – Filter 192
Disperse Phase 2
Dispersion Parameter 12
Disposable Cage 170
Distribution Parameter 14
– Function 3
Drag Coefficient 4
Dralon 61
Dry Scrubbing 149
Drum Type Washing Machine 171
Dry Absorption 131
Dust Accumulator 135
– Cake 107, 110
– Deposition 36
– Fineness 232
– Penetration 38

Ease of Maintenance 243
Electrofilter 135
Electronic Control 244
Electrostatic Chargeability 57
– Precipitator 157
Electrostatic Forces 28
Electrostatics 49
Emulsifiers 115
Envelope Filter 41
Environmental Protection 173

Equations of motion 28
Equilibrium 44
Equivalent Diameter of Settling Velocity 3
– Diameter 11
Explosion Protection 195, 247
Extensibility 103, 127

Face Velocity 1, 31, 40, 173
– – Rating 103
Factors 231
Fatigue 145
Feret Diameter 3
Fibre Fineness 59
– Web 91
Fibreglass 66
– Fabric 132, 134, 210
Filter Bags 36
– Cake 37
– – Resistance 49
– Characteristic Line 123
– Efficiency 163
– Elements 244
– Geometry 24
– Layout 43
– Operating Conditions 204
– Resistance 229, 261
Filtering Cycle 23
Filtration Velocity 227, 233
Fineness 3
Flame-Retardants 143
Flammability 142
Floor Seams 244
Flow Forces 2
Flue Gas Distribution 223
– – Dust Load 232
– – Flow Control 233
– – Temperature 233
Fluid Dynamic Behaviour 28
Fly Ash 131
Free Sieving Surface 87
Freeze Drying 259
Friction 115, 171
Froude Number 28

Gauss Normal Distribution 14
Gel Permeation Chromatography 127
Geometric Equivalent Diameter 3
– Standard Deviation 14
– Particle Measurement 3
Glassfibre Fabrics 66
Graphite 138
Gravity Effect 28

Hair Felt 94
Hamaker Constant 33

Hardened Contamination 172
Heat Insulation 256
High Pressure 181
– – Cleaning 258
– – Reverse Cleaning 105
Hostaflon® 65
Hot Gas Filtration 131, 173
Hybrid Mixture 137
Hydrated Aluminium Oxide 117
Hydrodynamic Factor 25
Hydrolysis 125

Ignition Energy 155
– Temperature 143
Impact Energy 33
– Resistance 252
Inconel 135
Indeces 231
Indicator of Cleaning Efficiency 49
Indirect Heat Exhange 132
Inertia Parameter 28
Influence of Cleaning 208
Initital Dust Cake 87
Initiation Pressure 231
Injector Characteristic 124
– Feed Point 124
Insertion Points 117
Interception 26
Internal Pressure 48
Investment Analysis 265

Jet Cleaned Dust Collector 108
– Cleaning 156, 173
– Filter 105, 108, 167, 168, 219, 230
Jet Stream 191

Kuwabara Equation 25

Lamb, Kuwabara 28
Lattice Effect 143
Layout Specifications 229
Lead Smelting 149
Length of Tear 144
Life 164
Light Scattering Particle Analyser 21
Limiting Particle Trajectory 26, 27
Line Filters 259
Logarithmic Normal Distribution 14
L.O.I. Value 142
Loop Strength 127
Low Pressure Cleaning 181, 259
– – Reverse Celaning 105, 157
Low Volume – High Velocity 152

Main Dimensions 3
Maintenance 172, 195, 243, 245
Market Trends 196
Martin Diameter 3
Mass Balance 23
– Concentration 40
– per Unit Area 24
– Velocity Ratio 123
Measurement Techniques 264
Measures of Fineness 6
Mechanical Cleaning 156
Median Adhesion Value 31
– Value 10
Medium Pressure 181, 258
Melt Spun 59
Metal Fibres 67, 138
– Fibre Fabric 173
– – Felt 134, 173
– – Addition 157
Mineral Fibres 58, 66
Minimum Ignition Temperature 136
– Spacing 33
Modal value 10
Model Filters 36, 37
Moments of Distribution 10
Mono Disperse 17
– Sized Particles 3
Multichamber Dust Collectors 50

Needled Nonwoven Felt 88
Needlefelts 23, 36, 56, 88, 100, 102, 103, 171
Needles 88
NOMEX® 64
Nonwovens 95
– by Swelling 95
– – Shrinking 95
– with Adhese Binders 95
Number of Insertions 90
Nylon 58

Off-line Cleaning 158, 221, 233
Oil and Coal Fired Steam Generation 157
On-line Cleaning 233
Operating Costs 265
– Temperature 164
Optical Projection Methods 3
Optimal Cleaning Conditions 49
Orlon 61
Oxidation 125

Parameters 27
Particle Concentration 24
– Distribution 2, 4
– Characteristics 2
– Movement 3

Index

- Shape 2
- Systems 2
- Trajectory 26
Parts Warranty 161
PE Felt 36
Peclet Number 25
Performance Warranty 161
Perlon 58
Permeability 36
Piezoelectric Sensor 204
Pinhole Effect 39
Plant Shut-Down 172
Pneumatic Cleaning 145
Pocket Cage 166
- Shaker Type Filter 185
Polyacrylonitrile 58, 61, 126
Polyamide 58
Polyester 58, 61, 126
Polymer Damage 65
Polyolefines 64
Polypropylene 64, 126
Polyurethane 126
Pore Volume 1, 89
Porosity 23
Potential Flow 28
Power Distribution according to Gaudin-Schuhmann 12
Pressure Pulse Slope 215
- Release Hatch 248, 250
- Curves 214
- Filter 141, 160, 181
- Gradient 193
- Release 252
- Resistant 252, 254
Primary Air 190
Probability Plot 14
Processing Filter 232
Projected Fibre Cross Section 24
PTFE 65, 126
Pulsating Cleaning Jet 230
Pulse Cleaning 46, 211
- Intervals 244
- Jet Cleaning 44
Pulse/Impact Conversion 193
PVC 58, 126
PYROTEX 132

Quartz Fibre 135

Rate of Pressure Rise 206
Recoil Chamber 255
Recycling 173
- Energy 132
Redox 146
Repellency 116

Residual Dust 263, 265
- Emission 264
- Filter Resistance 43
- Mass per Unit Area 47
- Pressure Drop 46
- - Loss 44
Reverse Flow Air Velocity 210
- Air Filter 230
- Flow Filter 181, 186
- Gas Celaning 181
- Jet Filter 181, 188
Reynold Number 28
Ring Jet 109
- - - Filter 183, 190
RRSB Distribution 14
Ryton 65

Safety Limit 139
Scolloping Effect 212
Secondary Air 190
Self Ignition Temperature 142
Separation 2, 19
- Forces 32
- in Fibre Assemblies 23
Service Life 155
- Temperature 61
Settling, Velocity 4, 6
Shaker Filter 105, 152, 181, 184, 210, 230
Shaking 155
Sieve Effect 24, 37
Single Bag Cleaning 108, 186, 193, 215
- Use Cartrige 170
Single Fibre Collection Efficiency 23
- - Separation 28
Solution Effect 127
Sound Insulation 256
Specific Air to Cloth Ratio 227
- Filter Cake Resistance 43
- Surface Area 17
- Total Cost 268
- Viscosity 128
Spunbonded 95
Stable Operating Conditions 41
Stainless Steel 135, 140
Standard Design Guide 230
- Deviation 11
Staple Fibre 82
- Length 56
Statement of Warranty 161
Static Electricity 136
Stochastic Motion 25
Stokes Law 4
Storage Filter 1
   Pressure 124
Suction Filter 181

Sum of Distribution 9, 20
Support Cage 167, 169, 245
− Fabric 89, 103, 110, 155
− Ring 169
Surface Filtration 114
− Resistance 138
Swelling 127
− Agent 149

TA Luft 98, 264
Tear Strength 127
TEFLON® 65, 158
Temperature Induced Damage 146
− Resistance 143
Tempering 130
Tensile Strength 103, 127
Test Comparisons 204
Tex 61
Third Dimension 106
Three Phases of Cleaning 192
Throat of the Bag 219
Top Lift Clean Gas 245
Total Collection Efficiency 18, 19
− Separator 232
Toyaflon® 65

Transport Mechanism 25
Trend 172
Tropical Factor 233

UHP-Process 147

Variance 11
Velocity of Blow-off 31
Venturi 190
− Jet 109
Vertical Strip 144

Warranties 160, 168
Washfastness 116, 143
Water Retention Capacity 56
Wearing Parts 245
Weave Pores 87
Weight of Filter Cake 43
Weighted Mean 11
Welding 165
Wire Mesh 169
Wool 57
Wool Felt 92

Yarn Pores 87

MIX
Papier aus verantwortungsvollen Quellen
Paper from responsible sources
FSC® C105338

If you have any concerns about our products,
you can contact us on
ProductSafety@springernature.com

In case Publisher is established outside the EU,
the EU authorized representative is:
**Springer Nature Customer Service Center GmbH
Europaplatz 3, 69115 Heidelberg, Germany**

Printed by Libri Plureos GmbH
in Hamburg, Germany